# Methods for Evaluating Plant Fungicides, Nematicides, and Bactericides

Prepared jointly by
The American Phytopathological Society
and the
Society of Nematologists

Published by
The American Phytopathological Society

**Editorial Committee**
Eldon I. Zehr, Chairman
George W. Bird
Kenneth D. Fisher
Kenneth D. Hickey
Fred H. Lewis
Roland F. Line
Samuel F. Rickard

ISBN: 0-89054-025-X
Library of Congress Catalog Card Number: 78-63414

The American Phytopathological Society
3340 Pilot Knob Road, St. Paul, Minnesota 55121

Printed in the United States of America

# Preface

Systematic testing of chemicals for controlling plant diseases is a young science. Many researchers who investigated and helped to introduce the first organic fungicides are still active plant pathologists. A large reservoir of knowledge about test procedures for chemicals has accumulated, but comparatively little of the information has been published. Thus, much of it is unavailable to newcomers in this area of investigation. The need to conserve the knowledge and experience developed over the past few decades is the motivating force of this book.

Such a need coincides with others that have become important in recent years. As costs of pesticide development rise, improving efficiency and proficiency in research with chemicals becomes increasingly important. How may current testing programs and procedures be made more efficient? Is greater or lesser uniformity in test methods needed? How can greenhouse and field tests be improved so that procedures and results are easily communicated and understood? Can nematologists and plant pathologists who investigate chemicals for nematode and plant disease control assist the regulatory agencies that must enforce the laws limiting pesticides? This book is a first step in an attempt to relate to these issues.

Individuals experienced in the subject matter wrote the chapters of this volume. At least two reviewers who are knowledgeable in the subject matter area and who consider the procedures to be adequate for the purposes stated have studied each contribution. The methods described, however, are not presented as practices that The American Phytopathological Society, the Society of Nematologists, or the editorial committee recommends. Our purpose is merely to outline the art of testing new chemicals for disease control, without suggesting how other individuals or institutions should use the procedures. The volume is intended as a reference for those who investigate chemicals as plant disease and nematode control agents, students, technical personnel, government officials, and others who may be interested in the procedures used to test new chemicals.

The editoral committee has solicited manuscripts from plant pathologists and nematologists in the United States and abroad that describe methods for testing new fungicides, nematicides, and bactericides. We are grateful to those who responded, and to others who offered support and encouragement.

The collection of papers in this book was made under the supervision and advice of the APS New Fungicide and Nematicide Data Committee and in cooperation with the APS Chemical Control Committee.

Eldon I. Zehr, Chairman

Committee Members

| | |
|---|---|
| George W. Bird | Fred H. Lewis |
| Kenneth D. Fisher | Roland F. Line |
| Kenneth D. Hickey | Samuel F. Rickard |

# Acknowledgments

We acknowledge the contributions of the E 35 Pesticides Committee of the American Society for Testing and Materials (ASTM). Some manuscripts for fungicide test procedures were prepared first under the auspices of this committee and are published here in modified form. The paper by Kenneth D. Fisher describes the background of the ASTM involvement in test methods for pesticides. We also acknowledge the inputs of the Joint Society of Nematology/ASTM E 35.16 Committee on Nematode Control, which was responsible for preparing the manuscripts on nematicide test procedures. Some of these papers are written in the format used in ASTM publications.

Manuscripts were judged on the basis of relevance of subject matter, reliability and thoroughness of procedures, organization, and clarity of writing. Two individual referees reviewed each manuscript; however, the entire Joint SON/ASTM E 35.16 Committee reviewed the manuscripts for nematicide test procedures. The committee chairman and a technical editor also edited each manuscript.

We thank the following individuals who served as members of the editorial review board for this publication:

Robert E. Baldwin
I. Fred Brown, Jr.

Robert C. Cetas
Duane L. Coyier

Harrison L. Dooley
Arthur W. Engelhard
Robert H. Fulton
Earl D. Hansing
Graydon C. Kingsland
Harold W. Lembright
Robert H. Littrell

Iain C. MacSwan
Dan Neely
John B. Rowell
Paul B. Shoemaker
Wayne R. Sitterly
Donald H. Smith

We also thank these additional persons who reviewed manuscripts for this publication: Steven V. Beer, Carlyle N. Clayton, Robert A. Conover, Alan L. Jones, Alan A. MacNab, R. Walker Miller, Charles C. Powell, Michael Szkolnik, Frederick L. Wellman, Charles E. Williamson, and Gayle L. Worf. The services of Julie Lewis as technical editor are much appreciated.

Eldon I. Zehr, Chairman
George W. Bird
Kenneth D. Fisher
Kenneth D. Hickey

Fred H. Lewis
Roland F. Line
Samuel F. Rickard

# Contents

PREFACE     iii

ACKNOWLEDGMENTS     iv

## I. PRELIMINARY CONSIDERATIONS

1. Fungicide-Nematicide Testing and Pesticide Registrations —*Joseph E. Elson*     1
2. Use of Statistics in Planning, Data Analysis, and Interpretation of Fungicide and Nematicide Tests—*Larry A. Nelson*     2

## II. LABORATORY AND GREENHOUSE PROCEDURES

3. A Cellophane-Transfer Bioassay to Detect Fungicides—*Dan Neely*     15
4. Methods for Monitoring Tolerance to Benomyl in *Venturia inaequalis, Monilinia* spp., *Cercospora* spp., and Selected Powdery Mildew Fungi—*K. S. Yoder*     18
5. Method for Screening Fungicides for Coryneum Blight Control Using Inoculated Detached Prunus Leaves—*H. H. Harder and N. S. Luepschen*     21
6. Procedures for Laboratory Evaluation of Thermal Powder and Thermal Tablet Fungicide Formulations—*Patricia L. Sanders and Herbert Cole, Jr.*     22
7. Greenhouse Evaluation of Seed Treatment Fungicides for Control of Sugar Beet Seedling Diseases—*L. D. Leach and J. D. MacDonald*     23
8. Greenhouse Procedures for Evaluation of Turfgrass Fungicides —*Patricia L. Sanders and Herbert Cole, Jr.*     25
9. Test Procedures for Fungicides and Bactericides Used to Control Foliar and Soilborne Pathogens of Ornamental Tropical Plants —*James F. Knauss*     27
10. Greenhouse Evaluation of Soil-Applied Fungicides for Fusarium Wilt of Chrysanthemum—*Arthur W. Engelhard*     30
11. Greenhouse Method for Screening Protective Fungicides for Apple Powdery Mildew—*H. L. Dooley*     32

## III. FIELD TEST PROCEDURES
### A. Tree Fruits and Nuts

12. Testing Chemicals in the Field for Apple Powdery Mildew Control in Colorado—*N. S. Luepschen*     35
13. Method for Field Evaluation of Fungicides for Apple Powdery Mildew Control—*K. D. Hickey*     37
14. Field Plot Tests of Apple Fungicides—*F. H. Lewis*     40
15. Field Test Procedures for Fungicides Used to Control Apple Diseases in South Carolina—*Eldon I. Zehr*     43
16. Techniques for Field Evaluation of Spray Materials to Control Fire Blight of Apple and Pear Blossoms—*Steven V. Beer*     46
17. Method of Testing Fungicides on Tart Cherries (*Prunus cerasus* L.)—*F. H. Lewis*     51
18. Laboratory and Field Test Procedures for Evaluation of Fungicides for Control of Brown Rot Diseases of Stone Fruits —*J. M. Ogawa et al.*     54
19. Field Evaluation of Fungicides Used for Control of Peach Leaf Curl—*H. L. Dooley and I. C. MacSwan*     57
20. Evaluation of Cytospora Canker Severity in French Prunes —*P. F. Bertrand and Harley English*     59

**B. Vegetable Crops**

21. Evaluation of Fungicides in the Field for Control of White Rot of Onion—*P. B. Adams*     61

22. Procedures for Field Testing Foliar Fungicides for Potato Late Blight Control—*R. C. Cetas et al.*     63

23. Field Evaluation of Protectant Fungicides for Controlling Celery Late Blight—*H. L. Dooley*     68

**C. Diseases of Tropical Plants**

24. Laboratory and Field Evaluation of Fungicides for Control of Coffee Berry Disease—*D. M. Okioga*     70

25. Laboratory and Field Evaluation of Fungicides for Control of Coffee Leaf Rust—*D. M. Okioga*     71

26. Screening Fungicides for Control of Blister Blight of Tea in Northeast India—*G. Satyanarayana, G. C. S. Barua, and K. C. Barua*     72

27. Procedures for Screening Fungicides for Control of Greasy Spot, Melanose, and Scab on Citrus Trees in Florida—*J. O. Whiteside*     73

28. Methods for Fungicidal Control of Fruit Diseases of Avocado —*H. T. Brodrick*     78

29. Methods for the Control of Anthracnose and Other Diseases of Mango—*H. T. Brodrick*     80

**D. Ornamentals and Turf**

30. Methods for Evaluating Fungicides for Controlling Powdery Mildew, Black Spot, and Rust of Roses—*H. L. Dooley*     82

31. Field Procedures for Evaluating Fungicides for Control of Pythium Blight of Turfgrasses—*Herbert Cole, Jr., C. G. Warren, and Patricia L. Sanders*     85

32. Field Evaluation of Fungicides for Control of Ascochyta Blight of Chrysanthemums—*Arthur W. Engelhard*     86

**E. Seed Treatments**

33. Techniques for Evaluating Seed-Treatment Fungicides—*Earl D. Hansing*     88

34. Screening Fungicides for Seed and Seedling Disease Control in Plug Mix Plantings—*Ronald M. Sonoda*     92

35. Control of Seed and Seedling Diseases of Cotton With Seed Fungicides—*Earl B. Minton and C. D. Ranney*     95

36. Field Evaluation of Fungicides as Soybean Seed Treatments —*N. G. Whitney*     97

**IV. NEMATICIDE TEST PROCEDURES**

37. Guidelines and Test Procedures for Nematicide Evaluation —*Kenneth D. Fisher*     99

38. Guidelines for Evaluating Nematicides in Greenhouses and Growth Chambers for Control of Root-Knot Nematodes—*C. P. DiSanzo et al.*     101

39. Evaluation of Nematicides for Systemic Eradication of Root-Knot Nematodes—*M. Thirugnanam*     103

40. Tests for Nematicidal Efficacy Using Larvae of *Heterodera schachtii*—*Arnold E. Steele*     105

41. Test Materials and Environmental Conditions for Field Evaluation of Nematode Control Agents—*A. W. Johnson et al.*     106

42. Site Selection Procedures for Field Evaluation of Nematode Control Agents—*G. W. Bird et al.*     108

43. Determining Nematode Population Responses to Control Agents —*K. R. Barker et al.*     114

44. Plant Responses in Evaluation of Nematode Control Agents —*C. C. Orr et al.*     126

45. Nematicide Evaluation as Affected by Nematode and Nematicide Interactions With Other Organisms—*G. B. Bergeson et al.*     128

46. Legal, Human, and Environmental Aspects of Nematode Control Chemicals—*John H. Wilson, Jr. et al.*     135

# Preliminary Considerations

## 1. Fungicide-Nematicide Testing and Pesticide Registrations

Joseph E. Elson

Associate professor and assistant coordinator, Office of IR-4, Cook College, Rutgers University, New Brunswick, NJ 08903

Results of field trials are useful in judging the effectiveness of pesticides under various conditions. In the United States, the Environmental Protection Agency (EPA) must register effective agents before they can be used commercially. An expensive, time-consuming process, registration is hard to justify economically except in the case of major or high–cash value commodities.

Interregional Research Project No. 4, a National Agricultural Program for Clearance of Pesticides for Minor or Specialty Uses (IR-4), was initiated to aid in the development of data that would establish tolerances and to assist in the registration of pesticides for minor or specialty uses. IR-4 interacts with state and federal agencies, industry, and the EPA. It is responsible for accumulating and evaluating available data; determining whether additional efficacy, phytotoxicity, and environmental data are needed; seeking out resources for producing these data where lacking; and advising industry of available information and eliciting their cooperation in label registration.

The EPA requires certain data to establish tolerances and to register product labels. Whenever possible, researchers should consider registration requirements when planning fungicide and nematicide trials. Taking extra time to make careful notes of phytotoxic effects or to collect samples for residue analysis could make the product available a season earlier than otherwise possible. The EPA requires efficacy, phytotoxicity, and residue data derived from the recommended rate of use and from twice the maximum recommended rate. Researchers who test chemicals to determine desirable rates have a unique advantage: they can compile data on various rates and thus assist the registration process.

To expedite registration, the EPA must have certain specific information. Listed below are items that are always required.

1. **Identity of the host, disease, and pathogen**—Knowing the name of the crop, including the crop variety, the accepted common name of the disease, and the biologic organism that is controlled is absolutely essential. Such information is basic to any disease control recommendation.

2. **Plot size**—A description of plot size provides a basis for evaluating the precision of the test.

3. **Number of replicates**—Data from four replicates are usually sufficient to obtain statistically sound information.

4. **Application timing**—The actual dates of application must be recorded. To assist in label preparation, the stage of plant growth should also be recorded where appropriate (eg, prebloom, bloom, first true leaf).

5. **Dosage**—Rates used must be representative of those suggested for labeling, plus twice the maximum dosage to be recommended. This enables EPA to set residue tolerances with a safety margin and to determine potential phytotoxic reactions.

6. **Formulation of the material**—The data must be obtained from plots treated with the same formulation as the one to be registered, because variations in carriers, emulsifiers, and other inert ingredients may affect the performance of the products.

7. **Method of application**—The method should specify such information as type of sprayer, nozzle type, pressure, dip method, and seed treatment. EPA reviewers will consider similarity to commercial equipment and methods.

8. **Efficacy**—Data on disease control should be recorded as a percentage based on comparison with untreated plots. Disease severity also should be noted. Specific information on inoculated plots is helpful.

9. **Yields**—Where appropriate, yield data should be taken to verify usefulness of the product. Qualitative

and quantitative data are recommended.

10. **Phytotoxicity**—Any abnormal growth of host plants should be recorded. The effect of temporary damage on final yields will be of concern to reviewers.

11. **Environmental conditions**—Local temperature and precipitation (irrigation) information should be available.

12. **Soil type**—The soil type should be identified in the case of soil-applied pesticides. Where appropriate, such information as soil moisture and pH should be recorded.

13. **Additional information**—Amount and time of application of pesticides other than the test product must also be recorded to assist evaluation of the data.

The collection of samples for residue analysis is cumbersome for researchers, but advisable if the pesticide may be needed for commercial use on a particular commodity.

The time for taking samples varies with individual crops, but normally it is done at the earliest stage of maturity. Plots should be sampled individually, and samples should not be combined. Usually, they should be stored frozen until shipped to an analytic laboratory.

Because the requirements of pesticide registration are complex, any data that the researcher can make available to the IR-4 Project are appreciated. Consideration of the registration requirements when testing fungicides and nematicides may make an effective pesticide available for use sooner than would be possible if new experiments had to be performed.

Any questions about how researchers can assist the IR-4 Project should be directed to your state IR-4 liaison representative, or to Dr. J. E. Elson, Office of IR-4, Cook College, Box 231, Rutgers University, New Brunswick, NJ 08903.

# 2. Use of Statistics in Planning, Data Analysis, and Interpretation of Fungicide and Nematicide Tests

**Larry A. Nelson**

Professor of statistics, North Carolina State University, Raleigh, NC 27607

Statistics is the science and art of applying logical principles to experimental design and technique and in collection and analysis of data to make valid (though uncertain) inductive inferences. Statistical analysis and inference procedures usually include measures of the degree of uncertainty. The proper use of statistics depends on the application of these principles to each of several phases of scientific investigation. Federer (3) listed the following steps in scientific investigation:

1. Formulation of questions to be answered and hypotheses to be tested.

2. A critical, logical analysis of the problem or problems to be raised.

3. Selection of a procedure for research.

4. Selection of suitable measuring instruments and control of the personal equation.

5. A complete analysis of the data and the interpretation of results in light of experimental conditions and hypotheses tested.

6. Preparation of a complete, correct, readable report of the experiment.

Statistics is not a substitute for the above process, but it does provide the tools for implementing some of these steps. The steps are not entirely statistical, but all except 1, 2, and 6 involve at least some statistical aspects. The statistician can be of assistance in asking questions that will sharpen the researcher's understanding and formulation of the problem (steps 1 and 2). He can also assist in preparing the tabular portion of the research report (step 6). Steps 3 and 5 contain the most highly statistical elements. Unfortunately, many researchers only consult statisticians for matters involved in step 5.

The purpose of this chapter is to review and clarify the basic statistical principles as they apply to each phase of research so that the reader might, through better understanding, improve the quality of his fungicide and nematicide tests. More reliable testing should facilitate the registration of chemicals and enhance the acceptance of research findings. Some of the suggestions given here cannot be found in statistical methods textbooks.

This chapter is divided into three sections corresponding to the important phases of statistical application: (i) planning, (ii) data and their analysis, (iii) interpretation and reporting of results. Various aspects of steps 1, 2, 3, and 4 are included in the planning discussion, whereas the data analysis and interpretation sections deal with aspects of step 5. A short discussion of reporting results is designed to help the reader in implementing step 6. A detailed flow chart of the aspects of experimentation to be discussed is given in Fig. 2.1.

## Planning

Planning is designed to assure that the treatments selected for the experiment most effectively will provide relevant comparisons, estimates, or hypothesis tests and that the analysis of data will be straightforward. Planning also helps to assure that the size of the experiment will be appropriate for the particular situation. Undersized experiments often result in failure to detect important differences. Oversized experiments can be costly. Planning also involves choosing an appropriate method of assigning the treatments to the plots, thus assuring unbiased estimates of means, an estimate of their reliability, and valid statistical tests.

**Role of statistician**—A researcher should consult a statistician before starting his tests to work out the design details for the experiment. The statistician often makes his most valuable contribution by asking questions that cause the researcher to reexamine all aspects of the problem, including his reasons for conducting the test. The test then can be tailored to the particular experimental setting so that it is efficient and yields a maximum amount of information with a minimum amount of effort and cost.

**Steps in planning**—The scientific method suggests that planning of research be done in steps. These include (i) defining the objectives, (ii) formulating a set of detailed specifications for the test, and (iii) determining specifically the method to be used in analysis of the data.

*Step 1: Defining objectives*—Objectives may be questions to be answered, hypotheses to be tested, specifications to be met, or effects to be estimated. Many researchers are overly ambitious and thus have objectives that are too broad and diffuse. On the other hand, the objectives may be so limited that experiments can be condensed into a single test.

The statement of objectives should be clear, concise, and specific. It should define the extent of the population over which the generalizations are to be made (eg, one farm, six counties, poorly drained soils). The results of the research are usually based on a sample (one or several experiments) that must be representative of the population to which the conclusions are to be applied.

*Step 2: Detailing specifications*—The detailed specifications or procedures for an experiment should be formalized in writing. The proposed field layout may be shown in a diagram drawn to scale. Some of the more important specifications are (i) underlying population for which inferences are to be made, (ii) characters to be measured, (iii) sites, (iv) treatments, (v) experimental material, (vi) experimental design, (vii) replication, (viii) experimental technique, (ix) design and randomization in a series of tests, (x) supplementary variables, and (xi) sampling procedures.

*Step 3: Determining the method of analysis*—The procedures for data analysis should be described in writing during the planning phase. References for statistical methods may be cited for details of analyses. Also useful is an outline of the sources of variation (eg, blocks, treatments, harvest, error) and their respective degrees of freedom.

A brief discussion of each important specification designated under step 2 above follows.

**Underlying population**—For every experimental situation, some target population exists to which the results apply. The population must be specified in the planning stage to sample adequately the range of its properties and to determine the most appropriate statistical methods for a given property distribution. For example, a farmer might conduct his own test to determine the optimal rate of nematicide needed for a particular field. The population would consist of all the nematodes (of various species) in that particular field. A single randomized block experiment located in that field would provide the needed information. On the other hand, an extension worker might wish to study methods of control of a particular pathogen on a statewide basis. To make valid inferences, he would need to conduct a series of experiments at different locations that would represent the range in conditions under which the pathogen occurs.

**Characters to be measured**—The variables that are submitted to statistical analyses are called the characters. The number of diseased plants per 10 ft of row and the proportion of plants that the root-knot nematode affects would be good examples. In some areas of research, certain standard characters have been established and information on them is obtained routinely. In other cases, the researcher is not certain which characters best reflect the effects of the controlled variables. Therefore, he measures a number of them to select one or more that will be useful for future studies.

The time at which characters are measured is important. In some cases, multiple measurements should be taken on characters throughout the season to sample climatic effects adequately.

**Sites**—The sites are drawn to represent the area about which inferences are to be made. Final selection of sites is made after careful screening of a group of prospective sites. Often these are chosen with the aim of representing certain environmental regimes. On other occasions, the sites may be selected randomly (ie, each site has an equal chance of being selected for the sample of sites). The area within a site should be as homogeneous as possible regarding such things as soil type, cropping history, slope, drainage, and fertility.

**Treatments**—Selection of treatments involves consideration of the following:

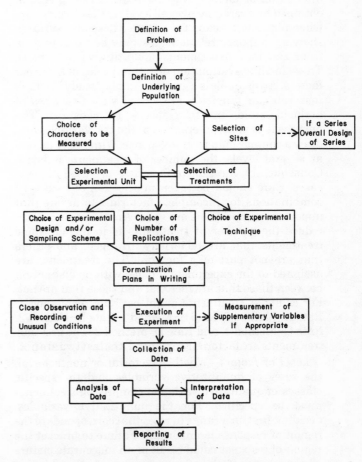

Fig. 2.1. Detailed flow chart of decisions and operational aspects of experimentation.

*Number of treatments*—This is determined to a large extent by the factors (and their levels) that will provide information on the problem being studied. A factor is a quantity under investigation in an experiment as a possible cause of variation. Its level is the intensity with which it is brought to bear. The upper limit of treatment number may be imposed by limited availability of experimental material or by the number of treatments that will give a reasonable block size (eg, 15 treatments).

*Function of the treatments*—The purpose of one type of experiment is to compare chemicals and spot the winner. Often the applications must be rate-specific to the chemicals. The treatments then consist of package combinations of chemicals and rates. The chemical and rate that perform best may then be recommended. Other purposes of experiments may be to determine the slope of a response to a single quantitative factor, to explore a response surface for the optimum combination of levels of more than one quantitative factor to obtain maximum yield, or to find an inflection point in a response curve. In these cases, it is important to choose levels of the factor or factors that will bracket the major region where response occurs and will give precise estimates of model parameters or optima and inflection points or both. In these cases, having the levels of the quantitative factor equally spaced may or may not be desirable.

*Factorial arrangement of treatments*—Treatments in factorial experiments are combinations of levels of two or more factors. An example is a $3 \times 2$ factorial with six treatment combinations of three levels of fungicide and two methods of application. One benefit of the factorial arrangement is the possibility of estimating the interaction of factors. Another is the added precision obtained by averaging over all levels of the other factors when obtaining means for a factor. Response surfaces showing response relationships may be fitted to data from factorial experiments (quantitative variables). These facilitate estimation of optimal rates of the input factors. An example is a growth chamber study in which four rates of fungicide are varied with five rates of herbicide, and growth response is measured. The purpose is to study interaction of the two variables and draw a contour map for the response. With many factors at several levels, the number of treatments is large. Consequently, incomplete factorials are used in some cases. Care should be taken in selection of treatment combinations for incomplete factorials to assure that important comparisons are valid. Another approach to reduce the number of treatments is to include some treatments that are factorially arranged, and others that are not part of a factorial. The treatments are assigned to the experimental plots, with no distinction between those that are factorial and those that are not. For example, treatments might be: Rate A–Method 1, Rate A–Method 2, Rate B–Method 1, Rate B–Method 2, Rate C–Method 1, and Rate D–Method 2. The first four treatments are factorially arranged; the last two are not.

*Level of factors*—Whether factorial or nonfactorial, the rates of quantitative variables and the specific classes of qualitative variables comprising treatments must be specified. For the quantitative variables, deciding first on a range in rates that corresponds to the region of response is important. Failure to bracket the region of response could result in an inaccurate picture of the shape of the response curve and could bias estimates of the optimal rate of the input factor. Once

the range in rates is established, one must decide how many treatment levels should be included within the range and how they should be spaced. In cases in which a curved response pattern is not likely in the population (linear response), only two levels, one at the low end and one at the high end of the range, would be sufficient. Usually the response pattern is not known, so at least one point in the middle of the range in addition to the end points is necessary to permit a check for curvature. In cases in which the shape of the curve is more complicated (eg, S-shaped curves), several levels are needed to estimate the number of parameters necessary to describe the curve adequately. The desire to keep the experiment within reasonable size places an upper limit on the number of levels of each factor.

*Use of controls*—These are often needed for a basis of comparison, especially if it is not known that other treatments will be effective. In factorial experiments in which none of the treatment combinations is an untreated check, a control can be randomized with the factorial treatments. Few situations call for a separate control for each treatment. The added precision gained from such an arrangement is usually offset by the additional experimental material required. Many experiments necessitate several types of controls. For example, in a nematicidal evaluation, if a chemical has both nematicidal and insecticidal properties, three checks should be included, ie, one with nematicide only, one with insecticide only, and one with neither nematicide nor insecticide. An analysis of variance for a $2 \times 2$ factorial can be conducted using data for the three checks and the chemical treatment. This permits testing for the separate effects of insecticide and nematicide and interaction of insecticide with nematicide. Similar approaches will be needed for chemicals that control several pathogens. Each check should allow for control of all but one pathogen. In addition, for one check, none of the pathogens should be controlled. This could result in the use of several controls in some cases, all of which are necessary to pinpoint the mechanism of the response to the chemical treatment. In other cases, no controls are needed. In fact, in tests involving airborne pathogens, the controls may be detrimental, because they supply spores that can contaminate the treated plots. In these cases, omitting the controls and using one of the chemical treatments as a standard for comparison is advisable.

**Experimental material**—Treatments are applied to experimental units that collectively are called the experimental material. Examples of experimental units are plots of land, trees, petri dishes, and pots of soil in a greenhouse.

Selection of the experimental units to be used in the experiment depends on the purpose of the experiment. If the experimental units (plots) are to be used as a medium on which treatments are compared, they should be as homogeneous as possible. That is, the factors that might influence response (eg, plant population, soil fertility and moisture, size of plant) are nearly constant. On the other hand, if the objective is to evaluate some property of the experimental units themselves (eg, average size of two different nematodes), less selection and more randomness is needed in choice of units. Selection may narrow the population to which the results apply.

In the field, the units are plots of land. The size and shape of plots must be determined, as well as the need for such items as border rows. These aspects of field plot technique have been described (7,8,11,18). Smith (14) reported some quantitative techniques for estimating optimal plot size from uniformity trial and cost data. He reported that within the range of one-fourth to four times optimal plot size, the efficiency of the experiment was not greatly different. Local conditions often call for modifications of the optimal size estimated by his procedure. The mechanical restrictions imposed by equipment and techniques used in the experimentation often dictate the size and shape of plots. In general, plots in most field experiments should be long and narrow, with the elongation in the direction of the gradient. Plots with this configuration provide minimum error and yet are convenient to handle with row-crop equipment. Plots in fungicide tests for foliar diseases where spores move freely over the borders are an exception. Square plots are preferable in this case, with only the center portion being used for experimental data. Square plots have proportionately less border than do long, narrow ones. The superiority of square plots over long, narrow ones would be greater when dispersal is at normal atmospheric turbulence than when it is at low turbulence.

Van der Plank (17) reported that mutual interference among plots under low air turbulence conditions can be decreased considerably by increasing the plot size. He gave some spore data for low turbulence conditions in which increasing the side of square plots from 1 to 10 m reduced mutual interference by more than half; increasing it from 10 to 100 m reduced mutual interference by more than five-sixths. When air was normally turbulent while spores dispersed, however, increasing the side of square plots from 1 to 100 m did not even halve mutual interference among plots.

Usually experience with a particular crop or pest species or both in one locality helps a researcher to understand the underlying population and the spatial and temporal distribution of the characters being studied. This experience should be useful in devising a suitable plot and a precise experimental arrangement that will provide unbiased estimates of treatment effects and be convenient operationally. The purpose of the experiment is important in plot size and shape determination. Breeding experiments may require completely different plot configurations than do experiments designed to compare the effects of chemical treatments.

Border areas (or guard rows) are used in cases in which the treatment imposed on one plot is expected to influence the adjacent plot. A correct judgment of the degree of this influence is important. Providing insufficient border area can produce interplot interference that may cause representational or cryptic error. An example is the movement of disease from unsprayed to sprayed plots. Information on the sprayed plots is not representative of what farmers who use spray would find, because the sprayed plots have been contaminated with disease and yield less than would uncontaminated plots.

Providing larger border areas than is necessary may require an excessive portion of the experimental area. Using at least 50 to 60% of the experimental area for border area is not uncommon. Even where spore movements may threaten mutual interference among plots, the border area may be reduced by keeping the range of disease small between the best and worst treatments (17). This implies that untreated plots must be omitted and that one chemical should be chosen as a standard for comparison.

Shoemaker (13) described three different designs to be used in experiments involving airborne pathogens. The first, designed for screening and ranking fungicides (protectants) under severe disease pressure, involves use of unsprayed zones bordering the sides of each treatment plot. Including the unsprayed check treatment with the fungicide treatments is optional. For the second design, protected (sprayed) buffer zones are placed between treatment plots and the unsprayed check is completely eliminated. Treatments are either protectants or eradicants. Stricter precautions against interference must be taken when some eradicants are included in the test. Such a design could still have some interference problems if the range in effectiveness of the materials is great. The third design that Shoemaker (13) mentioned involves only treatments that have been selected from preliminary testing and are candidates for recommendation. They do not differ substantially in their effectiveness to control disease. Plots should be large and data should be taken from their center. Unsprayed checks are omitted.

**Experimental design**—Several experimental designs are used extensively in many fields of research. The choice of an appropriate design for a specific situation involves a number of considerations, but the general rule is to keep it as simple as possible. A balanced design in which all treatment combinations have the same number of observations (usually replications) aids analysis and interpretation. Without this equal replication, analysis can be difficult and the resulting estimates poor. Imbalance is not a problem in the simple designs discussed below if the experimental design textbooks are followed. If the experimental plots are sampled and the number of samples varies from plot to plot, however, the balance will be lost and problems will result. Missing data due, for example, to environmental causes and mortality can also cause imbalance.

Given information on the variability pattern of the experimental material, the number and nature of treatments, and the experimental techniques employed, a particular design will usually emerge as the logical choice. The commonly used designs differ mainly in the way in which treatments are *randomly* assigned to the experimental units. Randomization means the process of assigning treatments to plots in such a way that all treatments have an equal chance of being assigned to a particular plot. Randomization provides assurance that a treatment is not continually favored or handicapped in various replications due to some extraneous source of variation, known or unknown. Randomization, like an insurance policy, protects against disturbances that could occur. Tables of random numbers or computers may be used as sources of random numbers. The commonly used experimental designs help to provide for control of known extraneous sources of variation by blocking. The basic difference among these designs is the number of restrictions on randomization, the restrictions being necessary because of the blocking.

With the *randomized complete block (RCB) design,* which is by far the most popular design, one restriction is placed on randomization. A complete set of treatments is randomized within each block (replication). An example of a randomized block design is shown in Fig. 2.2. The object is to block in such a way that although the blocks may differ from one another considerably, the units within each block are relatively uniform. Environmental sources of variation such as moisture and natural fertility of the soil often are a basis for blocking. In other cases, each run of a set of operations is a block, the runs being done at different times. Nearly square, compact blocks are preferred to long, narrow blocks in the field.

In addition to its precision, the RCB design is simple. The treatments are readily assigned to the experimental plots at random, and the field layout and analysis of data are simple. Furthermore, missing plot values may be readily estimated. This design accommodates a wide range of treatments and replications, although practical considerations place limits on both.

The *Latin square (LS) design* provides error control (blocking) in two directions. It is useful for high-precision experiments having from two to ten treatments, because the number of replications must equal the number of treatments. For squares having two or three treatments, several squares are usually needed to provide a reliable estimate of experimental error.

The LS design is not as simple to randomize as the RCB, because it has two restrictions on randomization: each row and each column must contain a complete set of treatments (Fig. 2.3), and the field layout could be more complex. Also, missing data cause more problems than with the RCB design.

The *split-plot (SP) design* is used in cases (i) in which the nature of the experimental material or mechanical aspects of the research require differential plot sizes for various factors or (ii) when more precision is desired to test some effects (subplot factors such as fungicide and interaction) than others (whole-plot factor such as crop variety), or both. It is also used for perennial experiments. In this case, years are the subplots. This design is commonly used in plant science subplots. This design is commonly used in plant science experimentation. In designing SP experiments, the factor requiring a more sensitive test is usually assigned to the subplot. Two separate randomizations are required, one for the whole plot and one for the subplot treatments within each whole plot (Fig. 2.4). If sampling is done within subplots, having constant numbers of samples to maintain balance is important.

The split-plot principle may be extended to more than one split (eg, split–split-plot design). In addition, some variants of the SP design involve stripping of plots for one factor over plots for a second factor (eg, split-block design). An example of a split-block design is shown in Fig. 2.5. These are usually used in cases in which mechanical factors prevent forming smaller plots within the whole plots (eg, spraying in one direction, plowing in another). Consultation with a statistician in planning a complex stripped design should assure that the experimental data can be analyzed.

The *completely randomized (CR) design*, although simple and flexible with regard to the number of replications within each treatment, does not control error variation through blocking. Therefore, it is not precise enough for most field experiments. It is used, however, for laboratory and greenhouse experiments. An example of a completely randomized design is shown in Fig. 2.6.

Cochran and Cox (1) have published an excellent detailed description of the above designs and a discussion of their use. This text also discusses more complicated designs (eg, lattice designs for experiments involving large numbers of treatments). Comprehensive experimental design bibliographies are available (4,5).

**Replication**—Replication assures an estimate of

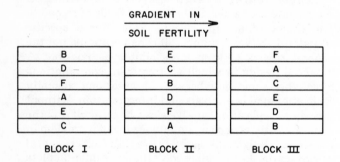

GRADIENT IN SOIL FERTILITY

BLOCK I / BLOCK II / BLOCK III

TREATMENTS: 6 VARIETIES: A,B,C,D,E, AND F
RANDOMLY ASSIGNED WITHIN EACH BLOCK

Fig. 2.2. Example of randomized complete block experiment in field to compare varieties for disease resistance.

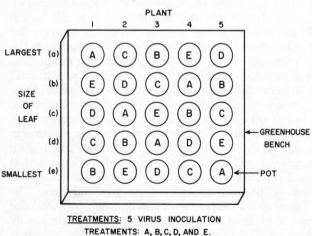

TREATMENTS: 5 VIRUS INOCULATION
TREATMENTS: A, B, C, D, AND E.
ROWS: 5 SIZES OF LEAF ON PLANT:
(a), (b), (c), (d), AND (e).
COLUMNS: 5 DIFFERENT PLANTS:
1, 2, 3, 4, AND 5.

Fig. 2.3. Example of Latin square experiment in greenhouse to compare virus inoculation treatments.

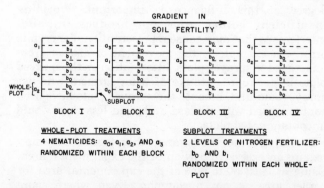

GRADIENT IN SOIL FERTILITY

BLOCK I / BLOCK II / BLOCK III / BLOCK IV

WHOLE-PLOT TREATMENTS
4 NEMATICIDES: $a_0$, $a_1$, $a_2$, AND $a_3$.
RANDOMIZED WITHIN EACH BLOCK

SUBPLOT TREATMENTS
2 LEVELS OF NITROGEN FERTILIZER:
$b_0$ AND $b_1$.
RANDOMIZED WITHIN EACH WHOLE-PLOT

Fig. 2.4. Example of split-plot experiment in field to compare four nematicides (whole-plot treatment) and two levels of nitrogen fertilizer (subplot treatment).

experimental error. Moreover, it is a simple means of increasing precision of estimates of means and the sensitivity of tests of significance. Thus, one is more apt to detect real differences among means. Beyond a certain number of replications, the precision benefits do not offset the cost of another replication. Table 2.1 in Cochran and Cox (1) may be used as a guide in choosing an appropriate number of replications if the coefficient of variation is available from previous experiments. The precision required, the magnitude of differences to be measured, and the significance level must also be specified. The table is accompanied by examples of its use. The amount of land, material, time, and money available for the study usually limits the number of replications. Single-replication experiments have been used in demonstrations to obtain preliminary indications but are not recommended for precise comparisons of treatments.

Repetition of experiments in several locations and years is another form of replication. Conclusions are usually better founded if based on data replicated over space and time. Interaction of treatments with environmental effects can be detected through repetition of this kind. For example, treatments may have a different response pattern in locations 1, 2, and 3 than in locations 4, 5, and 6. The presence of a significant interaction of treatments and sites might suggest that separate recommendations are in order for the two sets of locations.

**Experimental technique**—Statisticians have made one of their more important contributions to research programs by emphasizing that good technique increases the precision of experiments. Some ways of improving technique are to:

i. Write out procedures for conducting various phases of the experiment and a time schedule for their execution.

ii. Make all personnel dealing with the treatments, plots, and data aware of the various sources of error and the need for good technique.

iii. Apply the treatments uniformly.

iv. Exercise sufficient control over external influences so that every treatment produces its effect under controlled, comparable conditions. For example, having each of three men harvest an individual replication is better than having the three harvest all replications as a team. Still better is to have one person harvest all three replications. If it is impossible to control environmental conditions, readings on major environmental variables should be taken and used as covariables. (See sections on **design and randomization in a series of tests** and **supplementary variables**.)

v. Devise suitable unbiased measures of the effects of treatments.

vi. Prevent gross errors.

Statisticians sometimes use data from previous years to help to find changes in technique that might improve precision. In other cases, they design new experiments specifically to investigate ways of improving technique in a particular experiment. One type (preliminary experiments) is established to focus on limited aspects of technique. Another type involves superimposing a sampling and measurement study on existing experimental plots to facilitate estimation of relative sizes of various sources of variation (eg, plant-to-plant and leaf-to-leaf as compared with plot-to-plot variation).

All of the above designs assume proper randomization. Situations involving completely systematic arrangement of treatments are usually to be avoided. An example of improper randomization is the aerial application of fungicide treatments in long strips as shown in Fig. 2.7. A proper randomized arrangement (although more difficult and expensive to perform) is shown in Fig. 2.8. Use of the improper randomization results in an invalid test for treatments and a possible underestimation of experimental error.

Randomization can be a problem in split-plot experiments involving growth chambers. Often not enough chambers are available for replication of the temperature-humidity treatment (whole-plot factor). Changing the setting on temperature-humidity for random assignment of these treatments to various chambers is also difficult. As a result, a valid test for temperature-humidity does not exist. Multiple tests using different temperature-humidities in a chamber

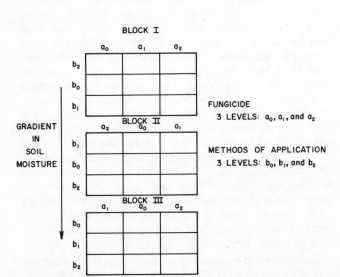

Fig. 2.5. Example of split-block experiment in field in which three levels of fungicide are stripped across three methods of application.

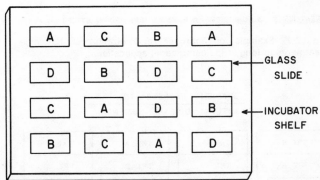

TREATMENTS: 4 FUNGICIDES: A, B, C, AND D APPLIED AS DROPS ON 16 GLASS SLIDES CONTAINING FUNGUS SPORES COMPLETELY AT RANDOM. GERMINATION COUNTS MADE FOLLOWING INCUBATION.

Fig. 2.6. Example of completely randomized experiment having four replications each of four fungicides assigned to glass slides containing fungus spores.

serve as a source of replication for the whole-plot factor, thus alleviating the problem.

The researcher must exercise judgment in implementing the randomization process. In some cases, certain possible randomizations are eliminated, and if one of these is obtained in the randomization process, a new set of random numbers is drawn. For example, if the original randomization of the three treatments in the four blocks of a randomized complete block design turned out to be 1, 2, 3; 1, 2, 3; 1, 2, 3; 1, 2, 3 for plots in the left-to-right positions, it would seem desirable to rerandomize to eliminate the systematic arrangement of the treatments. In other cases, some systematic arrangement of treatments is intentionally built into the experiment. In certain experiments designed for extension purposes, the treatments are arranged systematically in one replication so as to make the best demonstrational trial. For example, fungicide rates might be progressively increased from low to high from one side of the replication to the other. In another type of experiment, nematicides from each company might be placed together in groups within one replication. Treatments are then randomly assigned to the plots within each of the other replications. The increased facility for making visual comparisons of effects offsets the statistical disadvantages of this practice.

**Design and randomization in a series of tests**—Parallel design (common treatments and number of replications) and independent randomization of individual tests in a series of experiments facilitates the combined analysis of variance.

For a series of experiments, statistical models, which

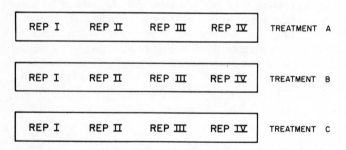

| REP I | REP II | REP III | REP IV | TREATMENT A |

| REP I | REP II | REP III | REP IV | TREATMENT B |

| REP I | REP II | REP III | REP IV | TREATMENT C |

TREATMENTS: 3 FUNGICIDES: A, B, AND C APPLIED IN STRIPS.

Fig. 2.7. Example of improper randomization of fungicide treatments in long strips using aerial application.

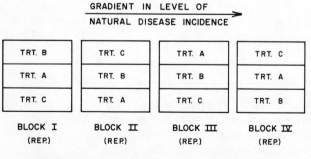

GRADIENT IN LEVEL OF
NATURAL DISEASE INCIDENCE →

| TRT. B | TRT. C | TRT. A | TRT. C |
| TRT. A | TRT. B | TRT. B | TRT. A |
| TRT. C | TRT. A | TRT. C | TRT. B |
| BLOCK I (REP.) | BLOCK II (REP.) | BLOCK III (REP.) | BLOCK IV (REP.) |

TREATMENTS: 3 FUNGICIDES: A, B, AND C
RANDOMLY ASSIGNED WITHIN EACH BLOCK

Fig. 2.8. Example of properly randomized arrangement for treatments of Fig. 2.7.

include all known sources of variation (ie, those due to variables deliberately controlled in the experiment and also uncontrolled variables), have recently been fitted to experimental data. The uncontrolled variables (rainfall, temperature, soil properties) primarily explain site-to-site variability, although they can also account for some intrasite variability. To develop an appropriate model, it is important to recognize those variables (both controlled and uncontrolled) that are related to the response and to measure them at an appropriate time during the experimental process. An example of the use of multiple linear regression for modeling such data that include controlled and uncontrolled variables was given by Laird and Cady (10). Some workers approach the problem using simulation models (systems science), but whether one should draw a sharp distinction between the two approaches is doubtful. In any case, success of modeling data from a group of experiments representing a range of environmental conditions depends on the ability to specify an appropriate model. With such a model, it should be possible to estimate the extent of the interaction between environmental and controlled factors. The model also allows a researcher to (i) synthesize the extant knowledge and understanding of a system, (ii) analyze critically the hypotheses about a system's functioning, (iii) identify fundamental constraints on a system's functioning, (iv) identify mechanisms to which a system's behavior is most sensitive, and (v) identify additional research needs.

**Supplementary variables**—Throughout the course of the experiment, the researcher should try to determine if environmental factors not controlled by the design are affecting the results. Readings on these supplementary variables should be recorded for possible use in statistical control. These variables are called covariables. They are used in an analysis of covariance for increasing the precision of an experiment and adjusting treatment means to a constant level of the covariable (eg, adjusting total plot yield for plant population where stands are uneven).

Estimating some covariables on a rather subjective scale may be necessary (eg, index of wind damage). Even with such subjectivity, precision may be improved.

**Sampling procedures**—Sampling is one of the most useful statistical procedures in nematological research. It may be done to estimate some character for an entire field or an individual plot within an experiment. Plot size for one important character may be larger than needed for supplementary characters. Expense and time may be saved if only a fraction of the plot is used to measure these characters. Also, this allows estimation of relative sizes of experimental and sampling errors, which is useful information for improving techniques for future experiments. For example, for estimating the percentage of plants affected by a disease, only 10 of 100 plants in a plot may be sampled. Yates and Zacopanay (19) studied the loss of information due to sampling relative to complete harvesting of a number of experiments on cereals. At a 6% sampling rate, the loss of information was 31.2%, or about one-third. Twelve percent sampling resulted in an information loss of 18% and 18% sampling resulted in a loss of 13%.

The objective in sampling is to estimate the value that would be obtained if all the individuals in the population were measured. The difference between the sample value and population value constitutes sampling error.

The best sampling procedures are those that give a small sampling error.

The unit in which actual measurement of a character is made is called the sampling unit. Some commonly used units are an insect, a 5-ft segment of a row of tobacco, a plant, a day (sampling time), and a 2-m$^2$ area of a grass plot. A good sampling unit must be easy to identify, easy to measure, and fairly uniform. Although several different units can satisfy the requirements given above, the best is one that will give the highest precision at low cost. Precise estimates are those with low variance. The cost is usually measured in labor, and low-cost estimates are those that save labor.

Once the sampling unit has been determined, consideration should be given to the population being sampled and how these units are distributed in it. How the samples are taken, as well as their number, depends on the distribution of units in the population. Visual observation or actual preliminary samplings may be used to study the distribution of units in the population.

The sampling plan specifies the sampling unit, sampling design, and sample size in measuring a character. Data from previous or ongoing experiments or sample surveys may be useful in developing an appropriate sampling plan. Also, experiments may be specifically designed for developing sampling plans. Gomez and Gomez (8) discussed at length the use of data from previous or ongoing experiments for developing sampling plans, taking their examples from the sampling of certain rice characters. They also described experiments for evaluating sampling plans.

A sampling design defines the manner in which sampling units (eg, sample row segments) are selected from the population (eg, plot). Three common sampling designs will be discussed.

In *simple random sampling,* each of the N population items has an equal chance of being selected in the sample of n sampling units. A simple procedure for taking such a sample is to number the items of the population from 1 to N and select, at random, n numbers in the range of 1 to N from a random number table. If we know nothing about the structure of the population other than its size, we cannot improve on simple random sampling.

In *stratified random sampling,* the population is first divided into s subsections and then a simple random sample of n sampling units is taken from each of these subsections. The subsections are referred to as strata.

Stratified sampling will increase the precision of estimates when there is more variation between units of different strata than within the same stratum. We never lose precision by stratifying, but we may gain considerable precision. Stratification is analogous to blocking in experimental design. The number of samples per stratum may either be constant or variable. Nothing is wrong with using variable size clusters as long as we know the chance of selection in each stratum and can adjust our estimator accordingly. Oversampling the more variable strata and undersampling the less variable strata is profitable. To maximize precision, the sampling fraction of each stratum should be proportional to the square root of the variance in that stratum. If the cost of an observation differs between one stratum and another, we need only replace the word "variance" in the rule given above by the ratio "variance/cost of an observation."

*Multistage or nested sampling* is a sampling design ordinarily used in sample surveys and in many sampling experiments in the biological sciences. For our illustration, we will assume that there are three tiers in the universe. From each of the $n_1$ levels of the top tier, A (eg, rows), $n_2$ random samples of B (eg, plants) are taken. Within each level of B, $n_3$ random samples of C (eg, leaves) are taken. Such a scheme permits estimation of the variance components due to the factors at the various tier levels (ie, rows, plants, and leaves). Considerably more information is available with this method than with a one-stage simple random sampling, and the cost is only a little higher. Using cost and variance estimates for units at the various tier levels, one can calculate the numbers of these units that should be used in future experiments (12).

**Size of sample**—There are several different ways to estimate the size of a sample required, depending on the sampling design and several other considerations. We have just mentioned one in connection with finding optimal sample sizes of various stages within a multistage sampling scheme. Perhaps the simplest case involves estimating a population average from a simple random sample. The investigator must specify both how closely he wishes to estimate the average and the chances he wishes to have of estimating it this closely. For example, assume that the mean number of diseased plants per row of plants is being estimated. The desired estimate is to be within 13 plants of the population average, with 99 chances out of 100 of being right. In other words, we want a 99% confidence interval whose half-length is 13 plants. The half-length of a 99% confidence interval is 2.6 times the standard error of the sample average (ie, Z = 2.6 from a table for the normal distribution). We therefore must fix sample size so that the standard error equals 13/2.6 = 5 plants (ie, $s_{\bar{x}}^2 = 25$ plants). Since the sampling variance is the population variance times (1/n), n must be made equal to the population variance divided by 25, ie, to .04 times the population variance. The estimate of the population variance is obtained either by previous experience, by making a small pilot survey, or by intelligent guesswork. The confidence statement is as follows:

$$\text{Probability}\ (\overline{X} - Z_{.01}\sigma_{\bar{x}} < \mu < \overline{X} + Z_{.01}\sigma_{\bar{x}}) = .99$$

where $\overline{X}$ is the sample mean, $Z_{.01}$ is the 1% tabular value of Z, $\sigma_{\bar{x}}$ is the population standard error of the mean, and $\mu$ is the population mean. The actual calculations proceed as follows:

$$\text{Let}\ Z_{.01}\sigma_{\bar{x}} = 13$$

$$2.6(\sigma/\sqrt{n}) = 13$$

$$(2.6)^2(\sigma^2/n) = 169$$

$$169n = 6.76\ \sigma^2$$

$$n = .04\ \sigma^2$$

Supposing that from previous studies in assessing the extent of disease incidence, it is known that $\sigma^2$ (measured from a number of different rows) is approximately 150. The sample size that should be taken then is (.04) (150) = 6 rows. The principal problem that can occur is that our estimate of $\sigma^2$ may not be accurate.

If one is dealing with attribute data such as proportion of leaves that are diseased, only one parameter, P, specifies the population completely. P for the example above is the population proportion of leaves that are diseased. Because P is usually not known, it is estimated by p, the sample proportion of leaves that are diseased. The estimated sampling variance of p, $p(1-p)/n$ cannot exceed $1/4n$ so that the standard error will not exceed the square root of this.

Thus, if we wish to estimate a proportion, P, to within ±.05 with 95% confidence, we must have .05 equal to two standard errors or a standard error of 0.025. Its square, the sampling variance, will be .000625, and this cannot exceed $1/4n$. Thus, the sample size (n) need be at most $1/([4] [.000625]) = 400$. This will be a sample large enough to achieve the specified aim, whatever the parameter P and whatever the statistic p.

Graybill and Kneebone (9) gave two procedures that are useful in determining sample size based on a knowledge of the coefficient of variation of the data, the acceptable length of confidence interval expressed as a percentage of the mean, and the risk level of type I error that one is willing to accept.

Federer (3) discussed a method for determining the number of samples that should be taken when sampling within experimental plots. Not only the sample size but also the method of sampling needs to be determined. Plant materials within a plot are often systematically sampled, although the first plant sampled is chosen at random. Thus, every fifth plant in a row of corn might be taken to constitute the sample. This method of sampling is more convenient and timesaving than sampling completely at random, and it does provide a representative sample of the entire plot.

### The Data and Their Analysis

**Data**—The data reflect not only treatment effects but also variation due to a number of other causes, known and unknown. A careful study of the data and appropriate analysis are necessary to separate the treatment effects (important) from the random effects (of little interest).

**Accuracy of data**—Two problems that continually arise in handling and reporting data are carrying more digits than are biologically measurable and accurate and, on the other hand, rounding data too much before analysis. The former can give the reader a false impression that the data were measured with a great deal of accuracy. A good discussion of the number of figures that should be carried in data is given in chapter 3 of Cochran and Cox (1). The rounding errors can cause

some extremely erroneous results in some regression analyses. The general rule when a series of calculations is to be performed is to round only after all the calculations have been performed.

**Data variation patterns**—Data variation patterns should be studied carefully prior to analysis of variance to find unusual values that might bias the conclusions about treatment effects. An easy way to identify outliers (unusual values) is to obtain the range in values for the various replications of each treatment. Extremely large or small ranges for one or two treatments indicate the presence of at least one outlier per treatment. In Table 2.1, the unusually large range of 12 for treatment 3 leads to further checking of 14 as a reading for treatment 3, block 2. The number 14 also would have been identified as unusual had the treatments within each block been ranked (from highest to lowest). The rankings for all other blocks are $1 > 2 > 3 \geqslant 4$, whereas that for block 2 is $3 > 2 > 4$. The rank ordering also serves a check for the possibility of a block times treatment interaction.

From 5 to 10% of the numbers in many data sets have been estimated to be in error due, for example, to misreading of instruments, copying errors, misplaced decimals, misidentification of treatment, and replication numbers. Other numbers reflect unusual environmental conditions that caused the biases in the numbers but that were not observed or recorded or both during the experiment. This emphasizes the need to take notes on unusual plot conditions such as flooding or wind damage. This will permit use of statistical techniques (eg, analysis of covariance or missing plot techniques) when aberrant observations are later identified in the data.

**Nonhomogeneity of data**—Many sets of data are obviously not homogeneous. Some treatments may have all zero values. Common sense and good judgment should be used in excluding data that have zero (or otherwise different) variance from the remainder of the data. Sometimes the untreated check should be omitted from the overall analysis of variance, because it has a higher or lower variance than the remainder of the treatments. Biological reasons often exist for expecting this variance to be different. A statistician should be consulted if any question arises about the homogeneity of variance in a data set. Otherwise, the analysis of variance may not reflect a true picture of the data if the error heterogeneity problem exists and is not rectified.

**Percentage data**—If the choice arises between conducting an analysis of variance on actual numbers of nematodes controlled and percentage of control, analyzing the actual numbers is usually better. Percentages based on widely varying denominators are apt to have different variances. A good way to handle the problem is to analyze and report the actual numbers controlled, and then include in the table of means percentage control figures calculated from means of the control data. No standard error would be reported with the percentage of control figures. In cases in which percentages are all based on the same divisor, percentages may be used in analysis, because conversion to percentages is equivalent to data coding.

**Data transformation**—Analysis of variance requires normality and equality of variances of the data. A slight departure from these requirements need not invalidate the results of analysis of variance, but drastic departure may require corrective action. This corrective

**TABLE 2.1. Use of Range to Inspect Data for Outliers**

| Treatment | Block 1 | Block 2 | Block 3 | Block 4 | Range |
|-----------|---|----|---|----|-------|
| 1 | 4[a] | 6 | 8 | 10 | 6 |
| 2 | 3 | 5 | 7 | 9 | 6 |
| 3 | 2 | 14 | 6 | 8 | 12 |
| 4 | 1 | 3 | 5 | 8 | 7 |

[a]Hypothetical data from randomized complete block experiment with four treatments and four blocks.

action usually consists of transforming the data from the original scale to a new one. The process is called transformation. Researchers should consult a statistician on questions concerning the need for data transformation and choice of an appropriate one. Common transformations for normalizing data and equalizing variances are angular or inverse sine transformation, square root transformation, and logarithmic transformation. For the angular transformation, the transformed data are angles ranging from 0 to 90 degrees that correspond to numbers in the original scale ranging from 0 to 1. For the square root transformation, each number in the transformed scale is the square root of a number in the original scale. When using a logarithmic transformation, the numbers that are analyzed are the logarithms of numbers on the original scale. Theoretical variances have been worked out for data on the various transformed scales, and these act as a basis for comparison to determine if the appropriate transformation has been made. Steel and Torrie (16) have discussed data transformation in detail.

The disadvantage of using a transformation is that the comparisons are made and reported on a scale that may not be familiar to the reader (eg, log, square root). Converting the means and standard errors back to the original scale for reporting and comparison purposes is not really valid. When transformations are made, common practice is to perform an analysis of variance on the original data as well as on the transformed data. Over a period of years, I have found a good correspondence between the results obtained in both cases from an interpretational viewpoint except in data sets having extreme deviations from the statistical assumptions.

**Analyses of data**—Subjecting data to statistical analyses before reporting them has several advantages. First, analysis is often necessary to separate the treatment effects from the random effects. Second, statistical analyses have the effect of reducing the bulk of the data so that their essential features may be presented in a concise fashion. Third, analyses provide standard errors that give a measure of precision of the experiment that generated the data. Statisticians now place more emphasis on estimating effects than on routine performance of tests of hypotheses. For example, it is not particularly informative to state that rates overall are significant in a nematicide rate test that had 0, 25, 50, and 75 kg/ha as treatments. Neither is it particularly helpful that the mean for 75 kg/ha is not significantly different from that for 50, but that the mean for 50 is significantly higher than the mean for 25, which in turn is significantly higher than the check mean. A more informative approach is to choose a realistic response model such as the quadratic polynomial ($\hat{Y} = b_0 + b_1 [Rate] + b_{11} [Rate]^2$) and find the parameter estimates ($b_0$, $b_1$, and $b_{11}$) through appropriate statistical techniques. Useful interpretations such as the anticipated yield without nematicide (ie, $b_0$), the initial response to nematicide (ie, $b_1$), and the optimal rate of nematicide to apply for maximum yield (ie, $-b_1/2b_{11}$) would be available. Another example is the comparison of two chemical fungicides. We are more interested in estimating the magnitude of their mean yield difference than whether this difference is statistically significant at some preselected probability level. Thus, we develop methods that provide unbiased estimates of the magnitude of the difference, because it is the basis for recommendations and economic analysis.

Specific analyses used in nematicide and fungicide tests are:

*Student's t-test*—This test is appropriate for a simple comparison of the means of two groups.

*Chi-square tests*—Applications of chi-square tests in nematological work are numerous. They are used with count data to test independence of factors. Chi-square is also used in breeding work to test hypotheses of specific genetic ratios. Chapters 17, 18, and 19 of Steel and Torrie (16) provide further information on the uses of chi-square.

*Analyses of variance*—Methods of analyses of variance for commonly used designs are reported in several texts (2,3,6,15,16) and should be routine to most researchers. This is probably the most common method of analysis of data.

*Regression*—This is used in survey or observational studies to relate the variation in a dependent variable to variation in one or more independent variables.

**Methods of computation**—Electronic computers perform routine calculations used in statistical analyses accurately and efficiently. They are invaluable in cases in which the designs are complicated or large numbers of characters measured at one or more locations need to be analyzed. At large computing centers, statistical system packages (eg, Biomedical Package [BMD], General Statistical Package [GENSTAT], International Mathematical and Statistical Libraries [IMSL], Statistical Analysis System [SAS], and Statistical Package for the Social Sciences [SPSS]) may be used with a minimum knowledge of programming and instructions to the computer. At small facilities, the researcher may need to devote considerable effort to writing computer programs for his own needs, and these will need to be checked for accuracy before routine use. If the computer is used, the researcher must study the data carefully before analysis, check for keypunch and programming errors, and understand the analysis that the computer is performing. Too often, data are placed on cards and analyzed on the computer before their patterns have been studied carefully.

In some cases, results from a more leisurely desk calculator analysis, which permits close scrutiny of the data, may be better. The disadvantages of desk calculator analysis are that the time and labor involved are usually considerably greater, and chance of error is greater than with a computer. If analyses are to be made solely on a desk calculator, an independent operator should check them.

## Interpretation and Reporting of Results

**Interpretation**—Interpretation of experimental results is one of the most important phases. It has theoretical aspects that are based on the mathematical laws of probability. Experimental observations are limited experiences that are carefully planned in advance and designed to form a secure basis of new knowledge. In the interpretation phase, the results of these planned experiences are extrapolated to the underlying population (eg, a five-county area). A degree of uncertainty is associated with this induction, but probabilities may be attached to our uncertainties.

Examples are those probabilities associated with the calculated F and t statistics in tests of significance.

Interpretation of the results of statistical analyses is also an art. Because the statistical methods books do not treat data interpretation in depth, some individuals have developed an intuitive faculty for this interpretation. This would account for the fact that two researchers might interpret the analysis of a set of data in different ways. Whether an intuitive or a more structured approach is used, much of the interpretation process involves careful study of variation patterns in the data. The results of statistical analyses and the conclusions should reflect trends seen in the data before the analysis is made.

The actual techniques used in interpretation vary with the purposes of the experiment and the nature of the treatments. Interpretational techniques for a few experimental situations are discussed below.

**Multiple comparisons**—Planned orthogonal comparisons are usually made if they make sense biologically. Orthogonal comparisons mean that the sums of squares of t minus 1 independent comparisons add to the total sum of squares for treatment with t minus 1 degrees of freedom. The biological basis for choosing a set of comparisons overrides the desire to keep all comparisons orthogonal. Selection of comparisons that are of interest from a biological viewpoint can usually best be accomplished through the cooperative efforts of a researcher and a statistician. Once a set of meaningful comparisons is derived, the statistician's responsibility is to provide standard errors

or multiple comparison criteria for the relevant comparisons.

Because of the confusion about the appropriate situations in which various multiple comparison procedures should be used, a brief description of four commonly used procedures is given below.

*Least significant difference (LSD)*—Because of its simplicity, the LSD is a widely used comparison procedure. Functionally, it requires small differences for significance and therefore has a high probability of rejecting false hypotheses. Errors are expressed as a percentage of the total number of comparisons made. The LSD has a fixed range, meaning that the same criterion is used for making all comparisons regardless of how close the means are with regard to their ranks. It is a poor test for comparing the largest and smallest means. Operationally, the LSD should be used only when treatments overall is significant.

*Tukey's-w procedure*—This fixed-range test is used in the same manner as the LSD, but requires larger differences between means for significance. Its rate of finding differences between means when they really do not exist is expressed on the basis of a percentage of experiments rather than a per-comparison basis as used in the LSD.

*Dunnett's procedure*—This is appropriate for comparing the mean of the untreated check with each of the other means individually (16).

*Duncan's new multiple range test*—When virtually no basis is available for making logical comparisons or subdividing treatments into groups biologically (eg, crop variety trials), multiple comparison procedures such as Duncan's new multiple range test (or Student-Newman-Keuls procedure) may be used. This is the only situation in which use of these multiple comparison procedures is valid. This explains why the LSD and Duncan's new multiple range test cannot be used interchangeably.

**Interpretation of factorial experiments**—Main effects and interactions should be tested as part of the analysis of data from factorial experiments. In addition, other aspects of the analysis of data from factorial experiments depend on whether the factors are quantitative or qualitative. If the treatments are a series of levels of a quantitative factor (eg, dosage of fungicide), curve fitting would show the overall relationship between the response means and the levels. A statistician should be asked to provide standard errors for the specific contrasts that are made in curve

**TABLE 2.2. Example of Complicated Table With Missing Interpretational Information**

| Days After Inoculation | Inoculum Levels (Egg Masses) | | | | |
|---|---|---|---|---|---|
| | 0 | 5 | 10 | 20 | 40 |
| | Mean Top Weights (g) | | | | |
| 7 | 2.33 | 2.51 | 3.18 | 2.49 | 7.36 |
| 14 | 6.30 | 11.69 | 8.21 | 11.72 | 9.51 |
| 21 | 22.29 | 14.30 | 13.58 | 14.14 | 15.63 |
| 42 | 24.84 | 25.62 | 20.74 | 25.69 | 21.34 |
| Inoculum level means | 13.94 | 13.53 | 11.43 | 13.51 | 13.46 |
| $LSD_{.05} = 7.11$ | | | | | |

**TABLE 2.3. Typical Well-Constructed Table**

| Treatment (Variety) | Disease Index[a] | Arc Sine Percent Defoliation[b] | Yield (ton/ha) | | | | Value/ha ($)[c] | Arc Sine Percent No. 1 Spotted[b] |
|---|---|---|---|---|---|---|---|---|
| | | | No. 1 | No. 2 | Culls | Total | | |
| Ohio No. 1 | 4.4 | 80 | 0.2 | 0.3 | 1.2 | 1.7 | 70 | 20 |
| Redwing | 3.3 | 42 | 2.1 | 1.8 | 1.1 | 5.0 | 600 | 5 |
| Golden Nugget | 1.8 | 11 | 2.6 | 2.2 | 1.3 | 6.1 | 740 | 2 |
| $LSD_{.05}$[d] | 1.2 | 21 | 0.4 | 0.9 | NS[e] | 1.1 | 100 | 12 |
| $LSD_{.01}$ | NS | 30 | 0.6 | 1.3 | NS | 1.5 | 150 | 20 |

[a] 1 = No evidence of disease, 5 = dead plant.
[b] 100 = Complete, 0 = none.
[c] Value of grades/ton: No. 1 = $200, No. 2 = $100.
[d] LSD = least significant difference.
[e] NS = not significant.

fitting (eg, linear, quadratic). Use of LSD or Duncan's new multiple range tests does not make sense in this case.

If the treatments are primarily qualitative or combinations of levels of quantitative and qualitative factors (but not factorially arranged), detailed information about the treatments should be used in constructing comparisons among means.

**Reporting of results**—The description of materials and methods should include all statistical aspects of the test, such as experimental design, number of replications, plot size, and data transformation.

An example of a statement describing the statistical aspects of a test as it would appear in the materials and methods section of a report is as follows: "Plots were single rows 15 m long on 100-cm centers; the design was a randomized complete block with three treatments and six replications. The percentage of disease and defoliation data were transformed to arc sine prior to analysis."

The most common error in describing experimental design and techniques is omitting important details. Few articles are specific enough with respect to techniques used. This makes interpreting the results difficult for the reader. A statistician should be consulted if there is any question as to how the description of the statistical aspects should be worded.

Most journals accept tables of means but not analysis of variance tables. Consequently, the author of a technical article must interpret the data by techniques, some of which have been described in this paper, and then summarize the interpretations in the narrative portion of the paper or report. Tables of means, accompanied by standard errors or multiple comparison criteria, should also be reported. These tables should be kept simple but clearly annotated. One of the most common faults in table preparation is attempting to report too much information (on too many factors) in one table. In complicated tables, the reader is often bewildered about which comparisons are valid with the multiple comparison criterion reported (eg, LSD).

To illustrate this problem, Table 2.2 summarizes results of a $4 \times 5$ factorial experiment of time by inoculum levels. How the single LSD should be used for comparing means is not clear to the reader. Both main effect and interaction means are reported, but there is only one LSD. To correct the problem, an LSD for two inoculum level means as well as one for two date means at the same inoculum level should be reported and labeled appropriately.

Table 2.3 is a simple, yet well-constructed table for a research report. The NS was used in place of an LSD in cases in which treatments overall was not significant at a specified level. This is to protect against error in statistical inferences and also to give the reader information on the results of the test of treatments. One variable (disease index) was significant at the 5% level but not at the 1% level. The footnoting is adequate and does not leave unanswered questions about the variables and the data.

## Summary

The logical statistical principles that apply to the design of experiments, experimental technique, and the collection and analysis of data were reviewed, and suggestions as to how experiments might be improved were made. The importance of involving a statistician in the planning phase of an experiment was emphasized. Detailed and explicit written plans should be prepared before an experiment is executed. Experimental designs should be as simple as possible to control error variation. Plots should be long and narrow unless interplot competition is a problem. In that case, they should be square. Blocks in a randomized block design should be square or nearly square, unless another shape maximizes interblock variation and minimizes intrablock variation. Data evaluation and interpretation involve careful scrutiny of data that desk or pocket calculator analysis permit. The advantages of electronic computers for analysis are efficiency and accuracy. Data analytic technique should be in keeping with the experimental design, data variation patterns, and knowledge of the concomitant variables that affect the character being studied. An overall test of treatments is usually just the first step in comparing treatments. Single-degree-of-freedom orthogonal comparisons are desirable if they make sense biologically. The results can then be reported in the text portion of the report. Tabular material in reports, on the other hand, usually consists only of means and their standard errors (or multiple comparison criteria). The appropriate multiple comparison procedure depends on the nature of the treatments and the comparisons among them that are required. The least significant difference (LSD) is a procedure for making planned comparisons that are suggested by the factor combinations comprising the treatments. On the other hand, Duncan's new multiple range test is valid only for cases in which little information about the treatments is available; thus, no definite planned comparisons are suggested. Duncan's new multiple range test has been widely used for making all possible comparisons among varieties in variety trials.

A well-written research report should contain in its methods section a detailed description of the experimental design, experimental technique, and methods used in the analysis of the data.

Tables of means should be simple yet well annotated. Attempts to include too much information in a single table usually cause the reader interpretational difficulties.

## Literature Cited

1. COCHRAN, W. G., and G. M. COX. 1962. Experimental designs. Ed. 2. Wiley, New York.
2. COX, D. R. 1958. Planning of experiments. Wiley, New York.
3. FEDERER, W. T. 1955. Experimental design, theory and application. Macmillan & Co., New York.
4. FEDERER, W. T., and L. N. BALAAM. 1972. Bibliography on experiment and treatment design pre-1968. Oliver and Boyd, Edinburgh.
5. FEDERER, W. T., and A. J. FEDERER. 1973. A study of statistical design publications from 1968 through 1971. Amer. Statist. 27: 160-163.
6. FISHER, R. A. 1951. The design of experiments. Ed. 6. Hafner, New York.
7. GOMEZ, K. A. 1972. Techniques for field experiments with rice: layout, sampling, sources of error. International Rice Research Institute. Los Banos, The Philippines.
8. GOMEZ, K. A., and A. A. GOMEZ. 1976. Statistical procedures for agricultural research. The International Rice Research Institute, Los Banos, The Philippines.

9. GRAYBILL, F. A., and W. R. KNEEBONE. 1959. Determining minimum populations for initial evaluation of breeding material. Agron. J. 51:4-6.

10. LAIRD, R. J., and F. B. CADY. 1969. Combined analysis of yield data from fertilizer experiments. Agron. J. 61:829-834.

11. LECLERG, E. L., W. H. LEONARD, and A. G. CLARK. 1962. Field plot technique. Ed 2. Burgess Publishing, Minneapolis.

12. MARCUSE, S. 1949. Optimum allocation and variance components in nested sampling with an application to chemical analysis. Biometrics 5:189-207.

13. SHOEMAKER, P. B. 1974. Fungicide testing: Some epidemiological and statistical considerations. Fungicide Nematicide Tests 29:1-3.

14. SMITH, H. F. 1938. An empirical law describing heterogeneity in the yields of agricultural crops. J. Agr. Sci. 28:1-23.

15. SNEDECOR, G. W., and COCHRAN, W. G. 1967. Statistical methods. Ed. 6. Iowa State University Press, Ames.

16. STEEL, R. G. D., and TORRIE, J. H. 1960. Principles and procedures of statistics. McGraw-Hill, New York.

17. VAN DER PLANK, J. E. 1963. Plant diseases: Epidemics and control. Academic Press, New York.

18. WISHART, M. A., and H. G. SANDERS. 1958. Principles and practice of field experimentation. Ed. 2. Technical Communication 18. Commonwealth Bureau of Plant Breeding and Genetics, Cambridge.

19. YATES, F., and I. ZACOPANAY. 1935. The estimation of the efficiency of sampling, with special reference to sampling for yield in cereal experiments. J. Agr. Sci. 25:545-577.

# SECTION II.

# Laboratory and Greenhouse Procedures

## 3. A Cellophane-Transfer Bioassay to Detect Fungicides

### Dan Neely

Plant pathologist, Section of Botany and Plant Pathology, Illinois Natural History Survey, Urbana, IL 61801

To be effective, protective fungicides must be sufficiently concentrated at an inoculation site to prevent infection by pathogenic fungi. The cellophane-transfer bioassay described in this chapter has been used to detect both protective foliar fungicides and systemic fungicides.

Bioassays offer advantages and disadvantages when compared with chemical procedures for detection of fungicides and fungitoxicity. The advantages include uniform experimental conditions for dissimilar fungicides, simplicity, speed, economy, and reproducibility. The disadvantages are that they measure chemical presence indirectly as the effect on living organisms, the test organism often is not a pathogen or is a pathogen different from the target organism, and the measurements of toxicity or inhibition are seldom standardized.

The cellophane-transfer technique has been used to determine fungistatic and fungicidal concentrations of fungicides in vitro (5), to measure fungicides deposited on leaves of crop plants and their resistance to artificial rainfall in the greenhouse (4), to determine the persistence of protective foliar fungicides when exposed to weathering and degradation agents in the field (3), and to measure the translocation of fungicides introduced into the sap stream of higher plants in the greenhouse and field (2).

### Technique

The cellophane (DuPont RGPD) used is 23 nm thick, transparent, and readily absorbent. It is air, dust, grease, and oil resistant. It is impermeable to dry gases but permeable to moist gases in proportion to their solubility in water.

Disks (5-mm diameter) were cut from folded sheets of the cellophane with a paper punch, placed in distilled water, and sterilized for 5 min in boiling water. The water was decanted, and the disks were placed on moist filter paper in a culture plate (9-cm diameter) and separated with forceps. Individual cellophane disks were transferred with forceps from the culture plate when needed.

Many fungi have been used successfully as test organisms. The eight plant pathogens used in these studies were *Stemphyllium sarcinaeforme* (Cav.) Wilts., *Curvularia trifolii* Parmelee and Luttrell, and *Alternaria* sp. (large spores); *Monilinia fructicola* (Wint.) Honey, (moderately large spores); *Fusarium moniliforme* Sheld., *Ceratocystis ulmi* (Buism.) C. Moreau, *Verticillium albo-atrum* Reinke and Berth.; and *Ceratocystis fagacearum* (Bretz) Hunt (small spores). The test fungi were grown in 9-cm culture plates on potato-dextrose agar (PDA) prepared from raw potatoes, at 25°C for 14, 21, or 28 days, depending upon the maturation of conidia. The colonies were flooded with sterile distilled water and agitated with a rubber policeman. The spore suspension was decanted into Erlenmeyer flasks, and the spore concentration was adjusted through dilution to these standards: large spores, 20 per field at 100× magnification; moderately large spores, 5 per field at 400×; small spores, 10 per field at 400×.

The spore suspensions were transferred to the cellophane disks with glass capillary tubing having a 0.85-mm inside diameter. Sufficient spore suspension was drawn into the tube to seed 18 disks. Each cellophane disk was lightly touched with the end of the capillary tube, depositing approximately 0.4 μl of the spore suspension.

**Fungicidal and fungistatic properties of chemicals**—A 100-ml aqueous stock solution or suspension of each test chemical was prepared, containing 1 g of active ingredient. All solutions or suspensions were prepared with deionized water passed through a 120-cm ion exchange column. Eight test concentrations, 10 ml each, were prepared in a dilution series of 500, 100, 20, 5, 1, 0.2, 0.04, and 0.008 μg/ml. Six concentrations (500 to 0.2 μg/ml) were used to test fungicidal properties and another six (20 to 0.008 μg/ml) to test fungistatic properties.

Coors white porcelain spot plates, each 112 mm long and 92 mm wide with 12 depressions 5 mm deep, contained the test concentrations. One filter paper disk (Schleicher and Schuell Co., No. 740E), 12.7-mm diameter, was placed into each depression. Drops of the desired test concentration were added to the filter-paper disk using a 5-ml pipette until a meniscus formed between the disk and the side of the depression in the plate. The disk was saturated with, but not floating in, liquid. The six concentrations for testing fungistatic activity were placed in order into six depressions on one half of the spot plate, and the six for testing fungicidal activity in the other six depressions. One spot plate was used for each test fungus.

Single cellophane disks were transferred with forceps from the culture plate and placed on top of each filter paper disk in the spot plate. Three cellophane disks were placed on each paper disk, beginning with the lowest and proceeding to the highest concentration. The forceps were washed at the end of each dilution series.

The spot plates were then stacked into a moist chamber. Glass desiccators or plastic containers lined with moist paper toweling served as moist chambers. An empty spot plate was placed on top of each stack. The cellophane disks were seeded with spore suspensions of the test fungus using a capillary tube as described earlier, and incubated at 25°C.

Two hours after seeding, the cellophane disks from the fungicidal portion of the spot plate were lifted with forceps, one at a time without inversion, and placed in lines on PDA in culture plates. The 18 disks from the six test concentrations were placed in one culture plate and incubated at 25°C. After the cellophane disks from the fungicidal portion of the plate were removed, the spot plate was returned to the moist chamber.

The fungistatic concentrations of each chemical were determined 20–24 hr after the cellophane disks were seeded. The cellophane disks were transferred from the filter paper to microscope slides, placed in six rows of three each, and covered with two 18-mm square cover slips. A drop of water was placed at the margin of each cover slip, and the fungus spores on each cellophane disk were examined. When 99% or more of the spores had not germinated or had germ tubes that were less than half the spore length, the test concentration was considered fungistatic.

The fungicidal concentrations of each chemical were determined after four days. If fungus growth was observed on any of the three disks, the test concentration was considered nonfungicidal; if no growth was observed, the test concentration was considered fungicidal. To determine precisely the fungicidal or fungistatic concentration, a second series was prepared using narrower concentration gradients.

**Deposition and tenacity of foliar protectant fungicides**—After fungistatic concentrations of test chemicals were determined, fungicides were applied to leaf surfaces to determine the fungicide dilution that would prevent germination of 99% ($ED_{99}$) of the conidia. The highest concentration tested was that which is commercially recommended for disease control, often 2.4 g/L (2 lb/100 gal). This concentration was prepared in tap water and diluted with distilled water to 50, 10, 4,

and 2% of the original concentration. Excised leaves or leaflets from the test plant were immersed in 400 ml of the test dilution, swirled for 5 sec, removed, and stored temporarily in a 9-cm culture plate.

Filter paper disks were placed in the depressions of Coors porcelain spot plates. Three drops of distilled water were added to each filter paper disk. A 1-cm square section of the treated leaf was placed on the filter paper disk. A cellophane disk, cut and sterilized as previously described, was placed on the treated leaf. A conidial suspension of the test organism was placed on the cellophane with a capillary tube. The spot plates were stacked in a moist chamber and incubated at 25°C for two days. The cellophane disks were then transferred to slides, and the spore germination determined. The deposition test was done four times.

The tenacity study tested the ability of fungistatic concentrations on leaves to withstand exposure to artificial rainfall. A 1-L suspension of each fungicide was prepared at a commercial concentration. An excised twig of a test plant was immersed in a fungicide mixture, swirled for 5 sec, removed, and held upright by inserting the cut stem in a flask containing distilled water. The fungicide on the leaves was allowed to dry for 1.5–2 hr.

Artificial rain was produced using a machine developed at the University of Illinois (1). Treated and nontreated plants in flasks were placed on a revolving tray in the raindrop machine and exposed to specific quantities of rainfall. All plants were exposed to 0.0, 2.5, 5.0, and 7.5 cm of rainfall. Fungicides that persisted after exposure to 7.5 cm of rainfall were applied to plants and exposed to 0, 5, 10, 15, 20, and 25 cm of rainfall. Fungicides persisting after exposure to 25 cm of rainfall were again applied to plants and exposed to 0.0, 7.5, 15.0, 22.5, 30.0, 37.5, 45.0, 52.5, and 60.0 cm of rainfall. After exposure to a specified quantity of rainfall, one leaflet or leaf section was removed from each plant and stored temporarily in a culture plate. The bioassay of leaf tissues then proceeded as previously described for deposition studies. The tenacity tests were done four times.

**Persistence of foliar protective fungicides**—Under field conditions, fungicides on leaves may be decomposed by various chemical, physical, and biological agents. The cellophane-transfer bioassay technique was used to detect fungicide residues on plants with differing leaf surface characteristics at varying times after field application.

After the fungistatic concentration ($ED_{99}$) was determined, a fungicide concentration common in commercial use, often 2.4 g/L (2 lb/100 gal), was prepared by mixing the fungicide in tap water. One twig of each test plant possessing mature leaves was immersed in the test fungicide, agitated slightly, and removed. One day after treatment and at weekly intervals thereafter, three disks of leaf tissue (9-mm diameter) were removed from a mature leaf with a paper punch. Disks were removed from the same leaf each time. The leaf disks were placed on Coors porcelain spot plates containing filter paper pads moistened with distilled water. A cellophane disk was placed on each leaf disk and seeded with a conidial suspension of the test organism. After 40 hr of incubation in a moist chamber at 24°C, the cellophane disks were transferred to microscope slides and examined for spore germination. The bioassay was terminated when the residual fungicide failed to inhibit 99% or more of conidial germination for two consecutive weeks.

**Translocation of systemic fungicides**—The cellophane-transfer technique was used to detect the presence of fungistatic concentrations of chemicals in the stem, branches, and leaf petioles of plants treated with fungicides through stem implantation, stem injection, root injection, and soil injection. To determine translocation in excised twigs of woody plants, the procedure described below was used.

Stock solutions or suspensions of 10,000 $\mu$g/ml of active ingredient were prepared for all compounds. Dilution to 500 $\mu$g/ml was completed on the day the test series began. Terminal twigs, 60–75 cm long with uniform diameter, were excised from field-grown trees, and the cut ends immersed in distilled water during transport to the laboratory. The twigs were pruned to 45–60 cm. The cut end was placed in a container with 100 ml of the test chemical. After 24 hr, the leaves were removed and three 1-cm long sections were cut from twigs 24–30 cm above the original level of the test chemical. The twig sections were placed in a quadrant-type culture plate containing distilled water and thick filter paper (Whatman No. 3). Closed forceps were used to punch three holes in the filter paper for each quadrant, and the three twig sections were placed upright in the holes. Four compounds were thus assayed in each culture plate.

A sterile disk of cellophane prepared as previously described was placed on top of each twig section. A spore suspension of the test organism was placed on the cellophane with a glass capillary tube, the culture plate cover replaced, and the plates incubated at 24°C for 24 hr. The cellophane disks were then transferred to microscope slides and examined for spore germination. If cellophane disks contained less than 1% of germinating spores (ED99), results were recorded as indicative of positive chemical translocation. Cellophane disks containing germinated spores but with germ tubes less than three times the length of the spore indicated probable chemical translocation. Cellophane disks containing mycelial colonies indicated lack of chemical translocation.

## Results

The cellophane-transfer bioassay technique is more sensitive than the seeded-agar bioassay technique in detecting fungistatic concentrations of chemicals on or in plants. In one test with 50 organic and 22 inorganic compounds, 41 were translocated through woody stem sections (2). At the concentration used (500 $\mu$g/ml) and with the eight woody species selected, the test materials were detected in 224 combinations by the cellophane-transfer and in only 95 combinations by the seeded-agar bioassays.

In tests to determine fungistatic and fungicidal concentrations, the small-spore fungi were more sensitive to broad-spectrum fungicides than were the moderately large- and large-spore fungi (5). Frequently the small spores failed to germinate in aqueous dilutions of chemicals one-fifth as concentrated as those in which the moderately large and large spores germinated. All commercial fungicides were fungistatic (ED99) at 20 $\mu$g/ml or less, many at 1 $\mu$g/ml, and a few at 0.04 $\mu$g/ml. Most commercial fungicides were fungicidal at 500 $\mu$g/ml or less, many at 20 $\mu$g/ml, and a few at 1 $\mu$g/ml. For some fungicides, the fungicidal concentration was 20 times that of the fungistatic concentration—a broad fungistatic range. For others, the range was much smaller; for dodine, the range was extremely narrow.

The specificity or the lack of toxicity of certain chemicals to certain fungi was evident. Cycloheximide was not toxic to *Ceratocystis ulmi,* for example. The presence of cellophane between fungus spores and a culture medium had little or no effect on the percentage of spore germination, although germ tube length was reduced by approximately 25%.

The cellophane-transfer technique was used successfully to determine the fungistatic concentrations deposited on the leaves of test plants (4). Most commercial fungicides effectively inhibited spore germination (ED99) on leaves when applied at one-fifth of the concentration commonly used in the field. Some fungicides inhibited spore germination at one-fiftieth of the field concentration. The deposition of chemicals on test plants varied inversely with the amount of pubescence on the leaf surface.

The tenacity of the deposit on leaves varied with the amount of rainfall (4). Many fungicides did not remain in fungistatic concentrations on the test plants following 2.5 cm of rainfall, while several remained after 30 cm. Only 2 of the 19 commercial fungicides, captafol and dithianon, persisted through 60 cm of rainfall on all test plants and prevented spore germination (ED99) of all test fungi. The tenacity of the fungicides usually varied directly with deposition. When the initial deposit was not more than twice the fungistatic deposit, the first few centimeters of rainfall usually reduced the deposit to less than a fungistatic concentration. Leaf pubescence affected fungicide tenacity. On pubescent leaves, less fungicide was deposited, but the deposit was more difficult to remove in simulated rainfall.

The cellophane-transfer bioassay detected the persistent residues of some fungicides that were sufficient to prevent germination of 99% or more of the conidia three months after application to plants in the field (3). Although a few commercial fungicides persisted in the field through weathering and degradation for less than one day, most persisted for one to three weeks and a few persisted for one to three months.

## Discussion

The cellophane-transfer technique can be readily adapted to determine the toxicity of a fungicide, the quantity deposited on the leaf surface, the tenacity to the leaf surface, and the resistance to weathering and degradation in the field. Comparable data from a number of dissimilar fungicides can be obtained rapidly. The technique can assist investigators in selecting fungicides to control diseases in the field. It can also be used to assist in choosing rates and frequency of application.

The cellophane-transfer technique is simple, quick, and gives reproducible results. No complicated equipment or costly chemicals are required, and most tests are completed within two days. Tests can be designed with sufficient replications for statistical analysis.

The cellophane-transfer technique is sensitive to

chemical concentration changes. The resulting fungistatic concentrations may appear higher than do those obtained using other procedures. The data, however, are based on an $ED_{99}$ and should be compared only with other $ED_{99}$ determinations. The cellophane is thin, the diffusion distance from the toxin to the fungus spores is short, and the dilution of the chemical by the seeding procedure is negligible.

## Literature Cited

1. CHOW, V. T., and T. E. HARBAUGH. 1965. Raindrop production for laboratory watershed experimentation. J. Geophys. Res. 70:6111-6119.
2. HIMELICK, E. B., and D. NEELY. 1965. Bioassay using cellophane to detect fungistatic activity of compounds translocated through the vascular system of trees. Plant Dis. Rep. 49:949-953.
3. NEELY, D. 1970. Persistence of foliar protective fungicides. Phytopathology 60:1583-1586.
4. NEELY, D. 1971. Deposition and tenacity of foliage protectant fungicides. Plant Dis. Rep. 55:898-902.
5. NEELY, D., and E. B. HIMELICK. 1966. Simultaneous determination of fungistatic and fungicidal properties of chemicals. Phytopathology 56:203-209.

# 4. Methods for Monitoring Tolerance to Benomyl in Venturia inaequalis, Monilinia spp., Cercospora spp., and Selected Powdery Mildew Fungi

## K. S. Yoder

Formerly research biologist, Biochemicals Department, Experimental Station, E. I. du Pont de Nemours & Co., Wilmington, DE 19898. Presently Department of Plant Pathology and Physiology, Winchester Fruit Research Laboratory, Virginia Polytechnic Institute and State University, 2500 Valley Ave., Winchester, VA 22601

Although resistance or tolerance of organisms to toxicants has been recognized for many years, only recently has the phenomenon reached economic significance with plant pathogenic fungi (1,3). The problem of fungicide tolerance was the subject of a recent symposium. Monitoring the fungal population for tolerance to preferred fungicides can reduce severe disease losses by enabling appropriate changes in the spray program to control the tolerant strain (7).

This chapter describes procedures for monitoring tolerance to benomyl (methyl 1-[butylcarbamoyl]-2-benzimidazole-carbamate) of *Venturia inaequalis* (Cke.) Wint. and *Podosphaera leucotricha* (Ell. & Ev.) Salm. on apples (*Malus sylvestris* Mill.), *Monilinia* spp. on stone fruits (*Prunus* spp.), *Cercospora apii* Fres. on celery (*Apium graveolens* L.), *Cercospora arachidicola* Hori on peanut (*Arachis hypogaea* L.), and *Erysiphe cichoracearum* DC. on cucurbits (*Cucumis* and *Cucurbita* spp.).

## Methods and Procedures

**General sampling procedures**—When fungicide tolerance is suspected as a cause of inadequate disease control, diseased material should be selected from well-sprayed areas with the severest infection. When the effects of altered spray programs on the proportion of tolerant and sensitive strains in the population are being investigated, however, samples should be randomly collected from large, replicated plots that are spaced to avoid interplot interference.

Collection of fresh, sporulating lesions will facilitate isolation and testing. Cross contamination, overheating, or delaying the processing of samples should be avoided. The isolated organism must be identified as the pathogen, perhaps through host inoculation studies.

**Apple scab, Venturia inaequalis**—*Sampling procedures*—Sporulating lesions on the youngest leaves or fruits are best for testing. For fruit in storage, small lesions that have not broken the cuticle are ideal. Leaf samples should be collected when the leaf surface is dry and placed into plastic bags without adding extra moisture. Refrigeration immediately after collection helps to prolong conidia viability and reduce contamination. The sample should include 20 or more leaves or fruit with lesions from different trees.

*Determination of tolerance*—In vitro tolerance to benomyl correlates closely with in vivo tolerance (Yoder, unpublished). *V. inaequalis* can be assayed on Difco potato-dextrose agar (PDA, 39 g/L + 5 g agar/L) acidified to pH 3.7 with eight drops (1.05 ml) of 85% lactic acid per liter and containing 5 mg/L of benomyl. This dosage provides approximately ten times the amount of toxicant required to inhibit sensitive strains under identical conditions. The lactic acid should be added after autoclaving when the medium has cooled to 60°C. For monitoring purposes, commercial formulations of benomyl may be added to PDA as a water stock suspension before autoclaving or as an ethanol stock suspension after autoclaving.

Divided plastic petri dishes containing benomyl-amended and nonamended PDA in the opposite compartments were used to simplify culture identification, record keeping, and fungal growth comparison. After the medium has solidified, the plates are returned to the sterile plastic sleeve in which they were purchased where they may be stored upside down for several months at ambient room temperature.

Because *V. inaequalis* is comparatively slow growing, care is required in the isolation technique. Sporulating lesions from unsterilized leaves or fruit are bisected with a sterile scalpel and excised with sterile forceps. Conidia are dislodged by rubbing one half of the excised lesion across the fungicide-amended agar and the other half across the control agar. By aligning the conidial streaks on opposite sides of the plate perpendicular to the center divider, treated and control conidia from six to eight lesions may be placed on a single plate. The concentration of conidia is greatest where the lesion first touches the agar surface. This initial contact should be made to accommodate microscopic examination.

Germination of conidia is observed after 24 hr at 16–19°C. Benomyl does not greatly reduce germination. Sensitive germ tubes grow only one or two cells in length, however, and appear distorted in the presence of the fungicide. Germ tubes from tolerant strains are uninhibited; growth on 5 mg/L of benomyl is the same as on nonamended medium. Since some conidia may have germinated on moist leaves before contact with the agar, a second transfer is needed to confirm tolerance. Spores that appear to be uninhibited are transferred as single-spore isolates onto sterile, benomyl-amended PDA plates for confirmation. A tolerant isolate will continue to grow and sporulate on 5 mg/L, but a sensitive isolate will not grow. Germinated conidia transferred with the agar block may grow from the top of the block without contacting the fungicide-amended PDA. Thus one must ascertain that the mycelium is in direct contact with the fungicide-amended PDA before confirming tolerance. Sporulation of the fungus in pure culture is desirable to confirm identity of the fungus.

In storage, scab lesions that have not broken the fruit cuticle may be surface sterilized by wiping lightly with a paper tissue soaked with 95% ethanol, bisected, and placed on divided plates as described, with the cut edge of the lesion in contact with the agar.

Conidia from a single, uncontaminated lesion are usually either all tolerant or all sensitive.

**Brown rot of stone fruits, Monilinia spp.**—*Sampling procedures*—Twenty or more infected fruits, blossoms, or twig cankers or a combination of these from different trees are collected in plastic bags and refrigerated immediately. Dry fruit mummies collected in paper bags are kept dry until processing.

*Determination of tolerance*—Sporulation of nonsporulating blossom infections and mummies may be induced by alternating daylight and dark conditions on 1.5% water agar. Cultures apparently will sporulate in variable day-length periods. Twig cankers are cut to 3 cm in length to include the canker margin (1 cm for the dead portion, 2 cm for the live portion), dipped for 2 sec in 95% ethanol, rinsed in sterile distilled water, dissected longitudinally, and plated on 1.5% water agar. A selective medium that Phillips and Harvey (8) described has also been useful for isolation from heavily contaminated samples.

Conidia from sporulating cultures or lesions are transferred with a sterile needle to fungicide-amended PDA prepared as described for *V. inaequalis*. Uninhibited growth on 5 mg of benomyl/L confirms tolerance, but sensitive germ tubes of *Monilinia* spp. may grow several cells in length before complete inhibition of sensitive strains occurs. Sporulation on 5 mg/L therefore is a more reliable indication of tolerance than is the percentage of germination or germ tube length. *M. fructicola* will sporulate within several days of the first transfer of conidia to PDA. *M. laxa* does not sporulate readily on the first transfer, but if an agar block from the first colony is transferred again to PDA, sporulation on the agar block will occur. To date, no tolerant *M. laxa* has been reported.

**Leaf spots of celery, peanut, and sugar beet, Cercospora spp.**—*Sampling procedures*—Fifty or more leaves with young, sporulating lesions from near the shoot apex are selected. Contaminated, senescent leaves near the soil surface should be avoided. Samples may be collected in plastic or paper bags, but crushing the sample may flatten the spores against the leaf surface and make isolation with a transfer needle more difficult. *Cercospora* spores are resistant to drying, so leaf samples may be either dried or refrigerated.

*Determination of tolerance*—Lesions without spores may be induced to sporulate by placing them in a humid atmosphere. The long conidia stand well above the lesion surface and can be readily transferred with a sterile needle to divided plates containing fungicide-amended PDA; the control PDA is prepared as described for *V. inaequalis*. Because *C. arachidicola* is slow growing, special care should be taken to touch only the tips of the conidia to avoid contamination in the transfer process.

The sensitive germ tube grows only a few cells in length on benomyl-amended PDA, but growth of tolerant isolates is uninhibited.

*C. beticola* and *C. apii* do not sporulate well on PDA, but will do so on transfer to a sugar beet leaf extract medium. The medium is prepared by steaming 16 g of dry weight or 130 g of fresh weight sugar beet leaves in 500 ml of distilled water for 20 min. The liquid is strained through four layers of cheesecloth, its total volume is brought to 500 ml, 10 g of agar is added, and it is autoclaved for 20 min. Cultures incubated 17 cm below 20-W General Electric cool white fluorescent tubes under 12-hr alternating light and dark periods at 25°C will begin to sporulate in about two days.

**Powdery mildews of cucurbit and apple, Erysiphe cichoracearum and Podosphaera leucotricha**—*Sampling procedures*—Twenty or more cucurbit leaves or apple shoots with lesions showing freshly sporulating areas are selected. The samples are placed in plastic bags and refrigerated until processing.

*Determination of tolerance*—To monitor for powdery mildew tolerance, susceptible plants must be isolated to maintain mildew-free conditions without the use of fungicides. Cucurbit or apple powdery mildews can be tested on young potted seedlings of susceptible varieties such as Straight Eight cucumber or Rome apple. The youngest fully expanded apple leaves are excised from seedlings or grafted stock and maintained throughout the incubation period by immersing the cut end of the petiole in a small vial containing a 3% (w/v) sucrose solution. Covering the opening of the vial with Parafilm "M" (American Can Company, Neenah, WI) retards evaporation from the vial and supports the excised plant parts. The sucrose solution is conveniently replenished with a narrow-spout laboratory wash bottle through a small slit in the Parafilm "M."

Cucumber foliage is sprayed to runoff with 50–100 mg/L of active benomyl ("Benlate" 50% WP). Because no tolerant apple powdery mildew has been reported, lower rates (20 mg of active ingredient/L) are used to assure adequate sensitivity of the test method.

Inoculum may be transferred from the disease sample by gently blowing across the sporulating surface toward the healthy plant or by preparing a spore suspension containing $5–10 \times 10^4$ conidia per milliliter of water and one drop of Tween 20 per 750 ml of water. Because the viability of conidia in the suspension decreases rapidly with time, such suspensions should be prepared immediately before spray inoculation with a DeVilbiss atomizer. After inoculation, the plants are incubated in translucent plastic containers covered with four layers of cheesecloth. Percentage of area infected and lesions per leaf are compared on treated plants and controls. Sensitive strains cause little or no infection on plants treated at the suggested rates. If a few lesions appear on the treated plants, reinoculation of treated plants from these lesions confirms tolerance or sensitivity of the organism.

## Discussion

The methods described are proved effective in detecting tolerance to benomyl with numerous samples of each organism except apple powdery mildew. No tolerant strain of the apple powdery mildew fungus has been detected. Related methods have been reported for *Cercospora* spp. (2,6), *M. fructicola* (4), and *V. inaequalis* (5). The methods reported here for *V. inaequalis* have also been used to test tolerance of peach and pecan scab fungi (*Cladosporium carpophilum* Theum. and *Fusicladium effusum* Wint.) and should also be adaptable to the pear scab fungus (*Venturia pyrina* Aderh.).

The in vitro rate of 5 µg/ml of benomyl is about ten times the amount required to inhibit sensitive strains of most fungi and can adequately differentiate tolerant and sensitive strains. Higher rates in vitro sometimes inhibit organisms tolerant to 5 µg/ml. Germination and initial germ tube development in sensitive strains are not greatly affected by 5 µg/ml of benomyl and therefore are poor indicators of tolerance. Extended germ tube development is a quick, fairly reliable criterion of tolerance if the possibility of pregermination is eliminated and if the spore is still visible for identification. Since numerous fungi are not inhibited, confirmation of the pathogenic organism is necessary.

In our experience, a relatively small sample size (20 lesions) may be used to determine whether ineffective control is due to high frequencies of tolerant propagules. Larger samples are needed to detect low frequencies of tolerant strains in population dynamics studies. If all the spores on one lesion arise from one parent spore, each lesion represents a sample size of one. Thus, one million viable spores from 20 lesions would represent a sample size of 20, not one million.

To have valuable predictive qualities, monitoring methods must detect tolerant pathogenic propagules at a frequency lower than that capable of causing visible disease losses beyond those expected in the absence of tolerant strains. This information must be available soon enough to prevent the organism from reaching economic significance before the control program can be altered to accommodate the tolerant strain.

Convenient mass screening techniques that are rapid, more predictive, and capable of detecting 1 tolerant spore in 1,000 need to be developed. Such techniques could involve mass trapping of airborne spores or composite spore suspensions plated onto highly selective media. Detection methods less sensitive than this may not forecast the threat of tolerance soon enough to avert serious economic loss. Because spontaneous mutations to benomyl tolerance may occur in some fungi at a frequency of 1 in $5 \times 10^7$ conidia (9), highly sensitive methods would likely detect spontaneous mutants if $10^7$ conidia were screened. Such information could be misleading, however, unless tolerant spores are tested for pathogenicity.

## Literature Cited

1. DEKKER, J. 1976. Acquired resistance to fungicides. Annu. Rev. Phytopathol. 14:405-428.
2. GEORGOPOULOS, S. G., and C. DOVAS. 1973. A serious outbreak of strains of *Cercospora beticola* resistant to benzimidazole fungicides in Northern Greece. Plant Dis. Rep. 57:321-324.
3. GEORGOPOULOS, S. G., and C. ZARACOVITIS. 1967. Tolerance of fungi to organic fungicides. Annu. Rev. Phytopathol. 5:109-130.
4. JONES, A. L., and G. R. EHRET. 1976. Isolation and characterization of benomyl-tolerant strains of *Monilinia fructicola*. Plant Dis. Rep. 60:765-769.
5. JONES, A. L., and R. J. WALKER. 1976. Tolerance of *Venturia inaequalis* to dodine and benzimidazole fungicides in Michigan. Plant Dis. Rep. 60:40-44.
6. LITTRELL, R. H. 1974. Tolerance in *Cercospora arachidicola* to benomyl and related fungicides. Phytopathology 64:1377-1378.
7. LITTRELL, R. H. 1976. Techniques of monitoring for resistance in plant pathogens. Proc. Am. Phytopathol. Soc. 3:90-96.
8. PHILLIPS, D. J., and J. M. HARVEY. 1975. Selective medium for detection of inoculum of *Monilinia* spp. on stone fruits. Phytopathology 65:1233-1236.
9. POLACH, F. J., and W. T. MOLIN. 1975. Benzimidazole-resistant mutant derived from a single ascospore culture of *Botryotinia fuckeliana*. Phytopathology 65:902-904.

# 5. Method for Screening Fungicides for Coryneum Blight Control Using Inoculated Detached Prunus Leaves

### H. H. Harder and N. S. Luepschen

Orchard Mesa Research Center, Colorado State University, Grand Junction, CO 81501. Colorado Agricultural Experiment Station Scientific Series Paper No. 1792. Mr. Harder is now with E. I. Du Pont, Biochemicals, Rt. 1, Fort Motte, SC 20950

Field incidence of Coryneum blight in Colorado differs from that found on the west coast, and long-accepted control measures are not always satisfactory (3). Due to the sporadic occurrence of Coryneum blight caused by *Coryneum carpophilum* (Lex.) Jauch, a method of culturing spores for controlled inoculations of *Prunus* foliage was developed for laboratory and greenhouse studies. Thus, controlled studies could be made on a natural substrate to determine the effects of different fungicides against the causal organism. Earlier work demonstrated that various fungicides incorporated into growth media differ in their ability to inhibit spore germination (2).

## Procedure

*C. carpophilum* was cultured in petri dishes containing 15–18 ml of peach agar. This medium consisted of 200 g of peach fruit homogenate and 40 g of Difco Bacto-Agar in 1,000 ml of deionized water. The fungus was incubated at room temperature (22°C) under ambient light conditions. Abundant viable spores were produced over the entire surface of the peach agar within 12 days. Mature spores were harvested by adding 20 ml of sterile deionized water to the petri dish and lightly brushing the culture with a small brush to dislodge the spores. Approximately one million spores were obtained per culture plate. These spores were then used for in vitro leaf inoculation tests.

Young apricot leaves (*Prunus armeniaca* L. 'Riland') were harvested in early June and held in loosely packed plastic bags at 1°C. The detached leaves were washed in cool tap water immediately before each test. Petri dishes were inverted, lined with No. 2 filter paper, and moistened with sterile deionized water. The detached apricot leaves were then placed in the prepared petri dishes. Additional moisture was provided daily.

A suspension of spores was atomized onto the upper surface of each leaf with a small spore dissemination tower. The tower consisted of two 1-gal tins brazed together end to end, with an atomizer mounted on the top and a space cut out at the bottom large enough to introduce a petri dish. The inoculated leaves were then incubated at 27°C with a light source set for 12 hr of alternate light and dark periods.

After 96 hr of incubation, red-bordered, light-centered lesions appeared on the leaf surface. Microscopic observation confirmed that germinated spores were associated with the red-bordered lesions. This method was used to evaluate potential chemicals for control of Coryneum blight (1).

In an evaluation of fungicide materials, the chemicals were applied to the individual apricot leaves before or after the spore inoculation step to evaluate protective and eradicative control potential. The leaves sprayed with protective fungicides were inoculated shortly after the spray film dried. For the eradicative spray treatments, the leaves were treated 24 hr after inoculation. All treatments were incubated at 27°C for 96 hr after inoculation. Coryneum blight was evaluated by counting the number of red lesions per leaf, using a ×5 hand magnifying lens. All data were subjected to a standard analysis of variance, the F test and LSD test being used to determine significant differences.

Table 5.1 gives the results obtained from two typical experiments in which ten detached leaves were used per treatment. These tests showed that protective spray treatments were uniformly better than postinoculation spray applications. Thiabendazole, dichlone, captan, and Citcop were the more effective materials.

Use of detached leaves as a natural host substrate for testing fungicide efficacy is suitable where field infection is absent or sporadic. It is more reliable in a screening process than is use of artificial media in vitro.

TABLE 5.1. Effectiveness of seven chemicals used as protective and eradicative sprays on detached Prunus leaves inoculated with Coryneum carpophilum

| Chemical, formulation, and rate per liter of water | Lesions per leaf[a] | |
| --- | --- | --- |
| | Protective | Eradicative |
| Thiabendazole, 60W, 1.2 g | 3.4 | 4.3 |
| Dichlone, 50W, 0.6 g | 3.3 | 6.4 |
| Citcop, 90L, 2,500 mg | 5.2 | 6.0 |
| Captan, 50W, 2.4 g | 5.0 | 7.3 |
| Tribasic copper sulfate, 54W, 3.6 g | 5.6 | 8.6 |
| Polyram, 80W, 1.2 g | 7.2 | 9.1 |
| Benomyl, 50W, 0.6 g | 8.9 | 14.1 |
| Unsprayed check | 8.4 | 10.0 |
| LSD 5% | 3.3 | 2.8 |

[a]Mean of ten leaves per treatment.

## Literature Cited

1. LUEPSCHEN, N. S., and H. H. HARDER. 1972. Control of Coryneum blight: Use of detached leaves as a tool for developing chemical controls. Colorado State University Experiment Station Progress Report 71-62.
2. LUEPSCHEN, N. S., K. G. ROHRBACH, C. M. CORE, and H. A. GILBERT. 1968. Coryneum blight infection and control studies. Colorado State University Experiment Station Progress Report 68-2.
3. WILSON, E. E. 1953. Coryneum blight of stone fruits. In: Plant Diseases, USDA Yearbook of Agriculture, 1953. pp. 705-710.

# 6. Procedures for Laboratory Evaluation of Thermal Powder and Thermal Tablet Fungicide Formulations

**Patricia L. Sanders and Herbert Cole, Jr.**

Research associate and professor, respectively, of plant pathology and the Pesticide Research Laboratory, The Pennsylvania State University, University Park, PA 16802

The procedures described herein have been developed to evaluate thermal powder and thermal tablet fungicide formulations for efficacy against spores of *Fusarium, Botrytis, Aspergillus,* and *Penicillium.* Using these procedures, we have tested thermal fungicide formulations for suppression of spore germination as a function of (i) fungal genus, (ii) distance from fungicide source, and (iii) physical orientation of the spore-bearing surface. We believe this method is appropriate for laboratory testing of thermal dust and tablet fungicide formulations against greenhouse foliar pathogens and pathogens of stored products where spores or mycelium or both are the major sources of inoculum. The procedures are easily done, require little space, and yield data within 72 hr.

## Fumigation Chamber Construction

A wood-framed, plastic-lined fumigation chamber, $1 \times 1 \times 2$ m, is used for testing. The chamber consists of four $1 \times 2$-m rectangular side frames and two $1 \times 1$-m square end frames built of $5 \times 5$-cm pine, with triangular wooden corner reinforcement. A $1 \times 2$-m piece of 0.63-cm plywood is fastened to one of the rectangular frames to form a rigid bottom for the chamber. Transparent polyethylene plastic of 4-mil thickness is stapled to the inside of each of the six wooden frames; the frames are fastened together with hooks and eyes to form a rectangular chamber. Such component construction allows easy disassembly, replacement of the plastic lining after each fumigation, and storage of the chamber when not in use.

## Procedure

Test fungi are grown at room temperature (about 21°C) on potato-dextrose agar slants for approximately two weeks. Spores are washed from the agar slants with about 30 ml of a filtered 2% gelatin solution in sterile distilled water. The resultant suspension of spores is decanted through a single thickness of paper tissue (Kimwipe) into a small beaker. Microscope slides, which have been cleaned in 95% ethyl alcohol and wiped dry, are dipped into the gelatin-spore suspension. The sides of a perforated wire test tube basket are used to hold the slides until they air dry. When the slides are completely dry, three replicate slides containing spores of each test fungus are placed immediately into a moist chamber, germinated, and used as untreated controls. The remainder of the test slides are placed at numbered locations in the fumigation chamber.

The method of placement of the spore-bearing test slides in the fumigation chamber depends on the intended use of the test fungicide. In evaluations of thermal fungicides designed for the treatment of storage areas, test slides are affixed to 30 numbered locations on the chamber walls for treatment. Data on variation in efficacy of fungicides against spores of various fungal genera and variation due to differences in distance from the fungicide source can be obtained by such slide placement. In evaluations of thermal fungicides intended for the treatment of greenhouse plants, test slides are placed in the fumigation chamber in a slide holder designed to simulate various leaf-angle orientations. This "artificial plant" is made by inserting round wooden dowels into holes drilled at various angles (horizontal, 45 degrees, and vertical) in a $5 \times 5$-cm piece of pine. Wooden clip clothespins are attached to the distal ends of each dowel to serve as slide holders. The dowels can be rotated so that the test slides present various surface angles between vertical and horizontal. Data on variation in efficacy of fungicides against spores of various fungal genera and variation due to differences in physical orientation of the spore-bearing surface can be obtained by such placement of test slides.

The amount of fungicide used for treatment is calculated on the basis of chamber volume converted from the manufacturer's recommended rate. The fungicide is placed at one end on the floor of the fumigation chamber equidistant from the sides, and ignited as prescribed by the manufacturer. Because a small quantity of fungicide is required for the 2-m³ volume of the chamber, difficulty may be encountered in obtaining sustained complete combustion after lighting. In such cases, the fungicide is completely burned by applying flame from a Bunsen burner to the fungicide through the partially opened chamber front. The fumigation chamber is immediately closed. After 24 hr, the chamber is opened and the test slides are removed and placed in a separate moist chamber for germination. After 48 hr of incubation, both treated and control slides are removed from the moist chambers and the percentage of germination is determined by microscopic examination.

## Collection and Use of Data

Counts are made on both sides of each test slide and recorded separately. One hundred spores, both germinated and ungerminated, are counted; the number germinated in each 100 spores is recorded as percentage of germination. Three such determinations are done at random locations on each side of all slides. Each determination is considered a replicate. Data obtained are subjected to analysis of variance and Duncan's modified least significance difference test.

Data from experiments in which test slides are affixed to the walls of the fumigation chamber provide a measure of the ability of the test fungicide to protect the treated area uniformly and effectively. Such experiments are appropriate for testing thermal formulations of fungicides that are intended for treatment of storage areas. Fungal species to be included in these experiments are those associated with damage to stored products.

Data from experiments in which test slides are held in the fumigation chamber on the artificial plant slide holder provide a measure of the ability of the test fungicide to contact and protect irregularly oriented surfaces effectively. Such experiments are appropriate for testing thermal formulations of fungicides intended for treatment of plants in an enclosed space such as a greenhouse. Fungal species to be included in these experiments are typically foliar pathogens of greenhouse plants.

These laboratory procedures are not a substitute for in situ testing of thermal fungicide formulations, but they can quickly provide quantitative data from which fungicide performance at various dosages and against various pathogens can be inferred. Such information is valuable in efficient planning of in situ evaluations of thermal fungicide formulations.

# 7. Greenhouse Evaluation of Seed Treatment Fungicides for Control of Sugar Beet Seedling Diseases

## L. D. Leach and J. D. MacDonald

Professor emeritus and assistant professor, respectively, Department of Plant Pathology, University of California, Davis, CA 95616

Damping-off of sugar beet (*Beta vulgaris* L.) seedlings is caused by several fungi. *Pythium ultimum* Trow, *P. aphanidermatum* (Edson) Fitzp., *P. debaryanum* Hesse, *Rhizoctonia solani* Kuhn, and *Aphanomyces cochlioides* Drechs. are all soilborne pathogens commonly associated with seedling disease of sugar beets (1). *Fusarium* spp. also can be involved, but because of the abundance of nonpathogenic forms that colonize the roots, isolates cannot always be assumed to be pathogenic. *Phoma betae* (Oud.) Fr. is a common seedborne pathogen of sugar beets that may be present on a large percentage of seed produced in moist climates (2).

While the ultimate test of seed-treatment fungicides is based on field performance, several factors often complicate evaluation of field results. Various fungi may be associated simultaneously with seedling diseases in the field, and fungicides frequently vary in their activity against different taxonomic groups of fungi. Moreover, excessive dosages of fungicides may have adverse effects on seedlings. Information relevant to specificity and phytotoxicity of seed-treatment fungicides can be obtained readily in the laboratory and greenhouse. This preliminary information is necessary for evaluation of fungicides in the field.

## Methods and Procedures

**Evaluation of phytotoxicity**—Fungicide treatments should be used at the manufacturer's recommended dosages, and should be applied to commercially processed, graded seed, using a variety adapted to the area. If the optimum dosage for use as a seed treatment is unknown, evaluating several rates may be necessary. The fungicides, however, must be tested first for any independent effects on seedling germination or growth at the rates to be used. This can be done by planting fungicide-treated seeds in flats of sterile or pasteurized soil.

Fungicides may be applied to seeds as dusts or spray suspensions in a revolving drum seed treater, or as a dust or slurry in a closed container. Four to six replications of each treatment dosage should be seeded into the pathogen-free soil in a randomized arrangement, with 25–50 seeds per foot of row (30.5 cm) in the flats. Seeds should be uniformly covered to a depth of 1 in. (2.5 cm), the soil moistened, and the flats incubated at 20–25°C.

Emerging seedlings should be counted daily for any evidence of reduced or delayed emergence. Delayed emergence is measured by calculating the mean emergence period (MEP). The MEP for each replication is determined by multiplying the number of seedlings that emerge each day (N) by the corresponding number of days since planting (D), and then dividing the sum of these products by the total number of seedlings that emerge (T).

$$MEP = \Sigma(N \cdot D)/T$$

In addition to changes in emergence characteristics, the seedlings should be observed for abnormal development. Seedlings should be harvested, washed, blotted dry, and weighed 20–24 days after planting.

With the above information, statistical comparisons of total seedling emergence, MEP, and seedling fresh weights can be made (3). A significant delay or reduction in emergence, or reduction in seedling weight, is

23

evidence of phytotoxicity. This information will identify dosages that can be safely used in tests for activity against specific pathogens.

**Protection from seedborne Phoma betae**—To evaluate fungicides for activity against *Phoma betae,* it is desirable to use a heavily infested seed lot. To identify an infested seed lot, samples of untreated seed should be assayed for *P. betae.* The water agar method (4,5) is relatively simple and effective. A shallow layer of 1.2% water agar is poured into 9-cm plastic petri dishes; 100 seed units are plated (5 per dish). An additional group of 100 seeds are plated in the same way following surface disinfestation in 0.5% NaOCl for 5 min. Seeds are then incubated at 20°C for seven days. The seeds are examined for fungal "holdfasts" (4,5) at intervals of three and seven days by inverting the dishes and observing them at a magnification of ×50–100 where the agar contacts the bottom of the dish. The number of seeds infested with *P. betae* is indicative of the level of infestation in the seed lot. The number of seeds with *P. betae* after disinfestation is an indication of the amount of seed with deep penetration of *P. betae* in the tissues. These values can be used to compare seed lots for the relative severity of *Phoma* infection (2).

After selecting a seed lot with a high level of *Phoma,* the effectiveness of seed-treatment fungicides can be evaluated by treating samples of the seed and planting them in pasteurized soil in the greenhouse. Nontreated and treated seeds should be planted in flats using a randomized design to allow statistical analysis of the results. Flats should be incubated at 12–15°C and watered as necessary to provide favorable conditions for seedling emergence and growth.

As infected seedlings are observed, they should be counted and removed at daily intervals. Removed seedlings should be washed and plated on water agar to determine if they were killed by *Phoma* or by other causes. Approximately 14 days after emergence, all seedlings should be harvested, washed free of soil, and examined for evidence of *Phoma* infection on the underground portion of the hypocotyls. Seedlings apparently infected by *Phoma* should be plated on water agar for confirmation.

The relative efficacy of candidate fungicides is measured as the improvement of emergence, or the relative freedom of seedlings from infection, in comparison with untreated controls.

**Protection from soilborne seedling pathogens**—Because several seed-treatment fungicides are selective in their protection, testing each fungicide in soils infested with a single pathogen is necessary before testing in soils infested with combinations of pathogens or in field soils where more than one pathogen is likely to occur. This allows identification of specificity and aids in selecting effective dosages.

The target pathogens should be maintained in pure culture and added to pasteurized soil using a method that provides consistent inoculum levels. One possible method involves incorporation of pathogen-infested washed vermiculite into the soils. Another method that has been successful with *P. ultimum, R. solani,* and *A. cochlioides* involves growing the fungi on nutrient agar and preparing an inoculum suspension by finely chopping the entire culture in water. This suspension is then mixed into the soil; the soil is moistened and allowed to incubate for several days at approximately 20°C. Soils should be remixed periodically to assure a uniform distribution of the inoculum before planting.

Seeds treated with fungicides that are candidates for use in these tests should be relatively free of *Phoma* to assure that only the soilborne pathogen is involved. Again, each treatment should be planted in rows 1 in. (2.5 cm) deep, randomized, and replicated several times. Daily counts of seedling emergence should be recorded, and dying seedlings noted, removed, and cultured in sterile water or on a suitable medium to confirm pathogen identity.

At the end of the experiment, the records should include the emergence per 100 seed units, the percentage of seedlings damped-off, the surviving seedlings per 100 seed units, and the pathogens identified from infected seedlings. Statistical comparisons of the treatments and controls will indicate which fungicides or dosages are effective against specific pathogens.

**Evaluation of seed treatment fungicides in field soils**—In preparation for field tests, a knowledge of the variety of pathogens present in the field soils is desirable. To determine this, soils from fields with a past history of severe seedling disease can be collected, screened, and transferred to flats in the greenhouse. Sugar beet seed that is relatively free of *Phoma* and not treated with fungicides should then be planted in the flats as previously described. During emergence, infected seedlings are removed and cultured for pathogen identification. Additional information concerning the variety of pathogens present in the soils can be procured by planting rows of seed treated with fungicides known to be effective against specific pathogens (ie, Dexon against *Pythium* spp. and PCNB against *Rhizoctonia*). These fungicides can be used singly or in combination.

Knowledge of the relative abundance of specific pathogens in each soil, and of the effectiveness of specific fungicides in controlling them, aids in selecting the particular seed-treatment fungicides, or combinations thereof, that have the best chance of controlling the seedling pathogens in the field. With this information, field trials can be conducted to test the fungicides under natural conditions and evaluation of the results should be simplified.

**Reporting test results**—A number of factors can influence the results of seed treatment tests. These factors must be described so that other workers can evaluate test results accurately. First among these is the need to identify the cultivar used. Sugar beet cultivars have been developed specifically for different geographic regions of the United States according to problems endemic to those areas. In addition, the processing grade of the seed should be indicated. The use of processed (seed that has been milled to remove superficial tissues) or nonprocessed seed can influence the efficacy of fungicides, as has been demonstrated with *Phoma betae* (2).

The planting date and mean temperature during seedling emergence should also be reported, since temperature can influence the activity of seedling pathogens. Failure to recognize this possibility could lead to erroneous conclusions about fungicide efficacy. A listing of pathogens involved, and their relative frequency of isolation from diseased seedlings, should also be reported.

A description of the plot design should be given, as well as the identification of statistical tests used in mean separation. Duncan's multiple range test is popular, but other tests are available (3) and may be better suited to the experimental design employed.

## Literature Cited

1. BENNETT, C. W., and L. D. LEACH. 1971. Diseases and their control. In JOHNSON, R. T., et al. (eds.) Advances in Sugar Beet Production: Principles and Practices. The Iowa State University Press, Ames, Iowa, pp. 223-285.
2. LEACH, L. D., and J. D. MACDONALD. 1976. Seed-borne Phoma betae as influenced by area of sugar beet production, seed processing and fungicidal seed treatments. J. Am. Soc. Sugar Beet Tech. 19:4-15.
3. LITTLE, T. M., and F. J. HILLS. 1978. Agricultural Experimentation: Design and Analysis. John Wiley and Sons, New York.
4. MANGAN, A. 1971. A new method for the detection of Pleospora bjoerlingii infection of sugar beet seed. Trans. Brit. Mycol. Soc. 57:169-172.
5. MANGAN, A. 1974. Detection of Pleospora bjoerlingii infection on sugar beet seed. Seed Sci. Technol. 2:343-348.

# 8. Greenhouse Procedures for Evaluation of Turfgrass Fungicides

### Patricia L. Sanders and Herbert Cole, Jr.

Research associate and professor, respectively, of plant pathology and the Pesticide Research Laboratory, The Pennsylvania State University, University Park, PA 16802

Our laboratory has used the greenhouse procedures described herein to evaluate protectant and systemic fungicides for phytotoxicity of, control of, and duration of efficacy (residual activity) against Sclerotinia dollar spot, Rhizoctonia brown patch, and Pythium blight on Penncross creeping bent grass (*Agrostis palustris* Huds.). With suitable modification, they may be appropriate for preliminary testing of fungicides for other fungus diseases of turfgrass that can be induced under greenhouse conditions. Testing can be done at any time of the year if temperature-controlled greenhouses or growth chambers are available. The data obtained are valuable in determining dosages and application methods that may provide maximum disease suppression, longest residual efficacy, and minimum phytotoxicity under field conditions. Such information provides a basis for more efficient field testing of fungicides.

## Test Plants

Penncross creeping bentgrass is the host that is usually used in our greenhouse tests. This cultivar is widely used in fine turf situations, and is easily managed under greenhouse conditions. As the need arises, other species and cultivars may be used. With ryegrasses, fescues, or bluegrasses, the seeding rate and emergence time vary. Seeding rates are converted from commercial rates (pounds per 1,000 ft$^2$) to grams per 78 cm$^2$ (the area of a 10-cm pot). In general, the seeding rate should be two to four times the equivalent commercial rate to allow for rapid establishment of a uniform, dense stand. The time between seeding and availability of grass for use depends on cultivar and species. Penncross bentgrass, ryegrasses, and fescues may be used three to four weeks after seeding. Certain bluegrass cultivars, however, may require five or more weeks from seeding to use.

In our procedure, Penncross creeping bentgrass is seeded in 10-cm–diameter plastic pots of steam-sterilized greenhouse soil (two parts soil, one part sand, one part peat). A seeding rate of 0.2 g/10-cm pot (equivalent to four times the recommended commercial rate) is used with Penncross to assure rapid establishment of a uniform, dense stand. After seeding, the pots are lightly top-dressed with a calcined clay topdressing to speed germination. Any turfgrass topdressing material may be used for this purpose, or the pots can be top-dressed with the sterilized soil mix. Throughout the experiment, the grass is regularly watered and maintained at a cutting height of about 2.5 cm above the soil surface. Chemical treatments may be applied three to four weeks after seeding.

## Fungicide Application

Fungicides may be tested as foliar sprays, soil drenches, or granular formulations. To obtain a wide spectrum of control effectiveness and phytotoxic manifestation, the test fungicide is applied at a range of dosages both above and below the recommended rate. Fungicide dosages and water diluents are calculated and applied to the pots of grass on an area basis at rates equivalent to the standard 93-ca (1,000 ft$^2$) dosages commonly used in turfgrass research and management. An active ingredient or product basis may be employed as appropriate.

The following are some sample calculations:

i. Area of a 10-cm pot:

$$\pi r^2 = 3.14 \times 25 \text{ cm}^2$$
$$= 78.5 \text{ cm}^2$$

ii. Dosage per 10-cm pot (78.5 cm$^2$):

$$2 \text{ oz}/1{,}000 \text{ ft}^2 = 56.7 \text{ g}/93 \text{ m}^2$$
$$= 56{,}700 \text{ mg}/930{,}000 \text{ cm}^2$$

$$\frac{56{,}700 \text{ mg}}{930{,}000 \text{ cm}^2} \times \frac{\text{X}}{78.5 \text{ cm}^2}$$

$$= 4.8 \text{ mg}/78.5 \text{ cm}^2$$

iii. Water dilution rate/10-cm pot (78.5 cm$^2$):

$$3 \text{ gal}/1{,}000 \text{ ft}^2 = 11.4 \text{ L}/93 \text{ m}^2$$
$$= 11{,}400 \text{ ml}/930{,}000 \text{ cm}^2$$

$$\frac{11{,}400 \text{ ml}}{930{,}000 \text{ cm}^2} \times \frac{\text{X}}{78.5 \text{ cm}^2}$$

$$= \sim 1 \text{ ml}/78.5 \text{ cm}^2$$

Since the method of application can greatly affect both the residual activity of a fungicide and the amount of phytotoxicity produced, any or all of the following methods of application may be used: (i) foliar spray at various water dilution rates, (ii) spray with or without previous wetting of the foliage, (iii) spray followed by irrigation of the foliage, (iv) soil drench, and (v) granular application. Fungicide sprays are applied with an air-pressure venturi-pickup atomizer sprayer that is powered by an electric pressure/vacuum pump.

All pots required for a given treatment block are sprayed simultaneously, but care must be taken so that each pot within the treatment block receives equivalent amounts of fungicide. Almost any pattern of spray that the investigator believes gives uniform coverage may be employed, but experience has shown that a circular pattern of spray on individual pots results in spray overlap in the center of the pot area. Therefore, a back-and-forth sweeping motion should be used instead.

Drenches are applied to pots on an individual basis at the appropriate water dilution rate. Granular fungicides may be applied on an individual pot or treatment block basis. Granulars are "diluted" with sterile sand to assure uniform coverage and applied in saltshaker fashion.

All pots of grass are treated with the test fungicides simultaneously. A standard fungicide of known effectiveness against the test pathogen is included for comparison. An untreated check is included in the test to verify the virulence of the test pathogen under the conditions of the experiment. All treatments are replicated at least three times. Factorial experimental designs with arrangement of the replications in a randomized block plan have proved most satisfactory for us.

## Inoculum Preparation, Inoculation, and Incubation

Virulent isolates of the fungus are grown for approximately one week on autoclaved rye grain. This medium consists of 50 g of rye grain, 1 g of CaCO$_3$, and 75 ml of water placed in 250-ml Erlenmeyer flasks. The flasks are stopped with cotton plugs and autoclaved for 30 min at 15 lb of pressure. Ten to 15 kernels of rye inoculum are placed in the center of the grass area in each pot. After inoculation, each pot is covered with a transparent polyethylene bag to maintain high humidity, and placed either on a shaded greenhouse bench or in a temperature-controlled growth chamber.

Temperature during incubation is maintained at the optimum for disease development and growth of the test pathogen (21–27°C for *Sclerotinia homoeocarpa,* 27–30°C for *Rhizoctonia solani,* and 30–36°C for *Pythium aphanidermatum*). These environmental conditions and the relatively large quantities of virulent inoculum provide a rigorous challenge to the fungicide treatments.

If the purpose is to test immediate suppression of symptoms by the test fungicide, inoculation is usually done within 24 hr after application. If, however, information is required on the duration of control provided by the fungicide, all pots of grass are treated simultaneously, and successive weekly inoculations are made. Initial inoculation of some of the treated plants is made within 24 hr after application of the fungicide. Inoculum is applied to a set of previously treated, noninoculated pots of grass at weekly intervals. Sequential weekly inoculations continue until the residual effect of the fungicide is no longer apparent. Testing protectant fungicides for duration of control usually requires two or three weekly inoculations after chemical treatment, but systemic fungicides may require six to eight weeks of testing after treatment before residual activity completely disappears.

## Disease Control Evaluation

The plastic covers are removed from inoculated pots and fungicide efficacy is evaluated about seven days after inoculation. This interval is determined by the rate of spread of the pathogen in the untreated check pots. When placed in the center of the pot of seedling grass, *S. homoeocarpa, P. aphanidermatum,* and *R. solani* produce foliar blight that advances radially toward the pot margins. Our usual procedure is to remove the plastic cover and terminate the test when the grass in the untreated check pots is completely blighted. At this time, all inoculated pots are evaluated individually for disease severity. A 0–10 visual rating scale is employed, which corresponds to the percentage of pot area blighted—0, no disease; 1, 1–10% blight; 2, 11–20% blight; 3, 21–30% blight; 4, 31–40% blight; 5, 41–50% blight; 6, 51–60% blight; 7, 61–70% blight; 8, 71–80% blight; 9, 81–90% blight; and 10, 91–100% blight, or essentially complete death of the foliage.

Data obtained from disease severity evaluations are subjected to analysis of variance (1) and Duncan's modified (Bayesian) least significant difference (DMLST) (2) tests. The DMLST was used instead of the older "new" multiple range test, because the use of the Bayes rule for symmetric multiple comparisons in the DMLST reduces the chances of type I and II errors in separation of means and determination of significant differences between treatments.

## Phytotoxicity Evaluation

Approximately one week after fungicide application, all treated, noninoculated grass is evaluated for phytotoxicity. Phytotoxic effects vary with the plant and the fungicide, and may be exhibited as stunt, thinning, discoloration, or chemical burn. Phytotoxicity evaluation is descriptive and is based on severity of symptoms and type and location of injury. An appropriate visual rating scale is employed.

## Correlation of Greenhouse Test Procedures With Field Test Results

The procedures described produce reliable, uniform data that are valuable for determining dosages and application methods for field use. The variable environmental conditions and differing physiologic states of plants encountered in field test situations, however, can influence the performance of fungicides. Greenhouse-grown seedling turf is usually more sensitive to phytotoxic substances than are mature field stands. One exception is clinical injury associated with senescent plants. Generally, however, if injury appears in greenhouse tests, one must be alert for similar results under field conditions.

In field evaluation of disease control, dosages that are effective in greenhouse pot tests are usually good in the field. A possible exception concerns systemic fungicides when thatch and organic matter interfere with their absorption from the soil. In this instance, greenhouse results are normally more optimistic than those obtained under field conditions.

Despite the foregoing limitations, preliminary greenhouse testing of fungicides has been invaluable in selecting dosages and application techniques to be used in field testing, and has enabled us to make more efficient use of time and space in our field tests.

### Literature Cited

1. STEEL, R. G., and J. H. TORRIE. 1960. Principles and procedures of statistics. McGraw-Hill Book Co., Inc., New York.
2. WALLER, R. A., and D. B. DUNCAN. 1969. A Bayes rule for the symmetric multiple comparisons problem. J. Amer. Stat. Ass. 64:1484-1503.

# 9. Test Procedures for Fungicides and Bactericides Used to Control Foliar and Soilborne Pathogens of Ornamental Tropical Plants

**James F. Knauss**

Agricultural Research Center–Apopka, Apopka, FL 32703

The ornamental tropical foliage plant industry produces more than 150 kinds of plants under climatic conditions favorable to development of diseases caused by foliar and soilborne plant pathogens. Economically significant host plant genera include species of *Aglaonema, Aphelandra, Areca, Brassaia, Calathea, Chamaedorea, Chrysalidocarpus, Crassula, Dieffenbachia, Dracaena, Ficus, Fittonia, Gynura, Hedera, Hoya, Kalanchoe, Maranta, Monstera, Nephrolepis, Peperomia, Philodendron, Pilea, Pleomele, Sansevieria, Scindapsus, Spathiphyllum,* and *Syngonium.* Fungal species belonging to the genera *Alternaria, Cephalosporium, Cercospora, Dactylaria, Fusarium, Helminthosporium, Rhizoctonia,* and *Phytophthora* are among the most common foliar fungal plant pathogens. Common bacterial pathogens causing leaf spotting and collapse, cutting and cane decay, and plant wilts are *Erwinia carotovora* (L. R. Jones) Holland, *E. chrysanthemi* Burkl. et al., *Pseudomonas* spp., *Xanthomonas dieffenbachiae* (McCull. & Pirone) Dows., and *Xanthomonas* spp. Soilborne fungal pathogens causing severe disease losses are *Phytophthora* and *Pythium* spp., *Rhizoctonia solani* Kuehn, and *Sclerotium rolfsii* Sacc. On occasion, species of *Fusarium* and *Sclerotinia* also cause economic disease losses. In Florida, severe disease losses from most of the previously mentioned pathogens can occur anytime during the year due to the strong influence of the subtropical climate on disease development.

The fungicide-bactericide evaluation program has been active at the Agricultural Research Center–Apopka for the past nine years. It was developed because of the almost complete lack of available published information on the efficacy and phytotoxicity of chemicals that control ornamental tropical foliage pathogens. When the program started in 1969, the cultural status of the industry in Florida was poor. Overhead irrigation, ground bed production, and slat shed structures were common. Severe disease caused by diverse pathogens was common, and the need for disease control chemicals extensive. The chemical control program was designed to obtain information in all three major pathogen areas so that growers could obtain disease control suggestions based on research and not on casual observation or hearsay. The methods developed and reported here were designed to provide broad-scale evaluations under severe disease pressure similar to that which occurs normally in commercial production.

A successful evaluation test of disease control compounds for ornamental tropical foliage plants requires:

i. Experimental host material of uniform quality, size, and age that is as pathogen-free as possible.

ii. Experimental soils, plant culture, and the like that simulate commercial horticultural practices.

iii. Quantitative and qualitative data determined throughout and at the termination of the test.

27

iv. Inoculation or infestation methods similar to those occurring naturally. Deliberate wounding of host tissue may produce unfair evaluations of otherwise effective control compounds.

v. Candidate compounds within the recommended shelf life, stored properly to maintain low humidity, and never more than two years old.

vi. The proper control treatments to determine the potential influence of factors other than those produced by test pathogens or candidate chemicals.

### Methods and Procedures

**Test site**—Most tests were conducted in a glass greenhouse with raised benches, pad and fan evaporative cooling, forced-air heating, temperatures of 18–27°C (winter) or 27–35°C (fall, spring, summer), and light intensities of 1,000–3,000 ft-c. Plants were watered carefully with a hose when necessary.

Some tests were conducted in a slat shed with 75% shade and 24–38°C temperatures. Elevated iron racks and hose watering were used for slat shed pot culture, and peat-amended ground beds and overhead irrigation for slat shed stock bed plantings.

**Test plants**—Plant material was selected from rooted and unrooted cuttings or seedlings, which were occasionally obtained from grower stocks. Whenever possible, cuttings taken from culture-indexed plants or seedlings produced at the Agricultural Research Center–Apopka were used in the tests. Selection of uniform plant material was especially important when tests included phytotoxicity evaluations.

**Test pathogens**—Whenever possible, the test pathogens were obtained from recent isolations of diseased host plant tissue. Where desirable, fungal stock cultures were stored at 15°C and transferred every two months. Stock bacterial cultures were also held at 15°C in sterile distilled water, with retrieval and subsequent identical storage repeated every 6–12 months. A preliminary evaluation of isolate pathogenicity was normally performed prior to each test, especially if the isolate had been stored or in culture several months.

Foliar fungal pathogens were grown in test tubes or disposable petri plates on media (eg, lima bean agar [LBA], potato-dextrose agar [PDA]) that would produce sufficient sporulation. Cultures were grown under cool-white fluorescent light (12 hr daily) at 23–28°C for one to three weeks. They were then initiated in petri plates by implanting in the center of the plate a disk taken from the advancing margin of a 24–72-hr-old culture. Test tube cultures were initiated by spreading a loop containing spores of the test pathogen over the tube's medium surface. Only with extremely slow-growing fungi, such as *Dactylaria humicola,* did it take longer than one week to produce spores suitable for inoculation. Spores were extracted by flooding the plate or tube medium surface with sterile deionized water, gently scraping the surface with a sterile loop, and straining the suspension through four layers of cheesecloth. The suspensions used always contained a large number of spores, but no attempt was made to standardize inoculum density.

In preparation for infestation of bacterial culture media, the bacterial test isolate was transferred several times at frequent intervals to PDA (*Xanthomonas* spp.) or LBA (*Erwinia* spp.) tubes. Once active growth was established, a loopful of a 24-hr-old culture was transferred to a flask containing either sterile nutrient broth or sterile yeast extract plus dextrose broth (10 g each/L). Flasks were allowed to incubate on a reciprocal shaker at 25–30°C for 4–10 hr (*Erwinia* spp.) or 16–20 hr (*Xanthomonas* spp.) to provide actively growing inoculum. For greenhouse studies, inoculum was extracted and prepared by taking a quantity of the appropriately aged culture, spinning in a centrifuge for 15 min, pouring off the supernatant, and resuspending the culture to the original volume with sterile deionized or sterile tap water. In studies conducted in the slat shed on plants growing in ground beds, the cultures were not centrifuged and resuspended but simply diluted 1:1 with plain tap water.

Culture media and the time and temperature of incubation used to prepare inoculum of soilborne pathogens are noted in Table 9.1.

The media in Table 9.1, except for potato-dextrose broth, were autoclaved for 1 hr at 121°C and 15 psi on two consecutive days before inoculation. Potato-dextrose broth was autoclaved once for 15 min. Cultures of *R. solani* and *S. rolfsii* were grown on PDA, and *Phytophthora* and *Pythium* spp. on LBA, in petri dishes at 27–30°C. An agar disk was cut with a sterile No. 2 size cork borer from the advancing margin of growth and added to the culture media (Table 9.1). The mixture was contained within 1,000-ml (millet seed, oatmeal-sand, wheat seed) or 250-ml (potato-dextrose broth) Erlenmeyer flasks. After incubation, the inoculum (with the exception of the broth cultures) was removed from the flasks, placed into a clean container, and thoroughly chopped and mixed prior to soil infestation. Mycelial mats grown on potato-dextrose broth were removed and fragmented in sterile deionized water (1 mat/100 ml of water) in a sterile Waring Blendor.

**Soil and fertilization**—Tests were performed in steam-pasteurized soil. The soil consisted by volume of 1–3 parts of German, Canadian, or horticultural Florida peat to 1 part coarse builder's sand. After pasteurization, 4.2 kg of dolomitic limestone and 0.6–1.8 kg of Perk, a minor element supplement,[1] were added per

**TABLE 9.1. Culture media and conditions used to produce inoculum of soilborne fungal pathogens in evaluations of soil fungicides**

| Pathogen | Culture media | Incubation Time (weeks) | Temp. (°C) |
|---|---|---|---|
| *Phytophthora* and *Pythium* spp. | 100 g Millet seed + 130 ml deionized water, or 40 g 100% Rolled oats + 200 g washed coarse builder's sand + 100 ml deionized water | 2–3 | 27–30 |
| *Sclerotium rolfsii* | 100 g Wheat seed + 175 ml deionized water | 2–3 | 30 |
| *Rhizoctonia solani* | Oatmeal-sand or wheat seed (as above), or | 2–3 | 27–30 |
| | Potato-dextrose broth as a stationary culture | 1–2 | 27–30 |

[1]Kerr-McGee Chemical Corp., Oklahoma City, OK. Containing percentage by wt: Cu, 0.23; Mn, 2.29; unchelated Fe, 3.44; chelated Fe, 0.23; S, 4.50; Mg, 9.17; Zn, 0.69; B, 0.023; and Mo, 0.002.

cubic meter of soil mix. Once established, plants or seedlings were fertilized every 7–14 days with water-soluble fertilizer containing 0.12 kg 20-20-20/100 L of water. The frequency of watering depended on plant size, plant need, and environmental conditions. Plants were never fertilized without receiving two previous applications of water.

In some slat shed studies, a 14-14-14 coated, slow-release fertilizer was used and applied to the surface of previously fumigated ground beds at the rate of 280 kg N-P-K/ha/year.

**Test chemicals**—Test compounds within prescribed shelf life, or less than two years old, were used. The current year's supply from unopened containers was used whenever possible.

Foliar fungicides and bactericides were applied in the greenhouse with either a 3.8- or 7.6-L stainless steel hand sprayer maintained at 25–40 psi. In slat shed tests, applications with bactericides were made at 125–200 psi using a 94.6-L power sprayer. In all foliar fungicide and bactericide tests, the adjuvant Plyac[2] was used. Plyac was never added to chlorothalonil sprays. With greenhouse tests, two identical sprayers were employed. This allowed thorough cleaning of one while the other was in use. Spray applications covered both the upper and lower leaf surface, usually on a seven-day schedule. Plants were sprayed one to three times before inoculation, and sometimes afterward.

In tests involving drench treatments with soil fungicides, the fungicide suspension or solution was made up in 7.6- to 94.6-L lots. Vigorous agitation was achieved by a hand-operated paddle or mechanical device. The volume of the fungicide suspension or solution applied was based on the soil surface area to be covered. Volume rates currently used for test applications correspond to 473,709, and 946 ml/929 cm$^2$ (1.0, 1.5, and 2.0 pt/ft$^2$) of soil surface area. In experiments with pots, the corresponding volume rates listed in Table 9.2 would apply.

Drench applications were made one to three days after soil infestation, depending on the growth rate of the pathogen and the type of host tissue being tested. Drenches were made within one day in all tests involving seedlings. Drenches to rooted and unrooted cuttings were made two to three days after infestation with *R. solani* or *S. rolfsii*. Applications to rooted cuttings were made one day after infestation with *Pythium* or *Phytophthora* spp.

Fungicide-amended soils were prepared in a thoroughly cleaned cement mixer by adding the required concentration of fungicide to 0.028 m$^3$ (1 ft$^3$) of pasteurized soil mix. Soils containing fungicides were used within one week after mixing.

**Experimental design**—All tests included the appropriate noninfested (or noninoculated) and untreated controls. In studies involving pots, soil was added to new plastic 10.2–15.2-cm diameter pots with one to four cuttings or plants and up to 20 seedlings per pot. Each treatment consisted of at least five pots, with one pot of each treatment set into a randomized block design within the experimental area. With potted plants, the only exceptions to the randomized block design were long-term foliar fungicide, bactericide, and

fungicide-insecticide phytotoxicity tests, in which six to eight weekly applications were used. In these cases, blocks containing one to several plant types, with one to several replicates within each block, were sprayed with the same compound and concentration. Spray shields were used to eliminate drift and three or more similar blocks were randomized throughout the experimental area.

In slat shed bactericide studies, ground bed plots were employed, with three or more per treatment and 5–20 test plants within each plot.

In soil fungicide evaluation studies, noninfested but chemically treated controls were included for phytotoxicity determinations. Each pot was placed on a steam-pasteurized wooden block to reduce the potential of cross-contamination.

Most tests of efficacy and phytotoxicity were repeated at least once before reaching a conclusion on the potential value of the treatment for use in the ornamental tropical foliage plant industry.

**Inoculation and disease development**—With foliar fungal and bacterial inoculations, test plants and their respective controls were first watered thoroughly, then placed under overhead mist irrigation (15 sec every 15 min) for several to 24 hr immediately before inoculation. A specific volume of spore or bacterial suspension inoculum was applied to the upper and lower leaf surfaces of the plants to be inoculated with a previously autoclaved atomizer. Noninoculated controls received a similar amount of sterile deionized water. To promote proper incubation after inoculation, moistened paper toweling was placed on the soil surface of each pot and the plant and pot covered with a new plastic bag secured to the pot with a rubber band. This unit was then placed under intermittent mist for two to seven (foliar fungal pathogens), one to three (*Erwinia* spp.), or four to ten (*Xanthomonas* spp.) days. This incubation procedure provided a continuous film of moisture on the foliage and foliar temperatures cool enough to promote fungal spore germination and fungal and bacterial ingress into the foliage. At the conclusion of the incubation period, the mist was discontinued, the bags removed, and the pots handled as stated previously.

In slat shed studies involving bacterial pathogens, plants were watered for 1 hr by overhead sprinklers immediately before inoculation. A uniform volume of bacterial suspension was applied to each treatment with a 7.6-L stainless steel hand sprayer. After inoculation, the overhead sprinklers were turned on for 1 min every 15–30 min, the frequency depending on

**TABLE 9.2.** Soil fungicide rates for drench applications to pots

| Pot size (cm) | Pot shape | Area[a] (cm$^2$) | Vol (ml) applied to achieve | | |
|---|---|---|---|---|---|
| | | | 946 ml/ 929 cm$^2$ | 709 ml/ 929 cm$^2$ | 473 ml/ 929 cm$^2$ |
| 10.2 | Square | 104.0 | 104 | 78 | 52 |
| 12.7 | Round | 126.5 | 128 | 96 | 64 |
| 15.2 | Round | 181.4 | 184 | 138 | 92 |
| 17.8 | Round | 248.7 | 252 | 189 | 126 |
| 20.3 | Round | 323.5 | 328 | 246 | 164 |
| 22.9 | Round | 411.6 | 416 | 312 | 208 |
| 25.4 | Round | 506.4 | 516 | 387 | 258 |

[a]Determined at rim of pot.

[2]Allied Chemical Corp., Morristown, NJ. Containing principal functioning agents emulsifiable A-C polyethylene and octylphenoxy polyethoxy ethanol.

environmental drying conditions. This schedule was in effect for 8 hr after inoculation and for 4 hr the following morning.

With tests involving soilborne fungal pathogens, infestation of pots with potato-dextrose broth cultures was accomplished by pipetting 25–50 ml of inoculum uniformly over the soil surface. With other culture media, a measured volume of inoculum (2.5–10 cm$^3$) was placed into a 2- to 3-cm hole made in the soil surface at the corner or center of each pot. Noninfested treatments received the same quantity of sterile media. Pots were then watered carefully by hose and fertilized as stated previously.

Data were taken periodically throughout and at the conclusion of the tests using one or more ratings based on severity of leaf blight (foliar fungal and bacterial diseases); the number of lesions produced (foliar fungal); the severity of stem rot, root rot, wilt, and collapse (soilborne and bacterial pathogens); and character and severity of foliar injury (phytotoxicity studies). Quantitative data taken at the conclusion of the tests included one or more of the following: fresh weight of roots and tops, height of tops, and number of leaves or nodes produced. Data were subjected, in most cases, to analysis of variance, with means compared by Duncan's multiple range test.

### Bibliography

ALFIERI, S. A., Jr., and J. F. KNAUSS. 1970. Southern blight of schefflera. Proc. Fla. State Hort. Soc. 83:432-435.

ALFIERI, S. A., Jr., and J. F. KNAUSS. 1972. Stem and leaf rot of peperomia incited by *Sclerotium rolfsii*. Proc. Fla. State Hort. Soc. 85:352-357.

KNAUSS, J. F. 1970. Ascochyta leaf spot, a new disease of leatherleaf fern, *Polystichum adiantiforme*. Proc. Trop. Reg. Amer. Soc. Hort. Sci. 14:272-279.

KNAUSS, J. F. 1971. Rhizoctonia blight of Florida Ruffle fern and its control. Plant Dis. Rep. 55:614-616.

KNAUSS, J. F. 1972. Description and control of Pythium root rot on two foliage plant species. Plant Dis. Rep. 56:211-215.

KNAUSS, J. F. 1972. Foliar blight of *Dionaea muscipula* incited by *Colletotrichum gloeosporioides*. Plant Dis. Rep. 56:391-393.

KNAUSS, J. F. 1972. Resistance of *Xanthomonas dieffenbachiae* isolates to streptomycin. Plant Dis. Rep. 56:394-397.

KNAUSS, J. F. 1972. Field evaluation of several soil fungicides for control of *Scindapsus aureus* cutting decay incited by *Pythium splendens*. Plant Dis. Rep. 56:1074-1077.

KNAUSS, J. F. 1973. Description and control of a cutting decay of two foliage species incited by *Rhizoctonia solani*. Plant Dis. Rep. 57:222-225.

KNAUSS, J. F. 1973. Rhizoctonia blight of syngonium. Proc. Fla. State Hort. Soc. 86:421-424.

KNAUSS, J. F. 1974. Nurelle, a new systemic fungicide for control of Phytophthora crown rot of *Peperomia obtusifolia*. Proc. Fla. State Hort. Soc. 87:522-528.

KNAUSS, J. F. 1974. Pyroxychlor, a new and apparently systemic fungicide for control of *Phytophthora palmivora*. Plant Dis. Rep. 58:1100-1104.

KNAUSS, J. F. 1975. Field evaluations for control of Phytophthora blight of petunia. Plant Dis. Rep. 59:673-675.

KNAUSS, J. F. 1975. Control of basal stem and root rot of Christmas cactus caused by *Pythium aphanidermatum* and *Phytophthora parasitica*. Proc. Fla. State Hort. Soc. 88:567-571.

KNAUSS, J. F., and S. A. ALFIERI, Jr. 1970. Dactylaria leaf spot, a new disease of *Philodendron oxycardium* Schott. Proc. Fla. State Hort. Soc. 83:441-444.

KNAUSS, J. F., and J. W. MILLER. 1972. Description and control of the rapid decay of *Scindapsus aureus* incited by *Erwinia carotovora*. Proc. Fla. State Hort. Soc. 85:348-352.

KNAUSS, J. F., and J. W. MILLER. 1974. Etiological aspects of bacterial blight of *Philodendron selloum* caused by *Erwinia chrysanthemi*. Phytopathology 64:1526-1528.

KNAUSS, J. F., D. B. McCONNELL, and E. HAWKINS. 1970. The safety of fungicides and fungicide-insecticide combinations for selected foliage plants. Fla. Foliage Grower 8:1-10.

KNAUSS, J. F., W. E. WATERS, and R. T. POOLE. 1971. The evaluation of bactericides and bactericide combinations for the control of bacterial leaf spot and tip burn of *Philodendron oxycardium* incited by *Xanthomonas dieffenbachiae*. Proc. Fla. State Hort. Soc. 84:423-428.

MARLATT, R. B., and J. F. KNAUSS. 1974. A new leaf disease of *Aechmea fasciata* caused by *Helminthosporium rostratum*. Plant Dis. Rep. 58:446-448.

# 10. Greenhouse Evaluation of Soil-Applied Fungicides for Fusarium Wilt of Chrysanthemum[1]

## Arthur W. Engelhard

Professor and plant pathologist, IFAS, University of Florida, Agricultural Research and Education Center, Bradenton, FL 33508

This method is intended for use in greenhouse evaluation of fungicides applied to the soil of potted plants for the control of Fusarium wilt of chrysanthemum.

### Test Fungus

The fungus suggested for use (because of its common occurrence) is *Fusarium oxysporum* f. sp. *chrysanthemi* Litt., Armst. and Armst. Another fungus, *F. oxysporum*

f. sp. *tracheiphilum* race 1 of cowpea, is also a known wilt pathogen of chrysanthemum, but is rarely found.

*F. oxysporum* f. sp. *chrysanthemi* sporulates readily when grown in petri plates on potato-dextrose agar (PDA) under continuous cool-white fluorescent lights at 27°C (80°F). Mycelial growth covers the plates completely within ten days. Sufficient spores are produced on ten plates to inoculate at least 60 pots. Conidia are harvested by flooding the petri plates with deionized or distilled water, rubbing the agar surface gently with a rubber policeman, and filtering the resulting suspension through cheesecloth. The culture

[1]Florida Agricultural Experiment Stations Journal Series No. 183.

should be derived from a single-spore isolate to increase genetic uniformity and insure a pure culture.

## Test Plants

Chrysanthemums should be planted one per 10-cm pot in a well-drained sterilized soil such as one composed of one-third peat and two-thirds sandy soil. The soil may be sterilized by steaming for 30 min at 82°C (180°F). The pH of the soil should range from 5.5 to 6.0. The plants should be fertilized with a balanced nutritional program that supplies the major and minor elements needed for good, vigorous plant growth. Nitrogen should be applied as ammonium nitrate. Using an ammonium source of nitrogen increases disease development, while an *all* nitrate source decreases it (2,4). A suggested fertilizer mix is 1.5 g of $NH_4NO_3$, 1.0 g of $NaH_2PO_4 \cdot H_2O$, 0.85 g of KCl, and 0.2 g of $MgSO_4 \cdot 7H_2O$ per liter. Apply 50 ml/10-cm pot twice weekly. Minor elements may be added to the soil mix, applied as a soil drench, or sprayed on the foliage.

Susceptible cultivars that may be used are Delaware and Yellow Delaware, although others also may be used (1). The most desirable plants for experimental purposes are young, vigorously growing plants.

## Inoculation and Incubation Procedure

The roots of a plant are injured by making a vertical cut with a knife into the soil 2 cm from the stem and 8 cm deep on opposite sides of the plant (2). Fusarium is a wound pathogen, and this procedure facilitates the infection process. The spore suspension is poured into the cuts. Good infection is obtained when at least ten million spores are applied in a 50-ml suspension to each 10-cm pot (200,000 spores/ml). Inoculation should be done 11 days after potting, which is one day after the second fungicide drench. Foliage symptoms (chlorosis, wilt) should become apparent on the controls about eight days later under ideal conditions for pathogenesis. Temperature is extremely important in the development of Fusarium wilt. It should not be allowed to drop below 24°C. Daytime temperature should be maintained within the 26–30°C range. No or few symptoms will develop at lower temperatures.

## Replication and Fungicide Application

All treatments, including the known standard fungicide and the water control, should be replicated a minimum of four times. Each replicate may consist of a single plant.

Fungicide drenches should be applied three and ten days after the plants are potted. The liquid should wet the soil mass. Usually about 50 ml/10-cm pot are required.

## Standard Fungicide

Benomyl 50W at 0.3 and 1.2 g/L (0.25 and 1.0 lb/100 gal) should be used as a standard treatment. When an ammonium source (ammonium nitrate, ammonium sulfate) of nitrogen is used as a sole nitrogen source in the nutritional program, disease should be severe at the 0.3-g/L rate, while control should be at a high level at the 1.2-g rate. The possibility of tolerance of the pathogen

should be investigated if benomyl ceases to give good control. This problem is overcome by using a nontolerant strain, and should not occur if the pathogen is maintained in sand culture in the absence of benomyl.

In research or more advanced testing, the relative effectiveness of a compound in controlling wilt may be evaluated using the integrated fungicide/lime/nitrate regime (2). The pH of the soil should be maintained at 6.5–7.0, and an all nitrate-nitrogen source (calcium, potassium, or sodium nitrate) substituted for the ammonium source. Complete control of symptoms is obtained when benomyl 50W is drenched on the soil at 0.3 g/L as previously outlined (2,3).

Woltz and Jones (5) have discussed the role of pH and nitrogen source in reducing wilt.

## Rating System

Fusarium wilt may be rated on a 0–5 basis as follows: 0, no visible leaf symptoms; 1, leaf chlorosis or vascular discoloration or both in one or more leaves; 2, chlorosis or vascular discoloration or both, plus epinasty of leaf (leaves) or stem or both; 3, leaf symptoms plus wilting; 4, leaf symptoms, wilting, and stunted growth of plant; and 5, dead plant.

Percentage of disease control is based on the formula:

Percentage of disease control =

$$\frac{\text{Mean disease} \quad \text{Mean disease}}{\text{rating in control} - \text{rating in treatment} \times 100}$$
$$\frac{}{\text{Mean disease rating in control}}$$

Additional rating indexes that may be employed are measuring the height of the plants, weighing the plants, and rating the flowering plants for commercial desirability either as cut flowers or as potted plants.

## Phytotoxicity

The crop should be carefully inspected at regular intervals for any indication of phytotoxicity. Phytotoxicity symptoms range in severity from death of the plants to marginal chlorosis or necrosis or both of the leaves, to plants that have no visible symptoms other than retarded growth. The kind and amount of phytotoxicity should be described and recorded.

## Statistical Analysis

The treatments should be analyzed by an appropriate method that may include an LSD or Duncan's multiple range rating system.

## Literature Cited

1. ENGELHARD, A. W., and S. S. WOLTZ. 1971. Fusarium wilt of chrysanthemum: Symptomatology and cultivar reactions. Proc. Fla. State Hort. Soc. 84:351-354.
2. ENGELHARD, A. W., and S. S. WOLTZ. 1973. Fusarium wilt of chrysanthemum: Complete control of symptoms with an integrated fungicide-lime-nitrate regime. Phytopathology 63:1256-1259.
3. ENGELHARD, A. W., and S. S. WOLTZ. 1973. Fusarium wilt of chrysanthemum: A new cultural-chemical control method. Fla. Flower Grower 10(9):1-3.

4. WOLTZ, S. S., and A. W. ENGELHARD. 1973. Fusarium wilt of chrysanthemum: Effect of nitrogen source and lime on disease development. Phytopathology 63:155-157.

5. WOLTZ, S. S., and J. P. JONES. 1973. Interactions in source of nitrogen fertilizer and liming procedure in the control of Fusarium wilt of tomato. HortScience 8:137-138.

# 11. Greenhouse Method for Screening Protective Fungicides for Apple Powdery Mildew

## H. L. Dooley

Environmental Protection Agency, Northwest Biological Investigations Station, 3320 Orchard Avenue, Corvallis, OR 97330

Several varieties of apple (*Malus sylvestris* Mill.), such as Gravenstein, Newtown, Jonathan, and Rome, are susceptible to powdery mildew. *Podosphaera leucotricha* (Ell. & Ev.) Salm., an obligate parasite, causes powdery mildew of apple and is a major problem in the western United States.

The method described in this chapter was designed to screen foliar-applied fungicides used for controlling powdery mildew of apple. The candidate fungicides selected by this method should be further evaluated in the field.

Advantages of the method are: (i) the test can be completed in five weeks, (ii) little test space is required, (iii) seedling trees can be used, (iv) low-cost equipment is used, (v) many fungicides can be evaluated, (vi) plants can be reused after eliminating disease by applying an eradicant or by severe pruning, (vii) small amounts of fungicide are required, (viii) the fungicide is challenged by the presence of the organism, and (ix) inoculum levels can be controlled. Disadvantages are: (i) timing of field applications cannot be evaluated, (ii) effects of the fungicide on fruit cannot be evaluated, and (iii) interaction of fungicides with temperature, relative humidity, light, and other variable environmental conditions escape evaluation and thus may cause effective chemicals in the greenhouse to be worthless in the field.

## Method

The fungus is maintained on susceptible Newtown apple seedlings in a separate greenhouse until needed as inoculum for a test. Heavily infected plants are introduced into the test greenhouse after the first sprays or dusts are applied and remain there throughout the experiment. Mature conidia are released from the infected plants daily with an air jet of 60–80 psi (414–552 KPa).

**Test plants**—Disease-free Newtown apple seedlings are used. Other seedlings may be used if they are susceptible to the fungus. Viable apple seeds are planted and germinated in a greenhouse flat containing sand or other suitable medium. After emergence, the seedlings are transplanted to 6-in. (152-mm) pots in potting media consisting of one part sand, one part loam, and one part peat moss.

Test seedlings should be grown in a heavily-shaded greenhouse. Supplemental light is provided for 10 hr by alternating warm- and cool-white fluorescent tubes spaced 6 in. apart over the bench. Greenhouse temperatures are maintained for 12-hr periods, the day being 21 ± 2°C and the night being 16 ± 2°C. Evaporative coolers may be used in areas of low humidity. Plants are maintained in good cultural condition by fertilizing every 21 days, or as needed, with an appropriate fertilizer.

Disease-free seedlings that are uniform in height and having five to six leaves are selected. Pruned seedlings with new, disease-free shoots of similar size may also be used. Their new shoots are susceptible to infection by *P. leucotricha*. The test begins with short seedlings, since they may grow 12-16 in. during the experiment.

**Procedure**—Plants are arranged on the greenhouse bench in a randomized block design according to Snedecor's randomly assorted digit table (5) or the equivalent. Each treatment should be replicated a minimum of four times. A standard fungicide and an untreated control should be included as treatments in each test.

The fungicide should be applied at weekly intervals and at least four times. Test plants are removed from the greenhouse, placed on a compound rotating turntable (3), and sprayed to runoff with a De Vilbiss paint spray gun with a No. 30 head using 25–30 psi (172–207 KPa) of pressure. Plants are dusted with a vacuum duster (1). Enough dust should be in the cup to cover the plant at 6 in. (152 mm Hg) of vacuum. After spraying or dusting, the plants are returned to the greenhouse.

Heavily mildewed plants are placed in the center row of the bench, between the second and third replications and after the first treatments have been applied. One infected plant should be used for every four test plants. Conidia should be discharged daily during the test with a 60–80 psi (414–552 KPa) air jet. The airblast is started from a different corner of the bench each day, making a complete circle around it. In four days, the conidia will have been blown from a different direction each time, thus allowing more uniform dispersal. This procedure must be continued throughout the test. Air currents from fans will also discharge conidia and can be used instead of pressurized air provided that allowance is made for equal distribution of inoculum to all plants.

A standard fungicide known to be effective, such as Parnon 1.3% emulsifiable concentrate used at the rate of 0.5 pt/100 gal of water (236.6 ml/378.5 L), would be acceptable. Untreated control plants (not sprayed or dusted) are necessary to determine the disease incidence and to provide a basis for treatment comparison. Both the standard and untreated controls should be replicated as treatments.

## Data Collection

Each plant is carefully inspected, and the incidence of powdery mildew on the foliage one week after the fourth spray application is visually rated. This is done subjectively by deciding whether the surface area is covered by more or less than 50% disease. If more than half of the plant is affected, one must decide whether its foliage is above or below the 75% disease level. If that disease incidence falls between 50 and 75%, the percentage is divided in half until the correct disease figure is determined. This procedure is also effective in calculating lower disease percentages.

The mean percentage of disease incidence for each treatment is found by adding the readings of all replications and dividing by the number of replications.

The mean percentage of disease control for each treatment is calculated with the following formula (2,4):

$$\text{Mean PDC} = \frac{\begin{array}{cc}\text{Mean} & \text{Mean} \\ \text{percentage} - \text{percentage} \\ \text{of DIC} & \text{of DIT}\end{array}}{\text{Mean percentage of DIC}} \times 100$$

where PDC is percentage of disease control; DIC, disease incidence in control; and DIT, disease incidence in treatment.

Phytotoxicity should be reported and described as to type and mean percentage if applicable. Common types of phytotoxicity are leaf burning, chlorosis, vein banding, and stunting.

## Results

Results should include the mean percentage of disease incidence that occurs in each treatment, standard, and check. The mean percentage of disease control for each treatment and standard is also required. The percentage of phytotoxicity and a description of its symptoms should be given.

Products are considered effective and ready for field testing if the mean percentage of disease control rating is 70 or more. Seventy percent disease control was chosen arbitrarily to allow for questionable products that may perform better under low inoculum pressures.

Results may be invalidated by insect infestations that may cause defoliation or leaf symptoms.

### Literature Cited

1. FARRAR, M. D., W. C. O'KANE, and H. W. SMITH. 1948. Vacuum dusting of insects and plants. J. Econ. Entomol. 41:647-648.
2. HORSFALL, J. G., and R. W. BARRATT. 1945. An improved grading system for measuring plant diseases. Phytopathology 35:655.
3. MC CALLAN, S. E. A., and R. H. WELLMAN. 1943. A greenhouse method of evaluating fungicides by means of tomato foliage diseases. Contrib. Boyce Thompson Inst. 13:93-134.
4. REDMAN, C. E., E. P. KING, and I. F. BROWN, JR. 1962. Tables for converting Barratt and Horsfall rating scores to estimate mean percentages. Eli Lilly and Company: Indianapolis, IN.
5. SNEDECOR, G. W., and W. G. COCHRAN. 1967. Statistical methods. Ed. 6. Iowa State University Press: Ames, IA.

# SECTION III.

# Field Test Procedures

# A. Tree Fruits and Nuts

## 12. Testing Chemicals in the Field for Apple Powdery Mildew Control in Colorado[1]

### N. S. Luepschen

Colorado State University, Orchard Mesa Research Center, Grand Junction, CO 81501

The procedure described in this chapter has been used successfully in the field for four seasons to evaluate the efficacy of fungicides against powdery mildew of apple trees (*Malus sylvestris* Mill.) caused by *Podosphaera leucotricha* (Ell. & Ev.) Salm. The method used is particularly suitable for the climate found in the western United States intermountain plateau: arid conditions (8–12 in. [20–51 cm] annual rainfall) requiring irrigation of orchards, hot summers, cold winters, and abundant sunshine. Powdery mildew is the main foliar disease affecting apples. The absence of apple scab allows us to omit all other sprays for diseases. One exception is control of fire blight whenever it exists. Our procedures have some aspects that are worthy of consideration even under humid conditions, including determining the mite inhibition properties of mildewcide materials.

### Technique

**Orchard site design and management**—Spray trials were conducted in ten-year-old orchard blocks of Red Rome and Jonathan apple trees located at Colorado State University, Orchard Mesa Research Center. The trees were on seedling rootstocks planted on 20 × 20-ft (6 × 6-m) spacing in a calcareous clay loam alluvium-type soil over porous, gravelly alluvium, which is typical of orchard soils found on the mesas in western Colorado mountain valleys. Irrigation was by open furrow or rill method, using three to five seasonal applications based

on soil moisture, crop load, and prevailing temperatures. Insecticide sprays were applied separately with a speed sprayer. They consisted of a delayed dormant spray (Diazinon plus oil) and four cover sprays (Guthion, Zolone, or Sevin) for codling moth, but no miticides until after mite counts were made. Weed control consisted of mowing the middles and spraying with herbicides (2,4-D and simazine) in the rows.

**Inoculum**—The powdery mildew fungus overwinters primarily as dormant mycelium in terminal buds. The leaves emerging from these buds usually become infected and bear conidia, which serve as inoculum for continuing cycles of infection on young leaves and fruits. The cleistothecial (sexual) stage has not been observed in Colorado.

A minimal powdery mildew fungicide program (one to three applications of Karathane) was used in these blocks the year before the test to allow sufficient inoculum buildup. Considerable inoculum develops on unsprayed check treatment trees and buffer row trees. Depending on weather conditions, unsprayed trees usually have 40–90% of the foliage infected with mildew. Mildew russeting of the fruit has been minimal.

**Experimental design and treatment application**—Replicated single-tree plots were used. A randomized block design of treatments in the row and buffer rows between treatment rows were used to negate spray drift due to prevailing east to west winds. Two six-row blocks each of Jonathan and Red Rome trees permit four complete tests each year. Each block contains 114 trees, or a 1.1-acre (0.45-ha) area. In practice, seven single-tree treatments were randomized within each

[1]Journal Series Paper No. 2139, Colorado Agricultural Experiment Station.

replicate, with replicates in sequence in test rows. Rows of nonsprayed trees bordered the test rows in the following pattern: rows two and five were treatment rows, and rows one, three, four, and six were border and buffer rows. Each test tree was flagged by color code for each treatment. The treatments were replicated five times.

Each tree was sprayed to the point of runoff using a Bean Spray-Miser handgun and a Hardie 150-gal, high-pressure sprayer at 350–400 psi (2.41–2.76 KPa). Test fungicides were applied at the dilute rate. In full foliage, about 4 gal (15.12 L) were applied per tree. Triton B-1956 was used as a spreader-sticker with all sprays at the rate of 4 oz/100 gal (120 ml/378 L) of water. Application of fungicides usually began at early pink stage, and was repeated at 14-day intervals (± two days) for a total of five to seven applications, to provide protection until terminal growth ceased.

**Treatments**—Karathane or wettable sulfur or both at the official recommended dosage was included each year as a standard for comparing the new experimental compounds or formulations. An unsprayed check tree was also included in each of the five replications. All fungicide treatments were applied separately from the insecticide program, which included all the trees in the block.

**Data collection**—The apple trees were evaluated for powdery mildew incidence in July. At this time, shoot growth had ceased, terminal buds were formed, and leaf tissue was mature. Daytime temperatures ranged from 90 to 115°F (32 to 46°C), with low relative humidity. Such conditions are unfavorable for further mildew development. Mite counts for two-spotted spider mites *(Tetranychus urticae)* were made at the same time or two to three weeks later.

Since concise mildew observations are difficult to make in full sunlight outdoors, ten current-season shoots were clipped at random from each tree with hand shears. They were cut from the outer circumference of the foliar canopy, placed in marked bags, and brought into the laboratory. The average number of leaves per shoot, which varies with tree vigor, was determined for the variety and block of trees used. The leaves were

visually examined for mildew mycelium or symptomatic leaf distortion caused by infection. The percentage of leaves infected per shoot was determined by counting the infected leaves and multiplying this number by the percentage factor. For example, with trees averaging 20 leaves per shoot, 8 infected leaves times 5 equals 40%; with 18 leaves per shoot, 8 infected leaves times 5.6 equals 44.8%. The ten readings per tree were averaged to determine the percentage of terminal shoot foliage infected for each tree.

Fruit samples for residue analyses (performed by sponsoring chemical companies) were harvested just before commercial picking time. Fruits sprayed with experimental compounds, or with unlabeled dosages of proprietary materials, were left to drop on the orchard floor and be disked under, or removed to a dump site. The fruit residue samples were placed into polyethylene bags, labeled, frozen at 0°F (−17.8°C) and later shipped by air freight in dry ice in insulated cartons to the respective residue laboratories.

For obtaining two-spotted spider mite counts, five to ten leaves were collected from each tree, loosely bagged, and examined under a binocular dissecting microscope in the laboratory. The lower leaf surface was scanned, and the number of mites recorded. An average number of mites per leaf was determined for each tree.

## Results and Discussion

Table 12.1, taken from 1975 results (4), illustrates a typical experiment with six spray treatments. Data presentation in tabulated form is similar to that used for Fungicide-Nematicide Tests Results (1), is uncomplicated enough for grower comprehension, yet presents a complete picture of the test. Chemicals are listed in terms of formulated product. Complete descriptions of the fungicides, including trade name, accepted common name or chemical name, amount of active ingredient, and manufacturer are listed in yearly progress reports (3,4,5).

The technique presented here provides reasonably accurate comparisons under field conditions. Results have been directly responsible for adding new disease control recommendations (2) as well as evaluating new compounds in cooperation with manufacturers. The data collection method is a severe test of the fungicide, since many of the infected leaves counted in the detailed laboratory evaluation would be overlooked in the field. Under commercial orchard conditions, the 75% control achieved with Benlate at the 6-oz (170-g) rate (Table 12.1) probably represents good control.

Weather variations affect test results. In years of low mildew incidence, finding significant differences between compounds or varying dosages is difficult. In such years the check trees may have only 10% of their leaves infected. Another variable concerns the overwintering inoculum in terminal buds, which is influenced by the previous season's spray treatment. To counter the effect of the previous season's program, single-tree plots need to be rerandomized each year, or the test block should be used only in alternate years. This allows an equalization of mildew inoculum in off years by using uniform or no treatment. For some purposes, however, a two- to three-year program of the same fungicide treatments may be desirable to observe long-range efficacy and to evaluate host response. For

**TABLE 12.1. Powdery mildew and two-spotted spider mite incidence in fungicide trials in 1975 with Red Rome apples**[a]

| Treatment | Rate per 100 gal (378.5 L) | Leaves infected July 31 (%) | Mites per leaf Aug. 20 |
|---|---|---|---|
| Unsprayed checks | | 52 a[b] | 60 a |
| Karathane 25 WD | 8 oz (226 g) | 26 b | 4 b |
| Benlate 50W | 6 oz (170 g) | 13 c | 3 b |
| Benlate 50W + oil | 2 oz + 1 qt (57 g + 946 ml) | 14 c | 4 b |
| Benlate 50W + Manzate 200 | 6 oz (170 g) 24 oz (680 g) | 12 c | 2 b |
| Sulfur 92 Wp | 6 lb (2.7 kg) | 7 c | 4 b |
| Sulfur, flowable | 3 pt (1.4 L) | 5 c | 4 b |

[a]Six applications were made at 14-day intervals beginning at pink stage.

[b]Values within each column followed by same letter are not significantly different at 5% level, Duncan's multiple range test.

example, a two-year Benlate and oil program appears to result in greener, heavier foliage (4). To observe short- and long-term treatment differences as they affect tree growth and crop yields, additional data have been or are planned to be taken on trunk circumference, length or weight of shoot growth, and size and yield of apples produced for each individual tree. Since phytotoxicity has been a minor problem, observations generally have been summarized in a few general statements, with no special detailed readings.

Another modification to improve reliability would be to decrease treatments to five or less, and to increase the number of replications within the test block. The 20×20-ft planting distance is satisfactory, since the tree canopies are kept about 3 ft (0.9 m) apart by moderate pruning. Use of semidwarf rootstocks allows more trees per acre, and perhaps provides buffer trees between each tree in the row as well as buffer rows on each side of treatment rows.

Including observations on spider mite suppression has resulted in a broader comprehension of the effect of fungicides on the orchard environment. A secondary beneficial role of mildew fungicides is seen in mite suppression where such populations are a major orchard problem.

The procedures outlined here are most applicable on experiment station facilities where commercial production can be simulated but commercial sales are not the first priority. Fungicide testing in grower orchards using speed sprayer applications to entire rows has also been done successfully using Mills' method (6).

## Literature Cited

1. AVERRE, C. W. (ed.). 1976. Fungicide and Nematicide Tests—Results of 1975. Vol. 31. American Phytopathological Society: St. Paul, MN.
2. HANTSBARGER, W. M. 1976. Fruit spray handbook. Colorado State University Extension Service: Fort Collins, CO.
3. LUEPSCHEN, N. S., H. H. HARDER, and L. E. DICKENS. 1974. Apple powdery mildew trials. Colorado State University Experiment Station Progress Report 74-2.
4. LUEPSCHEN, N. S., J. E. HETHERINGTON, F. J. STAHL. 1976. Powdery mildew and two-spotted spider mite control on apples: 1975 Results. Colorado State University Experiment Station Progress Report 76-1.
5. LUEPSCHEN, N. S., and R. D. MORRIS. 1975. Apple pesticide research for 1974: Fungicides, their effect on powdery mildew and two-spotted mites, and concentrate spraying for codling moth. Colorado State University Experiment Station Progress Report 75-1.
6. MILLS, W. D. 1959. Evaluations of fungicides in the field. In: HOLTON, C. S. (ed.). Plant pathology problems and progress 1908-1958. American Phytopathological Society, University of Wisconsin Press, Madison, pp. 257-261.

# 13. Method for Field Evaluation of Fungicides for Apple Powdery Mildew Control

## K. D. Hickey

Professor of plant pathology, The Pennsylvania State University Fruit Research Laboratory, Biglerville, PA 17307

Powdery mildew caused by the fungus *Podosphaera leucotricha* (Ell. & Ev.) Salm. is common on apple *(Malus sylvestris* Mill.) wherever this fruit is grown. The severity of the disease varies greatly with such factors as susceptibility of the cultivar, environment, magnitude of the fungus population, and kind and amount of fungicide used in the orchard. On susceptible cultivars such as Jonathan and Rome Beauty, use of an effective fungicide is often necessary to maintain healthy, productive trees. Other cultivars such as Delicious and Golden Delicious may need only a moderate amount of fungicide to maintain a satisfactory level of mildew control. Because of variations between orchards and production areas, research methods for testing mildew fungicides that are sufficiently flexible to meet the needs in each region are desirable.

The method described in this chapter is adaptable to tests using single or multiple-tree plots, depending on the number of usable trees in the orchard, disease pressure, number of treatments, amount of test compound available, type of application (dilute or concentrate), and type of program (protective or eradicative). Several parts of the procedure may be used in the evaluation of fungicides for control of other diseases.

## Technique

**Test site conditions**—The test site should be in an area with climatic conditions favorable for disease development, preferably one with a history of appreciable powdery mildew infection. The soil should be well drained and have enough fertility to assure average or above average growth of the tree. Fruits are useful, although not essential, to fungicide evaluation. Therefore, good air drainage is desirable to minimize the frost injury that often occurs in trees growing in low areas. Wind on high hilltop sites can interfere with proper fungicide application, leading to inadequate coverage or drift contamination from the various fungicide treatments being tested.

Cultivars of apple vary widely in susceptibility to powdery mildew. Only those that are of moderate to high susceptibility should be used. Cultivars of Jonathan, Rome Beauty, Cortland, Idared, Monroe, Baldwin, Gravenstein, Jonamac, Macoun, Stayman, Wayne, and others known to be susceptible are suitable for test plants. In some areas, cultivars of light to moderate susceptibility have been successfully used under high disease pressure, because lesions may be abundant and discrete. On highly susceptible cultivars, lesions often coalesce and leaves may be distorted. Trees should be planted in rows at spacings recommended for the particular rootstock-cultivar combination, or spaced at distances suitable to the application equipment. Trees should be at least in their fourth year of growth in the orchard site to provide an adequate number of shoots and acceptable fruit yield. Younger trees are susceptible, but will not produce enough uniform terminal shoot growth for good comparison. The trees should be sufficiently pruned and fertilized, and the orchard floor mowed. These practices allow for optimum tree growth and maximum penetration of spray droplets.

**Conditions favoring disease development**—The fungus is an obligate parasite that overwinters in infected buds on shoots (terminals) produced during the previous growing season. Fruiting or nonfruiting spurs may also be sites of overwintering, but most infected buds are located near the apex of the shoot. Infection on the first leaves that open results from conidia produced on these opening buds. Because opening of infected buds is slightly delayed, the first infections occur around the tight cluster stage of healthy bud development. Secondary spread results from inoculum produced by both primary and secondary infections.

According to Molnar (4), conidia of the mildew pathogen germinate and cause infections at temperatures between 4 and 30°C, when the relative humidity is between 70 and 100%. Leaves are susceptible to infection for only a few days after they emerge, but can be infected at any time if injured. Fruit infection occurs while the fruit is young—between pink bud stage and petal fall, or a few days later. The time of infection on developing buds, which carry the fungus over winter, is not well understood, but presumably occurs from midseason until the apical bud on the shoots has hardened off.

A high level of inoculum in the test site is essential for proper fungicide evaluation. Inoculum from infected shoots that develop from buds in which the fungus has overwintered is the most desirable type. Infected potted trees grown in the greenhouse and placed in the tree effectively provide inoculum. Excised infected shoots, either primary or secondary, also may be placed in the test trees in a container with water to prevent rapid wilting.

**Test procedure**—Treatments should be applied in a randomized block design according to Snedecor's randomly assorted digits table (5), or a similar random digit table or method. Sprayed plots consisting of single or multiple trees should be replicated four to six times. They should be arranged to prevent drift of spray droplets by leaving an unsprayed border row between sprayed rows or by having sufficient space between treated plots. By leaving a number of unsprayed trees in the test site, a heavier inoculum pressure can be obtained for both the present test and tests the following year. One of the replicated plots should serve as the unsprayed check to determine disease severity. In sites in which other diseases must be controlled, captan, or another fungicide ineffective against apple mildew, may be used instead of an unsprayed check. One of the plots should be treated with a commonly recommended fungicide to serve as a standard for comparison.

Treatments should be applied first at the tight cluster stage and continued at five- to seven-day intervals through bloom, and at 10- to 14-day intervals from petal fall through the shoot (terminal) growth period. Several application methods may be used, but a single one should be selected and used for the entire season for any specific test. Use of a high-pressure sprayer (21−35 kg/cm²) with handgun or spray-nozzle boom to spray the trees to the point of runoff (200−400 gal per acre [2,245−4,491 L/ha] for mature trees) is particularly desirable where small- to medium-size trees are used, or when the amount of test fungicide is limited.

## Data Determinations and Reporting

**Secondary mildew**—Visual estimates of disease incidence (3), or actual random counts of infected leaves and fruits, should be recorded for each plot at the end of the growing season. When disease incidence is moderate to light, or a critical evaluation of fungicide effectiveness is needed, all infected leaves should be counted. The count should include observations of all leaves on 15−20 terminals (shoots) per plot selected at random from all sides of the tree. Terminals that are of uniform length are suitable, because their growth spans the entire growth period. Short shoots that terminate growth early in the season should be avoided. In years when heavy prebloom infections occur, leaves on blossoming clusters may be used to measure early season effects. A disease severity level should be established at least for plots that are either unsprayed or sprayed with a wholly ineffective mildewcide by recording the number of lesions per infected leaf. Fruit infections should be recorded by observing 50−100 fruits per plot. Data on leaf and fruit infection should be expressed as percentage of disease incidence or disease control.

**Primary mildew**—The effect of treatment on the amount of bud infection and subsequent overwintering

**TABLE 13.1.** Incidence of powdery mildew on Jonathan apple treated with pyrazophos 30 EC 8.0 oz/100 gal dilute, as measured by percentage of leaves infected and Horsfall-Barratt rating system

| Fungicide | Rate per acre | Powdery mildew on Rome | |
| --- | --- | --- | --- |
| | | Leaves infected[a] (%) | Leaf area affected[b] (%) |
| DuPont DPX-10 | 50W 1.0 lb | 21.2 a[c] | 21.8 a |
| Topsin M, plus Thylate | 70W 0.5 lb 65W 3.0 lb | 29.3 b | 25.2 ab |
| Benlate, plus Polyram | 50W 0.5 lb 80W 4.0 lb | 37.5 c | 26.2 ab |
| Karathane, plus Cyprex | 25W 1.5 lb 65W 1.5 lb | 37.2 c | 28.6 bc |
| Cela-Merck W524 | 20 EC 40.0 oz | 42.1 cd | 34.9 d |
| No fungicide | | 70.3 e | 41.8 e |

[a]Determined on July 17, 1975, by observing all leaves on 20 terminals (shoots) per single-tree replicate (five replicates).
[b]Determined as in footnote a by using Horsfall-Barratt ratings (0−11), then converting to percentages.
[c]C = cover spray.
[d]Small letters indicate Duncan's multiple range groupings of treatments, which do not differ significantly at 0.05 level.

may be measured the following spring at the pink stage by counting the number of opening buds that are infected. It may be measured by one or all of the following three methods: First, the number of primary infected terminals (shoots) per tree that appear silver due to the surface mycelial growth may be determined. All terminals on small- to medium-sized trees should be observed, but only a portion (two quadrants) may be observed on large trees. Second, the percentage of buds infected or dead on 20 terminals per tree that are not obviously infected (nonsilvered) may be determined by observing the ten most distal buds starting at the apical bud. Mycelium and spores of the fungus are apparent on infected buds at this stage. Because infection cannot be assured in dead buds that remain unopened, a separate account of dead and infected buds should be kept. Third, the percentage of terminals (shoots) having infected buds may be determined by observing 100 terminals per tree at random.

**Phytotoxicity determinations**—Observations for leaf necrosis or chlorosis should be made several times during the growing season. Additionally, any phytotoxicity to floral parts or young fruit that may affect fruit set should be recorded. Evaluation of fruit for russeting or reduction in color intensity can be made by randomly observing 25–100 apples per tree on unharvested trees late in summer. Evaluation of harvested fruit is preferable, however. The data should be expressed as percentage of surface affected based on standard ratings systems such as the Horsfall-Barratt method (3) or other recognized standard grading systems for fruit. Effect on crop yield may be measured by total weight per tree, area, or individual fruit size and weight. Caution must be exercised in interpreting the results, since many factors such as previous crop size per tree, cropping method, or unrelated pest control programs can affect tree performance. Yield data should be obtained from the same orchard, with the same treatments repeated on the same trees for a minimum of three years. Also, unless the plots are large enough and replicated, the significance of powdery mildew control on yield will be difficult to assess.

**Data analysis**—The data should be analyzed for the variance among treatments using conventional analysis of variance methods (5). The percentages may be transformed using angular transformation or other suitable statistical treatments. Significance among treatments may be calculated using Duncan's multiple range test for significance (2) or Duncan's modified (Bayesian) least significant difference test (6).

**Data reporting**—This method is suitable for tests conducted under a number of conditions. Information that should be included in reporting results include (i) general conditions under which test was conducted, including type and number of trees used, disease pressure (eg, inoculum level, weather), (ii) test compound name, formulation, active ingredient or ingredients, and percentage of each, (iii) how the compound was applied, rate used, concentration (dilute or concentrate), spray volume per tree, equipment used, and number of applications, (iv) number of replicates, mean disease incidence, specifics on when and how the data were secured, level of control obtained, and relative effectiveness compared with standard treatments, and (v) type and amount of phytotoxicity.

## Results and Discussion

Researchers have used this evaluation method for many years. It has proved highly reliable in assessing fungicide performance when observations are made on 15–20 terminals (shoots) per tree and 4–6 replicates are used. The types of data obtained using this method are given in Tables 13.1, 13.2, and 13.3. Researchers have debated for several years the best method of determining disease incidence. Determining the percentage of leaves infected is laborious and time consuming, and fails to account for variation in disease intensity. Thus, counting the number of lesions per leaf is also necessary, but is unfeasible on a highly susceptible cultivar such as Jonathan because lesions often coalesce. Covey (1) has used rating systems with scales ranging from 0.1 to 1.0 or 1 to 5 with various mathematical treatments of the ratings and varying success and acceptance. The Horsfall-Barratt rating method (3) has been used, and I prefer it.

Tables 13.1 and 13.2 compare the results obtained in field experiments when disease incidence was determined by counting leaves showing at least one lesion, and by the Horsfall-Barratt rating system after

**TABLE 13.2.** Comparison of apple powdery mildew counts obtained by determining percentage of leaves infected on Horsfall-Barratt rating system on trees sprayed with various fungicides

| Fungicide | Amount per 100 gal dilute | Powdery mildew on leaves | | | |
| | | Jonathan | | Rome Beauty | |
| | | Leaves infected[a] (%) | Leaf area affected[b] (%) | Leaves infected[a] (%) | Leaf area affected[b] (%) |
|---|---|---|---|---|---|
| Elanco EL-222 | 12.5 EC 4.0 oz | 11.3 a[c] | 2.6 a | 7.3 a | 2.0 a |
| Karathane WD | 0.5 lb + Orthocide 50W 2.0 lb | 24.5 b | 8.0 a | 27.0 b | 10.5 b |
| Rohm & Haas 3928 | 50W 6.0 oz | 34.8 bc | 10.8 ab | 32.3 b | 14.8 b |
| NorAm SN 513 | 30 EC 9.0 oz | 44.8 c | 24.8 bc | 53.8 c | 33.8 c |
| Abbott ABG-2000 | 50W 3.0 oz | 63.5 de | 34.1 c | 77.4 d | 59.6 d |
| Abbott ABG-2000 | 50W 8.0 oz | 68.5 e | 36.1 c | 67.3 d | 51.4 d |
| No fungicide | | 84.8 f | 84.1 d | 90.8 e | 89.8 e |

[a]Determined by observing all leaves on 20 terminals (shoots) per single-tree replicate (four replicates) on August 5, 1975 (Jonathan), and August 8, 1975 (Rome).
[b]Determined as in footnote a by using Horsfall-Barratt ratings (0–11), then converting to percentages.
[c]Small letters indicate Duncan's multiple range groupings of treatments, which do not differ significantly at 0.05 level.

**TABLE 13.3. Comparison of apple powdery mildew counts determined by percentage of leaves infected or by severity rating system on trees sprayed with various fungicides using air-blast sprayer**

| Sprays (No.) | Timing | Powdery mildew on terminal leaves | |
| --- | --- | --- | --- |
| | | Leaves infected[a] (%) | Leaf area affected[b] (%) |
| 12 | 1/2 in. Green through seventh C[c] | 15.6 a[d] | 2.7 a |
| 10 | Pink through seventh C | 17.0 a | 3.0 a |
| 8 | Petal fall through ninth C | 35.2 b | 10.7 b |
| 6 | Bloom petal fall, first, second, fourth, sixth C | 27.0 b | 6.7 ab |
| 0 | | 81.6 c | 79.9 c |

[a]Determined by observing all leaves on 20 terminals (shoots) per single tree replicate (four replicates) on July 16, 1974.

[b]Determined as in footnote a and expressed as calculated loss of leaf area. Calculated by dividing percentage of mildew-infected terminals by severity factor based on scale of 1–5 (the greater the severity, the lower the rating).

[c]Small letters indicate Duncan's multiple range groupings of treatment, which do not differ significantly at 0.05 level.

converting the ratings to percentages. These data are similar in the percentage of leaves infected and the percentage of leaf surface actually affected by mildew. The rating system tends to show somewhat lower percentages, but it is more desirable because the estimated leaf surface diseased and the presence or absence of mildew on the leaf are both considered.

In a number of tests not reported here the Horsfall-Barratt method was easier to use than was the rating system used by Covey (1) and reported in Table 13.3. In the Horsfall-Barratt method, there are 12 unevenly divided categories, with narrow limits between categories on both ends of the scale and broad intervals in the middle. Therefore, the amount of diseased surface can more easily be assessed. The system is particularly effective on test compounds that allow a small number of lesions to become established, or those that have strong suppressive action on the lesions when used in several applications. The method appears to be adequate when used alone, but the actual percentage of leaves infected should be determined in addition to the rating method until enough experience has been gained to build confidence.

### Literature Cited

1. COVEY, R. P. 1976. Control of apple powdery mildew with fungicide sprays. Fungicide Nematicide Tests 31:17.
2. DUNCAN, D. B. 1955. Multiple-range and multiple F test. Biometrics 11:1-42.
3. HORSFALL, J. G., and J. W. BARRATT. 1945. An improved grading system for measuring plant diseases. Phytopathology 35:655 (abstr.).
4. MOLNAR, J. 1974. The effect of environmental factors on the germination of conidia of *Podosphaera leucotricha* (Ell. et. Eu.) Salm. Orchrana Rostlin 44:203-210.
5. SNEDECOR, G. W., and W. G. COCHRAN. 1967. Statistical methods. Ed. 6. Iowa State University Press: Ames, IA.
6. WALLER, R. A., and D. B. DUNCAN. 1969. Bayer rule for the symmetrical multiple comparison problem. J. Amer. Stat. Ass. 64:1484-1503.

# 14. Field Plot Tests of Apple Fungicides

## F. H. Lewis

Professor emeritus of plant pathology, The Pennsylvania State University Fruit Research Laboratory, Biglerville, PA 17307

A good apple fungicide must control disease over a wide range of conditions with an acceptable level of risk to the tree, the fruit, and the people who come in contact with the chemical. An evaluation of this ability begins in the laboratories and greenhouses and on the farms of chemical manufacturers; continues through agricultural experiment stations, government regulatory agencies, and commercial users; and ends only when no further interest in the fungicide exists. The evaluation is dynamic in that it is modified as additional data and observations accumulate. The methods are sufficiently refined so that, after two to four years of testing by experiment stations, probable performance of the fungicide under most commercial conditions is documented. The evaluation always remains an opinion, because a good fungicide will be used under diverse conditions of crops, cultivars, weather, mixtures, and methods of application.

Field performance of an apple fungicide is the end result of the interaction of nine major factors: (i) plant susceptibility, (ii) the fungus population, including size and dynamics, (iii) rainfall and temperature, particularly as they affect the frequency and severity of infection periods, (iv) tree growth, including nutritional status and the time when leaf growth ceases, (v) the characteristics of the fungicide, particularly its efficacy against specific fungi, (vi) the fungicide dosage or rate of application, (vii) the spray equipment, (viii) the spray timing, and (ix) the application method. Any one of these can affect the level of control obtained with each fungicide.

Evaluating each of these factors in one or a few field trials is rarely, if ever, possible nor is testing every new fungicide in all possible combinations essential. A choice must be made of the one variable to be introduced with each plot or treatment.

This chapter describes a method for comparing fungicides on apples (*Malus sylvestris* Mill.) under field conditions in which apple scab (caused by *Venturia inaequalis* [Cke.] Wint.) and powdery mildew

(*Podosphaera leucotricha* [Ell. & Ev.] Salm.) occur every year, and cedar-apple rust (*Gymnosporangium juniperi-virginianae* Schw.), quince rust (*Gymnosporangium clavipes* Cke. and Pk.), sooty blotch (*Gloeodes pomigena* [Schw.] Colby), flyspeck (*Schizothyrium pomi* [Mont. and Fr.] V. Arx), various fruit rots, and phytotoxicity may occur. The method minimizes labor and cost in testing about 30 fungicides and mixtures annually under conditions simulating those in commercial orchards in the Cumberland-Shenandoah Valley region of the eastern United States. Changes in procedure needed for evaluating low-volume sprays and other factors affecting fungicide performance also are described.

## Methods and Procedures

**Test site**—A site owned by Pennsylvania State University is used to accommodate legal limitations on pesticides and to maintain pest populations that are adequate to show differences in fungicide efficacy. The site is relatively frost-free and favorable for the growth of apple trees. The trees are reasonably uniform in size and vigor. The slope of the soil is less than 5% in most blocks, thus facilitating movement of equipment.

**Test plants**—Test plants are cultivars that are important in this area and throughout the United States. They are of an age and size that permit production of at least 1–2 bu of fruit per tree, plus removal of at least ten terminal shoots per tree for disease counts. They are either highly susceptible to disease or sufficiently so for the investigator's purposes.

The cultivar Rome Beauty is useful, because it is susceptible to all of the apple diseases common to the Cumberland-Shenandoah Valley region. Stayman Winesap is equally or slightly more susceptible to apple scab than is Rome Beauty, but it is less susceptible to powdery mildew and resistant to cedar-apple and quince rusts. The red sports of Delicious have been useful for studies on apple scab and phytotoxicity. The red sports in use in this laboratory, however, are immune to cedar-apple rust, and the frequency of scab and powdery mildew infection has been less than half that obtained on Rome Beauty. Quince rust is a minor problem that usually occurs only on the red sports of Delicious. These cultivars are satisfactory for studies on sooty blotch and flyspeck. Golden Delicious is preferred for evaluation of fruit russeting by fungicides, because the data are adequate to predict the relative danger of fruit russeting by fungicides for all commercially important cultivars. Russeting by fungicides differs in degree and importance on different cultivars, but the relative ranking of the fungicides is the same in nearly all cases. A notable exception is the russeting by captan of Ben Davis and Gano apples.

The preferred rootstock for test trees is Malling 7, but Malling Merton (MM) 106 and apple seedlings are also used. Trees on Malling 7 have been usable for fungicide trials at four years of age, and have been maintained at 12 ft (3.7 m) in height and diameter for 15–20 years of adequate fruit production. Trees on MM 106 are of desirable size and productivity, but tree losses due to crown rot (probably *Phytophthora* sp.) have been severe. Trees on apple seedling rootstocks are being used less, because they become too large for experimental purposes.

Greatest efficiency has been realized in blocks in which trees were planted in clumps or groups, with one tree of each of three cultivars planted in a single 12–14-in. (30–36-cm) hole and tree sites spaced 30 × 35 ft (9.1 × 10.7 m). Best results have been obtained with a combination of Rome Beauty, a red sport of Delicious, and Golden Delicious on Malling 7 rootstocks. Stayman Winesap on Malling 7 has been unsatisfactory in such plantings because of its rapid growth.

Other acceptable tree arrangements include (i) blocks of about 80 trees each, with single trees of a desirable cultivar planted 30 × 35 ft (9.1 × 10.7 m), (ii) a similar block with one or two additional cultivars grafted onto each tree during the second year in the orchard, and (iii) single trees of each of two to four cultivars planted 10 or 12 ft (3.1 or 3.7 m) apart in the row followed by a 30–50-ft (9.1–15.7-m) space for movement of equipment. The second and third arrangements have increased labor and losses from unsalable fruit while showing no material advantage over size-controlled trees planted in groups.

A block of trees four rows wide provides maximum flexibility in the movement of equipment and application of sprays. Orienting the tree rows perpendicular to the prevailing wind helps to minimize spray drift from plot to plot.

**Inoculum**—Inoculum of the apple scab and powdery mildew fungi is maintained at a moderate level by omitting fungicides on 10–25% of the trees in the orchard. Although the occurrence of quince rust, sooty blotch, and flyspeck is sporadic, no supplemental inoculum is provided. The location of the laboratory within a fruit district makes maintaining wild hosts of the apple disease fungi inadvisable.

Each year some 100 bu of overwintered apple leaves are collected in an abandoned McIntosh apple orchard and placed in experimental blocks within 24 hr after collection. This is done in April at the green-tip to delayed-dormant stage of growth. The leaves are placed in a 2 × 3-ft (0.61 × 0.91-m) area, and held in place for about four weeks by wire mesh. Leaf samples have indicated that the ascospores of the scab fungus mature and are discharged normally with this procedure.

One or more potted, greenhouse-grown, mildew-infected apple seedlings are tied to a limb or pole in the center of each test tree, usually during bloom, to provide additional inoculum of *P. leucotricha*.

Galls of the cedar-apple rust fungus are collected when the spore horns appear, soaked in water for 30 min, and distributed to small wire baskets suspended above the trees. Three large to six small galls are placed in each basket when it is raining and the forecast is for 12–24 hr of wetting at 53–65°F (11–18°C). This inoculation method is successful about one year in three.

**Equipment**—Equipment can be simple—one of the small commercial sprayers if tests include only a few treatments on small trees. The program at this laboratory, however, involves testing about 30 spray mixtures annually on trees from 8 to 15 ft (2.4 to 4.6 m) in height. Our equipment allows three men to apply one experimental treatment every 10 min or less, or two men to apply one treatment every 15 or 20 min. The sprayer has steel catwalks extending its full length on both sides and a 2-ft (61-cm) wide steel platform with a guard rail in the rear. It is mounted on a short-wheelbase truck with six forward speeds and mud and snow tires on the rear wheels. Major sprayer components include a steel tank divided into two 200-gal (757-L) compartments with steel tabs dividing each tank into 50-gal (189-L) increments, a 20-gal/min (76-L/min) pump designed to operate at 500–550 psi (3,500–3,800 KPa) of pressure, a motor to operate the pump and agitator paddles, 55 ft (16.8 m) of half-inch (1.27-cm) rubber spray hose, a trigger-type spray gun that delivers 12–15 gal (45–57 L) per minute, and two 2-in. (5.08-cm) pipes and lever valves arranged under the tanks and left catwalk to allow recirculation through the pump. The change from one tank to another can be made quickly while standing on the ground or on the left catwalk. A hinged steel step on the rear allows the person who applies the spray to ride from one replicate to another.

A truck-mounted nurse tank unit is used to haul water and provide all equipment necessary to wash and refill the tanks. Essential components include a 500-gal (1,893-L) tank, a centrifugal pump with a capacity of 100 gal (378 L) or more per minute, an air-cooled motor, a reinforced rubber discharge line 2 in. (5.08 cm) in diameter and 25 ft (7.6 m) in length, and a 1/2-in. (1.27-cm) garden hose 25-ft in length with a trigger-type nozzle.

Low-volume or concentrated sprays are applied with airblast sprayers capable of delivering 20,000 to 90,000 ft³ of air per minute. For spraying small trees, the preferred unit has a 100-gal (378-L) tank, a pump capable of operating at 300 psi (2,100 KPa), and an air delivery of 23,000 CFM at a velocity of 80 mi (128.7 km)/hr when the power takeoff on the tractor is operating at 600 rpm. This unit is mounted on a 60-hp tractor with independent control of the power takeoff speed and rate of travel.

**Planning the experiment**—All trials are planned and conducted in accordance with the principles of statistics (2). Each experiment has an objective and is carefully designed so that the results can be statistically analyzed. Preference is given to randomized block experiments with 10–12 treatments and a check replicated six times either on single trees or on groups of trees. Most products are tested at two to three rates ranging from 25% below optimum as indicated by the tests to that time to 25% above optimum. In trials in which little is known about the fungicides, the only variable is usually the fungicide or rate. When a fungicide is known to be ineffective against one of the diseases present, it is sometimes tested in combination with another fungicide.

When the fungicides are applied with airblast equipment, each replicate consists of two or more trees, with alternate rows serving as barriers to reduce drift between plots. Where two factors are varied, such as spray interval and dosage, strict adherence to a fixed time interval between sprays is considered essential. When problems such as the effects of fungicides on spores from visible lesions are studied, data are collected at fixed intervals (in series) even though the labor requirement is tripled or quadrupled.

**Application of sprays**—Most experimental fungicides are applied in dilute, high-pressure sprays until each tree begins to drip. Applications are made in the morning when wind movement is slow. This timing and the smallness of the trees minimize spray drift, the number of trees required for meaningful results, the amount of labor, and the amount of fungicide needed for

the trial. Fifty gallons (189 L) of spray mixture is more than enough for the six replicates of each treatment, but this amount is always mixed to avoid small errors in weighing the fungicides and measuring the quantity of water.

When low-volume sprays are used, the amount of fungicide needed per acre is based on a standard for dilute sprays (3). For example, good airblast equipment can provide excellent coverage of mature, 18–20-ft (5.5–6.1-m) apple trees using 400 gal (1,514 L) of dilute spray per acre. For low-volume sprays, fungicide rates per acre are reduced 20% because little runoff occurs and the size of the spray droplets is more uniform.

**Data collection**—Observations on phytotoxicity are made between spray applications and at harvest.

The most valuable indicator of disease control by a series of fungicide treatments on apple is a midsummer count of the leaves of shoots originating from vegetative terminal buds. Precautions are necessary on three points. First, in most situations the sample should not include shoots originating from buds in which the powdery mildew fungus overwintered, the count should be made to determine spread during the period when the fungicides were being applied. Second, the amount of disease on individual shoots from the same tree often varies by two to ten times the mean percentage of disease on the entire tree, so for data to be reliable, counts on at least 10–20 shoots per tree chosen for uniformity and location are required. Third, the maximum incidence of scab, powdery mildew, and cedar-apple rust rarely occurs on the same portion of the shoot. The most accurate record requires an examination of all unfolded leaves in the subsample.

Counts of the number of lesions sometimes provide additional insight into disease development and fungicide performance. Such counts rarely are justified, however. Data showing the percentage of diseased leaves are satisfactory in most cases with apple scab, cedar-apple rust, and powdery mildew. If mildew is present on more than 40–50% of the leaves, a rating system may be useful.

Errors are likely if one person attempts to count two or three diseases on a shoot before the data are recorded. Either one of two methods is used to avoid this problem: One person may count only one disease before it is recorded, but a preferred procedure is to use mimeographed record sheets that show simulated terminal shoots having short leaf petioles on alternate sides of the shoot. As the leaves are examined, the diseases on each are recorded. The record sheets provide an opportunity to determine the approximate time of infection if the data are coupled with records of leaf development, periods of leaf wetness as determined with a De Wit Leaf Wetness Recorder (Valley Stream Farm, Orono, Ont.), and temperature as recorded by a thermograph.

Determination of the percentage of spur leaves with scab has been useful when heavy early-season infection occurred. Such counts are made in most instances on 20 spurs per tree using only one type of spur (eg, with or without flowers, fruit, or growth from a lateral bud).

Data on the effects of fungicides on apple fruit are considered essential. For the cultivars used in our tests, the amount of disease on the fruit frequently has been less than on the leaves. Differences between fungicides have been harder to determine. Since major emphasis has been placed on leaf counts, the size of the fruit sample has been held to a minimum, usually 50 apples per tree.

Data on fruit finish are obtained by observation and by placing the apples in a sample from each tree into classes based on russeting or color or both.

In some tests, sooty blotch and flyspeck appear late in the season, so a count at or near harvest is adequate. Since the fungicide residue on the trees is a major factor in the control of sooty blotch and flyspeck, a series of counts should begin as soon as the diseases appear, and continue until harvest. Otherwise, the fungicide might be listed as ineffective when, in fact, control was excellent until spraying ceased one to three months before harvest.

**Statistical procedure**—Data are compiled and a percentage of infection for each treatment is determined. For statistical purposes, the figures often are converted to equivalent measurements of angle, square root of $X + 0.5$, or other suitable transformation (2). In most situations, comparing each treatment with all others in the test is desirable. Therefore, Duncan's multiple range test usually is used. Other statistical procedures are acceptable if properly used and reported.

### Discussion

Field plot tests of apple fungicides can be accurate if each step in the procedure is carefully examined and all sources of variation minimized except those deliberately introduced. Small differences of 2–6% can be measured. For example, in one test (1), scab varied from 0.4 to 9.9%, and a mean of 1.3% diseased leaves differed significantly from 5.1%. Powdery mildew varied from 1.2 to 4.6%, and 1.8% differed significantly from 4.6%. Cedar-apple rust varied from zero to 1.5%, and a mean of 0.2% differed significantly from 1.4%. Larger mean differences were needed in other tests in which more variation was present within treatments.

An attempt to obtain maximum accuracy in field trials is expected to reduce the number of trials required to document the probable performance of the fungicides under a wide range of commercial conditions.

Four factors are especially important in conducting a good field trial of fungicides on apple. First, insofar as possible, plan the experiment and collect the data to provide information on both the fungicide and the disease (eg, time and severity of infection, effects of the fungicides after infection, spray timing). Second, do everything possible to eliminate variation among the trees and in mixing and applying the sprays. Third, in making a decision regarding type and size of leaf and fruit samples, consider the high variability that occurs within such samples. Fourth, statistically analyze the data. This is helpful to the worker and important in convincing others that the test was sufficiently accurate for small differences to be meaningful.

### Literature Cited

1. LEWIS, F. H. 1973. Apple scab, powdery mildew, rust. Fungicide Nematicide Tests 28:26-27.
2. STEEL, R. G. D., and J. H. TORRIE. 1960. Principles and procedures of statistics. McGraw-Hill Book Company: New York, pp. 1-481.
3. TETRAULT, R. C., D. ASQUITH, W. M. BODE, C. M. RITTER, G. M. GREENE, D. H. PETERSEN, F. H. LEWIS, and D. R. DAUM. 1976. Tree fruit production guide. The Pennsylvania State University College of Agriculture: University Park. p. 1-98.

# 15. Field Test Procedures for Fungicides Used to Control Apple Diseases in South Carolina

### Eldon I. Zehr

Associate professor, Department of Plant Pathology and Physiology, Clemson University, Clemson, SC 29631

Contribution No. 1433 of the South Carolina Agricultural Experiment Station. Published with approval of the director

Apples (*Malus sylvestris* Mill.) are subject to many destructive diseases that may cause economic losses of fruits, leaves, branches, or entire trees. Resistant cultivars, exclusion or eradication of pathogens from orchard sites, or use of cultural practices that interfere with life cycles of pathogens can prevent or minimize some losses. Certain diseases, however, have resisted control by other than chemical means. In South Carolina, such diseases and their causal agents are scab (*Venturia inaequalis* [Cke.] Wint.), powdery mildew (*Podosphaera leucotricha* [Ell. & Ev.] Salm.), black rot (*Physalospora obtusa* [Schw.] Cke.), bitter rot (*Glomerella cingulata* [Ston.] Spauld. & Schrenk), flyspeck (*Microthyriella rubi* Petr.), sooty blotch (*Gloeodes pomigena* [Schw.] Colby), cedar-apple rust (*Gymnosporangium juniperi-virginianae* Schw.), quince rust (*Gymnosporangium clavipes* Cke. & Pk.), and fire blight (*Erwinia amylovora* [Burrill] Winslow, et al.). Losses due to these diseases can be kept relatively low with chemicals now available, but the search continues for better treatments effective at lower rates and with fewer applications.

In South Carolina, a research program for chemical control of apple diseases has been in progress since 1970.

The objectives of this research are to (i) evaluate in field trials the efficacy of new fungicides for control of apple diseases, (ii) improve the effectiveness and efficient use of commercial fungicides for disease control, and (iii) evaluate effects of variable environmental conditions relative to fungicide performance. In this research, certain procedures have been developed that are suitable for use in small-scale field testing programs in which these objectives apply. Details of these procedures are described in this chapter. More general descriptions have been published previously (5–7).

## Methods and Procedures

**Test plot**—A site on the Clemson University Experiment Station near Pendleton, SC, was chosen, based on the need to maintain an orchard site in carefully controlled conditions so that (i) residues of experimental fungicides would not threaten wildlife, humans, or domestic animals and (ii) cultural practices could be managed carefully for maximum information on fungicide performance. The soil was a well-drained Cecil clay loam of sufficient depth and structure to support vigorous growth of apple trees. Air drainage promoted rapid drying of fungicide suspensions after they were applied and minimized injury due to late-season freezes. Before planting, the site was subsoiled to a depth of 18–24 in. (46–60 cm) to rupture subsurface hardpans.

The cultivars selected were Red Delicious, Golden Delicious, and Jonathan propagated on Malling 7 rootstock. Red Delicious trees are susceptible to scab, black rot, bitter rot, and quince rust, and therefore are acceptable as test plants for control of these diseases. Red Delicious trees also are susceptible to sooty blotch and flyspeck, but background color sometimes interferes with evaluation of fungicide performance for control of these diseases. Golden Delicious trees are susceptible to cedar-apple rust, and fruit color and sensitivity make this cultivar excellent for evaluating fungicides for fruit decay, flyspeck, sooty blotch, and phytotoxic symptoms such as surface russeting. Jonathan apple trees are susceptible to fire blight and powdery mildew, and also may be used for evaluating efficacy of control of fruit decay, sooty blotch, and flyspeck.

Trees were planted on 15-ft (4.6-m) spacing in rows 20 ft (6.1 m) apart. Spacing in and between rows was similar to that in commercial apple orchards in South Carolina where this rootstock is used. Trees were pruned to promote an open habit of growth to facilitate thorough spray coverage by penetration of spray droplets.

To control soil erosion, fescue sod was maintained between orchard rows. Paraquat (1,1′-dimethyl-4,4′-dipyridinium dichloride) and dichlobenil (2,6-dichloro-benzonitrile) were applied in the tree row for weed control. Insecticides recommended in South Carolina (1) were applied by airblast sprayer or mixed with test fungicides in the spray tank. Fertilization was done according to customary commercial recommendations in South Carolina.

To evaluate the effectiveness of test materials, using trees that are at least 5 years old is usually necessary to obtain data on control of fruit diseases. Experiments may begin earlier for foliar diseases such as powdery mildew, scab, black rot (frogeye leaf spot), or cedar-apple rust.

**Plot design**—For most fungicide tests, a randomized complete block was used. Because the orchard was planted on the contour, however, a completely randomized design was used for some experiments. Four replicates of two trees each was customary, but for some experiments, five single-tree replicates were used. The number of replicates and the size of plots were based on subjective estimates of the number required for an adequate statistical comparison. The number of replicates varied according to number of trees available, uniformity of inoculum in the orchard, amount of plant material or fruits required for data collection, number of treatments, and uniformity of the physical and natural environment.

**Disease organism**—Because natural inoculum for the test fungi was abundant, trees were not inoculated artificially. Portions of the orchard not used for testing were sprayed with minimal fungicide applications so as to maintain abundant inoculum in the orchard. Use of herbicides for weed control contributed to accumulation of leaf litter, which added further to inoculum levels in the orchard. Cedar trees were allowed to grow in the vicinity of the orchard to provide inoculum for rust development.

**Controls**—Because each experiment contained new materials for control of apple diseases, trees that received no fungicide were always included. This practice provided a measure of disease severity in the test orchard, and a base line for measuring disease control and phytotoxicity by experimental materials. It probably contributed somewhat, however, to uneven distribution of inoculum in the test site as Shoemaker (2) and van der Plank (4) have discussed.

Fungicides that are used by many apple growers in commercial orchards in South Carolina served as the standard for comparison. In most tests the standard was captan, dodine, or a mixture thereof. Captafol or benomyl was used as a standard in some experiments.

**Fungicide application**—A John Bean Spartan® sprayer with a 30-gal tank and pump operating at 200–300 psi (2,100–3,200 KPa) was used routinely for spray applications. Spray applications were made with a single-nozzle handgun. This equipment was suitable for dilute sprays (ie, application to the point of runoff), but was not satisfactory for more concentrated suspensions. Therefore, all test materials were applied as a dilute spray.

The frequency of spray applications depended on the objectives of each experiment. When the objective was to simulate or compare with a standard orchard spray program, applications usually were at six- to ten-day intervals through bloom and at 14-day intervals thereafter. For experiments in which minimal spray applications were made, spray intervals were extended or applications were based on rainfall or infection periods. Applications usually began when the first green leaf tissue appeared in early spring.

**Data collection**—By 1976, powdery mildew had not become established in the test orchard, but other common diseases were prevalent. Fire blight, though prevalent, was not evaluated in experiments through 1976. For other diseases, data on disease control were collected three times yearly—in late May, in midsummer, and at harvest. Data collected in May were for early season scab control, rust infection, and frogeye leafspot caused by *Physalospora obtusa*. At that time, leaves on fruit spurs were fully developed and had been exposed to early-season infections. One hundred spur leaves per tree were counted at random around each tree at a height of 1–2 m from ground level. The percentage of leaves infected was recorded for each tree. No attempt was made to determine the percentage of leaf surface infected or fungicide effects on sporulation of the fungi.

Midsummer evaluations were made three to five weeks before harvest. Data were collected on the percentage of terminal leaves infected with scab. No attempt was made to determine the percentage of leaf surface infected. Ten terminal shoots per tree were taken at random, and ten leaves were counted per shoot. Because the shoots were still elongating, the four to six youngest leaves usually were omitted in data collection.

Data on fruit infection were collected when fruits were ripe—mid or late August in South Carolina. One hundred fruits per tree were harvested at random and examined for scab, rust, fruit decay, sooty blotch, and flyspeck.

Phytotoxicity was determined on the basis of leaf injury or abnormal growth throughout the growing season, and on russeting or other fruit injury at harvest. Fruit russet was evaluated by estimating the percentage of fruit surface russeted on 25 fruits per tree (Golden Delicious cultivar) at harvest.

Results of each experiment were subjected to statistical analysis. Because each treatment was to be compared with all other treatments in the test, Duncan's multiple range test usually was used. Other tests, such as the least significant difference (LSD), were used infrequently.

## Results and Discussion

The procedures described in this chapter have been satisfactory in testing experimental chemicals for apple disease control, especially in comparisons of chemicals that have potential for further development as fungicides. A relatively large number of chemicals can be tested under careful statistical control without large expenditures for equipment, labor, and loss of fruit due to lack of clearance of the experimental materials. Some limitations exist, however. First, dilute sprays have been superseded by low-volume (concentrated) sprays in the apple industry, so one must use other methods if needed to study any apparent discrepancies in fungicide performance in small plots versus general orchard use. Second, the equipment described is not large enough for spraying apple trees that are taller than 5 m. Third, minimal spray programs in the orchard contributed to a higher level of inoculum than would occur in most commercial orchard situations. The high inoculum level was the result of a deliberate attempt to create severe disease situations in the orchard and was necessary for evaluation of new fungicides, but it may have subjected test materials to more severe disease situations than would be likely to occur in commercial orchards.

Some improvements could have been made in this procedure, especially in design of the test orchard. The tree spacing (4.6 × 6.1 m) was too close and resulted either in leaving adjacent trees unsprayed (which further increased the inoculum level) or limited spraying to periods when air movement was nearly calm. A minimum tree spacing for experimental purposes should be 8 × 8 m.

In selecting the test orchard site, greater allowance should have been made for wind movement. Orchards planted on hilltops or on windward slopes will have greater problems of spray drift during application. Site selection should include consideration of shelter by topography or windbreaks from the prevailing wind.

Data collection methods have not included yield data for different treatments. The need for yield data was underscored by a recent article (3) that reported yield reduction of apples when oil was added to benomyl. In use of new chemicals, or in new uses of labeled products, potential effects on crop yield as well as disease control and phytotoxicity should be considered.

The procedures for evaluating fungicide performance were chosen to measure efficacy and phytotoxicity at three different times during the season. The results with these procedures generally were satisfactory for this purpose, but usefulness of these methods may be limited in other situations. When terminal leaves were examined for scab in midseason, the youngest leaves were omitted because they often were free of scab lesions even when not sprayed, especially when dry weather preceded the examination. This omission would not be acceptable in test plots where powdery mildew occurs. Perhaps in many situations the best measure is to examine all leaves on the shoot after terminal elongation stops.

## Literature Cited

1. MILLER, R. W., and D. K. POLLET. 1976. Commercial apple spray schedule. Information Card 110, Cooperative Extension Service, Clemson University.
2. SHOEMAKER, P. B. 1974. Fungicide testing: Some epidemiological and statistical considerations. Fungicide Nematicide Tests 29: 1-3.
3. SPOTTS, R. A., F. R. HALL, D. C. FERREE, and B. M. JONES. 1975. Yield reduction in apples related to benomyl-oil spray applications. Plant Dis. Rep. 59: 541-543.
4. VAN DER PLANK, J. E. 1963. Plant diseases: Epidemics and control. Academic Press: New York.
5. ZEHR, E. I. 1974. Apple scab, bitter rot, sooty blotch, flyspeck. Fungicide Nematicide Tests 29: 29.
6. ZEHR, E. I. 1975. Apple scab, frogeye, cedar-apple rust, sooty blotch, flyspeck, leaf blotch. Fungicide Nematicide Tests 30: 36-37.
7. ZEHR, E. I. 1976. Fungicide tests for control of apple diseases in South Carolina, 1975. Fungicide Nematicide Tests 31: 50.

# 16. Techniques for Field Evaluation of Spray Materials to Control Fire Blight of Apple and Pear Blossoms[1]

**Steven V. Beer**

Department of Plant Pathology, Cornell University, Ithaca, NY 14853

Fire blight of pear (*Pyrus communis* L.) and apple (*Malus pumila* Miller), which is caused by *Erwinia amylovora* (Burrill) Winslow et al., is devastating in some years and of little consequence in others (23). When fire blight is severe, the current season's crop is reduced in proportion to the percentage of blossoms infected (Beer, unpublished). If fire blight lesions initiated as blossom infections extend into older tree parts, structural damage reduces fruit production for at least several seasons following the year of infection. In severe cases, particularly in young orchards, trees may be killed or rendered permanently unproductive (5,16,28).

Infections that originate in primary blossoms are considered most important, because they reduce yields, kill major limbs, and provide inoculum for later secondary blossom infection and shoot infection (11,16,28). Specific measures aimed at reducing blossom blight incidence have been more successful and cost-effective than have measures designed to control vegetative shoot infection (8). Absence of blossom blight in areas where the disease is endemic may be explained by lack of sufficient inoculum (10,18), unfavorable environmental conditions before (26) or during and after bloom (10,23), nonsusceptibiligy of blossoms (32), or a combination of these factors. Because blossoms function as infection courts for only a short time, blossom infection will not occur unless conditions favorable for production and dissemination of inoculum and for development of infection prevail during this critical time.

Evaluation of new materials or techniques for control of blossom blight often has been disappointing because of the sporadic occurrence of the disease. Test plots commonly are located in commercial orchards in which success depends on the amount of natural infection that develops. Many experiments have failed under natural orchard conditions, because infection was insufficient or nonuniform in distribution.

Techniques involving artificial inoculation of orchard trees with *E. amylovora* were developed to increase the amount of fire blight that occurs under natural conditions (7). These techniques have been improved over a four-year period and have been used in western New York to evaluate standard and experimental spray materials for blossom blight control (3,4,6). Modifications of the techniques presented in this chapter should enable those in other areas to evaluate materials or methods reliably.

## Methods

**Test site**—The test site should be suitable for the growth of apple or pear or both with respect to soil and climate. Ideally, the test site should be located in an area apart from centers of commercial pome fruit production so that the anticipated extensive disease on the test site does not threaten commercial plantings. This is not required, however, if the precautions described below are taken. If the test site is dry during bloom, overhead irrigation facilities should be available to establish higher moisture conditions that favor development of the disease.

**Test plants**—Any fire blight-susceptible apple (2) or pear (25) cultivar can be used. Combination plantings of several cultivars are desirable to insure cross-pollination. If pollination occurs, infected blossoms are more likely to remain in clusters and lead to cluster infection. Idared, Rhode Island Greening, and Twenty Ounce apple cultivars and Bartlett and Bosc pear cultivars have been used with success in New York. Less susceptible apple cultivars such as McIntosh and Spartan also have been used successfully (Beer and Abdel-Rahman, unpublished). Preliminary data suggest that when such cultivars are used, more inoculum is required to produce a given level of infection, and damage to tree structure is less severe.

Trees propagated on size-controlling rootstocks are most desirable, because they bloom at a young age and are easy to manipulate. For apple, the clonal rootstocks Malling (M) 7, Malling Merton (MM) 104, and MM 106 are suggested. Rootstocks M 9 and M 26 should not be used, because their high susceptibility may lead to early tree death (14,20). Trees propagated on seedling understocks are acceptable, but they take longer to flower.

Trees should not be more than 3 m (10 ft) high, and should be spaced more widely than they normally are for commercial fruit production. Wider spacing allows for easier maneuverability of orchard machinery and reduces the effect of spray drift. If individual plots include sets of contiguous trees (see experimental design below), tree spacing within rows may be comparable to, or even less than, commercial practice.

**Test plot management**—Recommended commercial orchard maintenance practices (fertilization; cultivation; pruning; insect, disease, rodent, and weed control) should be used for the test site. Where feasible, a sod cover crop should be maintained to facilitate access to the test site when soil is wet. Test plots should be kept free of natural fire blight (5) so that inoculum levels will be uniform throughout the test orchard. If test trees are to be used repeatedly, at least one year's growth should be allowed between test years.

[1]Mention of a trademarked or proprietary product does not constitute endorsement or warranty of the product by the author, Cornell University, or The American Phytopathological Society, nor does mention imply approval of the product to the exclusion of other products that also may be suitable.

All pome fruit plantings within 400 m (1/4 mi) of the test site should be protected from fire blight by a rigorous protective spray program. Frequent patrols of nearby pome fruit plantings should be made while infection is developing in the test orchard (5).

**Experimental design**—Several considerations influence experimental design. The number of blossom clusters per tree and the uniformity of growing conditions in the test plot are of prime importance. As variability in the test plot and the degree of resolution desired between test treatments increase, the number of replicates required for a valid test increases. More replicates may be required for evaluation of materials of moderate to high efficacy for which small differences between treatments are anticipated than for tests of materials of unknown efficacy.

Individual blossom clusters generally are examined for infection. The smaller the number of blossom clusters per tree, the larger the number of trees needed in test plots. If the number of blossom clusters is limited or if the precision of the test is important, individual blossoms (within clusters) can be examined for symptoms. This is rarely done, however, because of the time and expertise required. Examination of individual blossoms must be timed properly to distinguish direct infections (from artificial inoculation) from those that develop via the spur base and peduncle from adjacent blossoms.

Candidate materials are applied to individual trees or sets of contiguous trees in the orchard. In small-scale tests in the greenhouse or field, individual branches, or even clusters, might be used if adequate separation of treatments is ensured.

The randomized complete block design is the most commonly used, but for special purposes, split plots, split-split plots, Latin squares, or completely randomized blocks are acceptable (21). Plots within blocks may consist of individual trees or sets of contiguous trees. If candidate materials are applied by airblast sprayer, plots of several trees facilitate uniform spray application. Buffer trees between plots or trees that are treated but not evaluated improve the precision of the test. Depending on the variability within the test site and the degree of precision desired, randomized complete blocks should be replicated at least twice, but preferably four or more times.

Interplot interference (30) normally would not be expected to be important in tests for control of primary blossom blight, since blossoms are susceptible for only a few days after inoculation. During this period, only small increases in inoculum on check trees would be expected and dissemination of *E. amylovora* from these to sprayed trees probably would be minimal.

**Check treatments**—A check (no chemical) treatment should be included in all tests. Check plots should be sprayed with water at the same time that test plots are sprayed with materials to be evaluated. Water-sprayed plots are preferred to unsprayed plots, because the carrier water used for spray application may affect redistribution of inoculum and development of infection. Check plots provide an index of the maximum amount of infection to be expected from a totally noneffective material.

A standard treatment that is expected to provide good disease control should be included in all tests. A commercially available formulation of streptomycin sulfate (such as Agri-Strep Type D, Merck & Co., Inc., Rahway, NJ) is suggested. By including streptomycin-treated plots in the test, the efficacy of candidate materials can be compared with an accepted standard fire blight control material.

**Pathogen**—Successful evaluation of materials depends on uniform disease pressure throughout the test plot. This may be accomplished by artificial inoculation with *E. amylovora* at an appropriate time (4,7). Any virulent single-cell or single-colony isolate of *E. amylovora* that is sensitive to streptomycin may be used. The isolate should be sensitive to streptomycin so that fire blight infection in adjacent orchards may be controlled by application of the antibiotic.

The pathogenicity and sensitivity to streptomycin of the candidate isolate must be tested before use. Streptomycin sensitivity may be tested by plating on nutrient agar (Difco Laboratories, Detroit, MI) that has been amended with 0, 10, and 100 mg of streptomycin sulfate per liter (9). The plates are then incubated for several days at 25–27°C. A strain of *E. amylovora* known to be resistant to 100 ppm of streptomycin should be included in the test as a check. Such strains have been isolated from orchards in California, Washington, and Oregon (13,22,24,28). Only strains of *E. amylovora* that fail to grow on medium containing 10 mg/L of streptomycin sulfate should be used for inoculation of orchard trees.

The pathogenicity of the *E. amylovora* isolate may be checked by several techniques. Succulent shoots of greenhouse-grown apple or pear seedlings or cultivars (2) or susceptible ornamental species such as *Cotoneaster* and *Pyracantha* (29) may be inoculated with the aid of a hypodermic needle and syringe. Etiolated apple or pear seedlings (27) also may be used. Several candidate isolates and a nonpathogen (as a check) should be included in pathogenicity tests.

**Preparation of inoculum**—Because the amount of fire blight that develops in inoculated blossoms is proportional to the concentration of inoculum applied (7,10), a uniform inoculum level must be used. As the ideal time for inoculation generally cannot be predicted more than one day in advance, a technique was developed that permits the preparation of inoculum of known concentration on short notice from frozen *E. amylovora* suspensions.

The selected *E. amylovora* isolate is cultured in "Media 523" broth (19) in 800-ml Kjeldahl flasks. Each flask, containing 400 ml of broth, is seeded with 1 ml of a 24-hr "Media 523" broth culture of *E. amylovora;* or with a suspension of cells in 50m$M$ potassium phosphate buffer, pH 6.5, made from a slant culture. The cultures are incubated at 25–29°C on a wrist-action shaker adjusted to provide vigorous mixing with aeration.

After incubation, the bacteria are handled as aseptically as possible. All containers to be used subsequently are autoclaved or rinsed with 70% ethanol and allowed to dry before use. *E. amylovora* cells are harvested during the log phase of growth (18–22 hr) by centrifugation at $10,000 \times g$ for 10 min at 4°C. The supernatant liquid is decanted carefully, and the pellets are resuspended in a milk suspending medium at 10–25% of the original medium volume. Suspending medium is prepared by rehydrating 200 g of instant nonfat dry milk (Carnation Co., Los Angeles, CA) in 1 L of distilled water. After vigorous shaking to dissolve it, the milk is filtered through glass wool and then pasteurized in an autoclave for 13 min at 115°C (75.8 kPa, 11 psi) (33). The resuspended pellets are pooled and mixed to equalize concentration, dispensed into polyethylene containers of 10–500-ml capacity, capped, and frozen at −20°C.

After at least one day in the freezer, samples of each inoculum batch are tested for viable cell concentration. Immediately after thawing at room temperature, the inoculum preparation and dilutions thereof are maintained in an ice bath. Dilutions of the inoculum concentrate are made in chilled 50mM potassium phosphate buffer, pH 6.5, and plated in triplicate on Crosse and Goodman's medium (15). Seeded plates are incubated (inverted) at 27–29°C for 48–72 hr. All colonies that exhibit the characteristic *E. amylovora* cratering on this medium are counted to establish viable cell concentration.

Inoculum should be maintained at −20°C until used. Large quantities of inoculum may be transported in a Styrofoam chest with solid carbon dioxide. For lesser quantities or for short trips, mixtures of ice and sodium chloride are effective for keeping inoculum frozen. Inoculum is prepared in the field by thawing containers in air or in a water bath at 25°C with shaking. After thawing, inoculum concentrates should be maintained in a ice bath to prevent changes in cell number.

The water used to dilute inoculum to field concentration should not contain residual chlorine or any other components that are lethal to the bacteria. If water quality is questionable, a sample should be tested for its effect on *E. amylovora* viability. Buffered distilled or deionized water, or buffered tap water with a final potassium phosphate buffer concentration of 5mM, pH 6.5, may be used. No more concentrate should be diluted to the application concentration than can be used in 1 hr.

**Inoculum concentration**—Other factors being equal, the level of infection is proportional to the number of viable *E. amylovora* cells applied to blossoms (7,10). If the inoculum preparation contains too few viable cells, infection is insufficient to test candidate materials properly. If inoculum is too concentrated, so much infection may occur that spray materials may be overpowered. Weather conditions following inoculation, blossom age, and cultivar also influence the amount of infection resulting from a given inoculum dose (Beer, unpublished; 1). Because weather conditions cannot be predicted reliably, choosing an inoculum concentration that normally results in a relatively large amount of infection is preferable. Table 16.1 lists the percentage of cluster infection that resulted from disease production tests on apple over a four-year period in western New York. An inoculum concentration of $1 \times 10^6$ cells/ml is suggested for initial trial. The most desirable inoculum level depends on individual orchard and climatic conditions.

**Inoculation techniques**—Appropriately diluted inoculum may be applied to blossoms by various techniques. For small-scale studies, individual blossoms may be inoculated by placing a small drop of inoculum in blossoms with a repeating pipet (1,10). For large-scale orchard tests, application by low-pressure sprayer is more efficient. Applying inoculum to individual blossom clusters to the point of runoff with a cone-jet tipped wand (Spraying Systems Inc., Wheaton, IL) has been successful (3,4,6). Inoculum is pressurized (21–28 kPa, 30–40 psi) with a power takeoff-driven nylon roller pump (Hypro Inc., model C-6100, St. Paul, MN). Low pressure is used to reduce contamination of adjacent trees or orchards. In areas isolated from other pome fruits, inoculum might be applied by airblast sprayer, but limited survival of *E. amylovora* in small droplets has been reported (31). The container used for inoculum should be thoroughly clean, to prevent inactivation of the pathogen by residues of previously used compounds.

**Application of materials to be tested**—Pesticides may be applied to test trees by several means. Dilute applications may be made either by handgun at 1,725–3,100 kPa (250–450 psi) to the point of runoff or with an airblast sprayer calibrated for complete tree wetting without runoff. Concentrate sprays may be applied with a low-volume sprayer such as the Kinkelder model P50 (Marwald, Ltd., Burlington, Ont.) at a rate

**TABLE 16.1. Effect of Erwinia amylovora concentration on percentage of fire blight blossom cluster infection of apple[a]**

| Inoculum concentration[b] | Year, cultivar, and percentage of blossom cluster infection | | | | | |
|---|---|---|---|---|---|---|
| | 1973 | 1974 | 1975 | | 1976 | |
| | Idared[c] | Idared[c] | Idared[d] | Idared[e] | Idared[c] | Twenty Ounce[c] |
| 0 | 0 | 16 | 1 | 1 | 8 | 23 |
| $5 \times 10^2$ | 0 | ... | 2 | 1 | ... | ... |
| $1 \times 10^3$ | ... | 37 | ... | ... | 4 | 30 |
| $1 \times 10^4$ | ... | ... | 31 | ... | 12 | 34 |
| $5 \times 10^4$ | 4 | ... | ... | 10 | ... | ... |
| $1 \times 10^5$ | ... | 79 | ... | ... | 21 | 54 |
| $1 \times 10^6$ | ... | ... | ... | ... | 35 | 85 |
| $5 \times 10^6$ | ... | 88 | 61 | ... | ... | ... |
| $1 \times 10^7$ | ... | ... | ... | 41 | ... | ... |
| $5 \times 10^7$ | 44 | ... | ... | ... | ... | ... |
| $1 \times 10^8$ | ... | ... | ... | ... | 58 | 95 |

[a]All tests were done in same orchard in Huron, Wayne County, NY.
[b]Number of viable *E. amylovora* cells per milliliter applied to clusters of *Malus pumila* with low-pressure sprayer.
[c]Inoculations made at 80–100% full bloom.
[d]Inoculations made at 50% full bloom.
[e]Inoculations made around 25% petal fall.

between 560 and 1,120 L/ha (60 and 120 gal/acre) (3,4). If low-volume sprayers are used, accurate calibration of the sprayer delivery system is particularly important and should be checked with water prior to spray application (17).

**Timing of inoculation and spray application**—The relative timing of spray material application and inoculation determines the type of evaluation obtained. Trees may be sprayed before inoculation to test protective action of a spray material, or after inoculation to test eradicative action of materials. In preliminary tests of a new material, a simple combination test in which sprays are applied both before and after inoculation may be worthwhile. Such tests quickly indicate whether more extensive and costly testing is justified.

For a test of protective action, materials should be applied one to three days before inoculation and when 40–70% of the blossoms are fully open (12). Some of the open blossoms that will be inoculated subsequently, however, would not have been sprayed. Therefore, full control of the disease cannot be expected by nonsystemic materials.

Materials to be tested for eradicative action should be applied one to three days after inoculation but before 20% petal fall. Inoculation should be done when at least 50–70% of the blossoms are open. Separate check plots and plots treated with a standard material must be included for each day of inoculation, because weather conditions and blossom age affect the amount of disease that develops.

**Data collection**—The efficacy of spray materials is based on the percentage of blighted blossom clusters on treated trees versus the percentage of blighted clusters on check (water-sprayed) trees. Control should be evaluated only after the pathogen has progressed into the cluster base and cluster leaves exhibit symptoms (wilting or black or brown necrosis or both). Evaluations should be made, however, before individual cluster infections coalesce and become indistinguishable. Under western New York conditions, the proper stage for evaluation is reached 20–30 days after inoculation. At that time, uninfected fruits are approximately 1–2 cm (0.4–0.8 in.) in diameter.

Extreme care must be exercised in examining clusters to determine if they are infected. Sometimes vegetative shoots or clusters exhibit symptoms at the same time as blossom clusters. Therefore, all infections must be examined closely to determine the source of infection. Blossom clusters have small infected or healthy fruit or peduncle scars on the cluster base. Noninfected clusters must be examined for peduncle scars to determine if the structure had been a blossom cluster or vegetative cluster. Although most blossoms that become infected remain attached to the cluster base, some abscise before *E. amylovora* enters the base. Clusters with symptomatic cluster leaves have been considered infected; those that only have necrotic peduncle scars have been considered healthy.

All clusters (up to a maximum of 400) on each test tree should be examined. On trees bearing more than 400 clusters, representative limbs of the tree are examined. Whole limbs in each tree quadrant are selected at random, and up to 100 blossom clusters on the selected limbs are examined.

**Phytotoxicity**—Trees treated with candidate materials should be examined frequently for evidence of phytotoxicity. Within two days after application, sprayed blossoms should be examined for evidence of petal burning or necrosis. Later, foliage and young fruit should be examined for evidence of chlorosis, necrosis, or abnormal morphology. In all cases, symptoms on trees treated with candidate materials should be compared with water-sprayed check trees.

**Data analysis**—Test data should be analyzed statistically by procedures suitable to the experimental design. Conversion of the raw data to the proportion of infected clusters per tree (a decimal between 0 and 1.0) followed by transformation to the arc sine proportion is usually necessary before statistical analysis, because small proportions of infection (0–0.1) often result from highly efficacious materials. The analysis of variance and separation of treatment means by Duncan's new multiple range test (at the 5% level) is then performed on the transformed proportions (21).

While statistical analysis is performed on infected cluster data, the effect of each treatment is interpreted more easily when these data are transformed to percent of disease control (PDC) than when they are not. This transformation is accomplished by the following formula:

$$ PDC = \frac{(DI_{ck} - DI_{tr})}{DI_{ck}} \times 100 $$

where $DI_{ck}$ is mean disease incidence in water-sprayed check plots and $DI_{tr}$ is mean disease incidence in treated plots. The effect of this transformation is to relate the efficacy of a candidate material to that of water. When PDC equals 100, infection is not present in spray material-treated plots; when PDC equals 0, spray material-treated plots have the same amount of infection as the water-sprayed check plots.

**Reporting test results**—Each investigator is responsible to his institution and industrial cooperators to report the results of tests of candidate materials. Reports to manufacturers affect future development of candidate materials and may be used in preparing applications for registration. Test reports are also useful in making recommendations to growers and to research specialists working on fire blight or other diseases of bacterial etiology.

Reports should be written as concisely and precisely as the results of any other research. The following information should be included in a complete report:

i. Name or code number of material tested, including trade name, chemical name, or common name or all three if established and not confidential.

ii. Formulation name and percentage of each active ingredient and manufacturer of formulation.

iii. Composition of spray mixture, including the amount of water carrier, candidate material, and any adjuvants or other pesticides.

iv. Rate of application, preferably expressed as amount of spray mixture or active ingredient or both applied per hectare, or acre, or per tree.

v. Dates that applications were made and tree growth stage (12) on each application date.

vi. Method of application, including a brief description of the application equipment.

vii. Location of test site.

viii. Experimental design, including the number of replications, distance between rows, spacing of trees within the row, number of treated trees per plot, and number of any buffer trees between plots.

ix. Cultivar or cultivars used, rootstocks, and tree age.

x. Date and method of inoculation, with details of inoculum concentration, origin of isolate, preparation of inoculum, and inoculum application.

xi. Number and dates of irrigation treatments, including the approximate amount of water applied and a description of the application equipment used.

xii. Brief summary of the environmental conditions during the growing season from the start of growth until evaluations are made. Detailed information should be provided concerning temperature, relative humidity, and rainfall for the period starting with the first spray or inoculation through the data collection period.

xiii. Other pesticides used in the test block, including rates and dates of application.

xiv. Mean percentage of clusters infected in candidate-treated, water-sprayed, and streptomycin-sprayed plots.

xv. Percentage of disease control.

xvi. Statistical analysis of data and methods used.

xvii. Type and severity of phytotoxicity.

xviii. A brief written evaluation and interpretation of the data to indicate the confidence that the investigator has in the results.

### Acknowledgments

The technique described could not have been developed without the expert technical assistance provided by James T. Bruno, James F. W. Eve, Jr., John L. Norelli, and Dan C. Opgenorth. I am grateful also to Freer Fruit Farms Inc., North Huron, NY, for their excellent cooperation, which facilitated development of this technique in an orchard leased by Cornell University from the corporation. Special thanks are extended to Frank Freer for his sincere interest in the project and routine maintenance of the test site. I gratefully acknowledge financial or material assistance or both provided during development and evaluation of the technique by Cornell University; Agway, Inc.; Cities Service Corporation; Kalo Laboratories, Inc.; Kocide Chemical Corporation; Marwald, Ltd.; Merck & Company Inc.; Uniroyal, Inc.; and 3M Company.

### Literature Cited

1. ALDWINCKLE, H. S., and S. V. BEER. 1976. Nutrient status of apple blossoms and their susceptibility to fire blight. Ann. Appl. Biol. 82:159-163.

2. ALDWINCKLE, H. S., and J. L. PRECZEWSKI. 1976. Reaction of terminal shoots of apple cultivars to invasion by Erwinia amylovora. Phytopathology 66:1439-1444.

3. BEER, S. V. 1976. Control of fire blight with fixed copper. Fungicide Nematicide Tests. 31:14-15.

4. BEER, S. V. 1976. Fire blight control with streptomycin sprays and adjuvants at different application volumes. Plant Dis. Rep. 60:541-544.

5. BEER, S. V. 1976. Fire blight—its nature and control. NY State Coll. Agr. Life Sci. Info. Bull. 100.

6. BEER, S. V. 1976. Preliminary tests of an experimental compound for the control of fire blight. Fungicide Nematicide Tests 31:14.

7. BEER, S. V. 1974. Production of fire blight of apple under field conditions. Phytopathology 64:578 (abstr.).

8. BEER, S. V., and H. S. ALDWINCKLE. 1974. Fire blight in New York State. NY Food Life Sci. Quart. 7(1):16-19.

9. BEER, S. V., and J. L. NORELLI. 1976. Streptomycin-resistant Erwinia amylovora not found in western New York pear and apple orchards. Plant Dis. Rep. 60:624-626.

10. BEER, S. V., and J. L. NORELLI. 1975. Factors affecting fire blight infection and Erwinia amylovora populations in pome-fruit blossoms. Proc. Am. Phytopathol. Soc. 2:95 (abstr.).

11. BEER, S. V., and D. C. OPGENORTH. 1976. Erwinia amylovora on fire blight canker surfaces and blossoms in relation to disease occurrence. Phytopathology 66:317-322.

12. CHAPMAN, P. J., and G. CATLIN. 1976. Growth stages in fruit trees—From dormant to fruit set. NY Food Life Sci. Bull. 58.

13. COYIER, D. L., and R. P. COVEY. 1975. Tolerance of Erwinia amylovora to streptomycin sulfate in Oregon and Washington. Plant Dis. Rep. 59:849-852.

14. CUMMINS, J. N., and H. S. ALDWINCKLE. 1973. Fire blight susceptibility of fruiting trees of some apple rootstock clones. HortScience 8:176-178.

15. CROSSE, J. E., and R. N. GOODMAN. 1973. A selective medium for and a definitive colony characteristic of Erwinia amylovora. Phytopathology 63:1425-1426.

16. EDEN-GREEN, S. J., and E. BILLING. 1974. Fireblight. Rev. Plant Pathol. 53:353-365.

17. FISHER, R. W., and W. HIKICHI. 1971. Orchard sprayers—A guide for Ontario growers. Ont. Dept. Agr. Food. Pub. 373.

18. HILDEBRAND, E. M. 1937. Infectivity of the fire blight organism. Phytopathology 27:850-852.

19. KADO, C. I., and M. G. HESKETT. 1970. Selective media for isolation of Agrobacterium, Corynebacterium, Erwinia, Pseudomonas and Xanthomonas. Phytopathology 60:969-976.

20. KEIL, H. L., and T. VAN DER ZWET. 1975. Fire blight susceptibility of dwarfing apple rootstocks. Fruit Var. J. 29:30-33.

21. LITTLE, T. M., and F. J. HILLS. 1972. Statistical methods in agricultural research. Agricultural Extension, University of California, Davis and Berkeley.

22. MILLER, T. D., and M. N. SCHROTH. 1972. Monitoring the epiphytic population of Erwinia amylovora on pear with a selective medium. Phytopathology 62:1175-1182.

23. MILLS, W. D. 1955. Fire blight development on apple in western New York. Plant Dis. Rep. 39:206-207.

24. MOLLER, W. J., J. A. BEUTEL, W. O. REIL, and B. G. ZOLLER. 1972. Fire blight resistance to streptomycin in California. Phytopathology 62:779 (abstr.).

25. OITTO, W. A., T. VAN DER ZWET, and H. F. BROOKS. 1970. Rating of pear cultivars for resistance to fire blight. HortScience 5:474-476.

26. POWELL, D. 1963. Prebloom freezing as a factor in the occurrence of the blossom blight phase of fire blight of apples. Trans. Ill. State Hort. Soc. 97:144-148.

27. RITCHIE, D. F., and E. J. KLOS. 1974. A laboratory method of testing pathogenicity of suspected Erwinia amylovora isolates. Plant Dis. Rep. 58:181-183.

28. SCHROTH, M. N., S. V. THOMSON, D. C. HILDEBRAND, and W. J. MOLLER. 1974. Epidemiology and control of fire blight. Annu. Rev. Phytopathol. 12:389-412.

29. SEEMULLER, E. A., and S. V. BEER. 1976. Absence of cell wall polysaccharide degradation by Erwinia amylovora. Phytopathology 66:433-436.

30. SHOEMAKER, P. B. 1974. Fungicide testing: Some epidemiological and statistical considerations. Fungicide Nematicide Tests 29:1-3.

31. SOUTHEY, R. F. W., and G. J. HARPER. 1971. The survival of Erwinia amylovora in airborne particles: Tests in the laboratory and in open air. J. Appl. Bacteriol. 34:547-556.

32. THOMSON, S. V., M. N. SCHROTH, W. J. MOLLER, and W. O. REIL. 1975. Occurrence of fire blight of pears in relation to weather and epiphytic populations of Erwinia amylovora. Phytopathology 65:353-358.

33. WEISS, F. A. 1957. Maintenance and preservation of cultures. In PELCZAR, M. J., Jr. Manual of Microbiological Methods. McGraw-Hill: New York, pp. 99-119.

# 17. Method of Testing Fungicides on Tart Cherries (Prunus cerasus L.)

## F. H. Lewis

Professor emeritus of plant pathology, The Pennsylvania State University Fruit Research Laboratory, Biglerville, PA 17307

The commercial tart cherry industry is based primarily on the cultivar Montmorency. This cultivar is highly susceptible to cherry leaf spot caused by *Coccomyces hiemalis* Higgins, which can defoliate the trees within a month after bloom. It also is susceptible to powdery mildew caused by *Podosphaera oxyacanthae* (DC.) d By., which receives little attention unless it becomes unusually prevalent. Brown rot caused by *Monilinia fructicola* (Wint.) Honey, is important in Pennsylvania only on frost-injured flowers or during the last half of long, wet harvest seasons. In some areas, brown rot is caused by both *M. fructicola* (Wint.) Honey and M. *laxa* (Ader. and Ruhl.) Honey; the problem can become serious. Fruit rots caused by *Alternaria, Glomerella,* and *Botrytis* spp. occur where contributing factors such as hail favor the fungi.

Fungicides affect tart cherry trees both directly and indirectly. Directly, they may be phytotoxic and cause fruit injury, leaf necrosis, and defoliation. They may affect fruit skin toughness, which often is associated with ease of separation of the fruit from the pedicel, and the amount of scald that occurs during transportation and cooling. They may affect fruit color, size, and soluble solids. Odors due to fungicides may be present in canned cherries. Indirectly, fungicides may affect fruit yields, color, and maturity through their control of cherry leaf spot and the defoliation caused by it.

Other factors may be important in choosing a fungicide for use on tart cherries. If lead arsenate is used for control of cherry maggots (*Rhagoletis cingulata* Loew and *R. fausta* Osten Sacken), a fungicide such as captan or folpet should be used to reduce arsenical injury to the fruit pedicels. Allergic reactions to some fungicides, usually manifested as an itch, have occurred infrequently in harvest workers.

The quantity of fungicides used on tart cherries is insufficient to justify large expenditures on the part of industry in a search for new products. Therefore, fungicide testing on tart cherries is part of a large program aimed at products that could be applied to many crops. If a fungicide can be marketed for use on apples, for example, the manufacturer may be interested in obtaining labeling that meets U.S. government regulations for use of the compound on cherries.

This chapter describes a method for comparing a series of fungicides and mixtures for use on tart cherries to determine their effects on disease control, phytotoxicity, and fruit quality. Details of spray equipment and other items applicable to both apples and cherries are described in chapter 14.

## Methods

**Test site**—The site should be relatively frost-free, sheltered from the prevailing wind, and suitable for growth of cherry trees. Tart cherries are more susceptible to frost and wind injury than are apples or peaches.

Ideally, the organization sponsoring the work should own or lease the site. Trees in commercial orchards rarely are planted to permit the most desirable plot layout. Some of the fungicides may not be labeled for commercial use, so the fruit cannot be sold. Few growers are willing to cooperate in experimental work that tends to complicate harvesting operations.

**Test plants**—Preferred test plants are 5 to 12 years old, nearly uniform in size, free of obvious symptoms of virus diseases, and expected to yield 50 to 200 lb of fruit. Older trees can be used, however, if the experiment is designed with the expectation that the tree-to-tree variation will be fairly large.

Ring spot and yellows are so common in tart cherry trees that are more than 10–12 years old that using such trees in fungicide trials may be necessary. The ring spot virus usually begins to spread soon after the trees begin to bloom, with spread reaching a peak in eight- or ten-year-old trees and declining as infection nears 100%. Shock symptoms during the first year after infection may cause so much fruit and leaf injury that the trees are unusable for fungicide trials. After the first year, the trees usually show only mild leaf symptoms of ring spot and may be used for fungicide trials. Yellows appears after ring spot, and may cause severe defoliation and a reduced crop of large fruits. The results obtained in fungicide trials have been applicable to both healthy and virus-diseased trees. Thus, the problem is one of statistical error in the test.

The spacing of trees for use in fungicide trials is important in allowing movement of equipment and avoidance of spray drift. A tree spacing of 20 ft (6.1 m) on the square may be adequate for handgun applications of test fungicides to single-tree replicates if the trees are kept small by pruning and care is taken to minimize drift. Spacing at 30–35 ft (9.1–10.7 m) on the square is desirable if enough land is available. Another type of planting is one with three or four trees spaced 15–20 ft (4.6–6.1 m) apart in the row, with a 35–50-ft (10.7–15.2-m) space for movement of equipment. The rows in such a block should be 30–35 ft apart to minimize spray drift. Other spacing is suitable if the sprays are to be applied from equipment moving the full length of each row. My preference is for trees 20 ft (6.1 m) apart in rows 35 ft (10.7 m) apart.

**Maintenance of inoculum**—Where epidemics of cherry leaf spot have occurred year after year, inoculum is more than adequate for fungicide trials. If the amount of inoculum is low, cultivation should be delayed until the fungus spreads into the tree from the old leaves on the ground. High populations of all fungi causing cherry diseases are favored by not spraying 10–25% of the trees. After the appearance of leaf spot in unsprayed trees, one or more applications of fungicide may be necessary to avoid complete defoliation and death of the trees.

**Equipment**—Application of experimental fungicide treatments requires equipment adequate to apply the sprays uniformly during times when wind conditions are apt to be relatively stable. The equipment that we chose for such work is described in chapter 14.

**Planning the experiment**—Preference is given to randomized, complete-block experiments with up to 12 treatments and a check that is handled like other treatments except for omission of fungicides. Single-tree replicates are preferred if the trees show no symptoms of virus infection, but multiple-tree replicates are advisable in any situation in which tree-to-tree variation is apt to be high (abandoned orchards, weak trees, trees older than 10–12 years, trees defoliated early in the previous year).

One treatment should be a standard fungicide that has little effect on fruit quality and is cleared for use on tart cherries—captan, folpet, or captafol is preferred. Each experimental fungicide should be used at two or three rates. Only one factor should be varied with each treatment or plot unless satisfactory results are known to be obtainable only with a certain mixture. In that case, the mixture is treated as a single fungicide. Insecticides should be included with each fungicide, and should be the same throughout the experiment. In most cases, the interval between sprays should be the same as that suggested for commercial orchards. If, however, the fungicide is promising in all respects except for appearing weak in leaf spot control, it should be tested at all rates with at least two time intervals between sprays.

**Application of sprays**—High-pressure, dilute sprays should be applied at the amount necessary to cause the trees to begin to drip. The method of application (eg, handgun, spray boom, airblast equipment) is not important if all plots are treated alike, the plot design is suitable for the type of equipment, and the method of application is reported. If low-volume sprays are used, the fungicide rate per acre for mature trees should be based on the amount that is required in a standard amount of dilute spray, usually 300, sometimes 400, gal. The rate per acre can be reduced on short trees as compared with mature trees.

**Data collection**—Symptoms of phytotoxicity should be described and their severity estimated at intervals necessary to show the full extent of the injury. Leaf spot and powdery mildew counts should be made in a manner that reveals the percentage and severity of infection and any effects that the fungicide has on spore production on the lesions. Observations and counts should be timed to detect differences in the rate of defoliation following leaf spot infection.

The control of cherry leaf spot should be measured by (i) the amount of leaf abscission caused by the disease and (ii) the presence of disease on leaves remaining on the tree. The labor involved in counting the number of lesions is not justified in most experiments, but the reader may wish to study the reports of Ehlers and Moore (1,2).

One method of counting leaf spot infection and defoliation involves choosing four to eight terminal shoots distributed around the tree at a height of about 5 ft (1.2–1.5 m). Prior to loss of the first infected leaves, a tag 3–4 in. (7.6–10.2 cm) long is placed at the base of each shoot. A small tag is placed above the last unfolded leaf on each shoot, or if most of the terminal growth has

been completed, the tip bud and all immature folded leaves are pinched out. When the tags are placed, the number of leaves and the number of diseased leaves on each terminal are recorded. The count is repeated as needed to show the progress of the disease. If the counts are made at fixed intervals of two to four weeks, the data can be used to measure fungicide performance and the effects of various factors on disease incidence.

Modification of the above procedure may be desirable when terminal growth is rapid at the time the first count is made and an accurate record of late-season disease spread is wanted. The above instructions must be followed first. At each subsequent collection of data the number of leaves that have unfurled above the tag is counted. The number of diseased leaves and leaf scars above the tag is recorded, assuming that each leaf scar represents a leaf that dropped because of infection. The data from the tip of the terminal are added to that obtained from the tagged area at the base of the terminal; percentages are calculated on the basis of the total number of leaves that have unfurled at the time the count is made.

For less detailed counts, one method is to count the number of leaves present, the number of diseased leaves, and the number of leaf scars, assuming that in the absence of phytotoxicity, leaf fall was due to leaf spot. One problem with such counts is that leaf scars at the base of the terminal are small and closely spaced, making an accurate count difficult.

Powdery mildew control can be evaluated on the mean percentage of leaves with mildew on 10–20 shoots per tree. These shoots are selected in a uniform manner from the periphery of the tree at a height of about 10 ft (3 m). Mildew often is most abundant in the treetops.

Arsenical injury or dry stem should be counted if present near the middle of the harvesting season. Small branches are selected at intervals around the tree. The number of cherries with dead pedicels are recorded, including all cherries on each branch for a total of 200 or more fruits per tree. The data are calculated as the percentage of fruits with dry stem.

Most fruit quality measurements are made on samples of 2 lb (907 g) picked from each tree soon after harvesting begins. Equal samples may be taken on successive dates. Samples taken at random around the tree at a height of 4–5 ft (1.2–1.5 m) are preferred, but this sampling method may underestimate the number of poorly colored cherries on thick trees. Another method is picking fruits from triangular areas extending 2 ft (0.6 m) or more into the tree and using at least two such areas per tree or replicate.

The samples from each tree or replicate should be placed in a 12-in. (30-cm) plate with a small label showing the treatment and replicate numbers. All samples from each treatment are placed in a row across a laboratory table. This should be repeated for each treatment until all samples are exposed. Three workers can rearrange the samples until all agree on the relative ranking of the treatments based on the average color in all samples. This color ranking has been more useful than have mathematical values that measure intensity of colors with a Hunter color-difference meter (3–5).

The size of the cherries is determined by counting the number of fruits in a 2-lb (907 g) sample from each tree or replicate. With multiple-tree replicates on variable trees, a sample from each tree is desirable. The data may be expressed as the mean number of cherries per pound. The percentage of soluble solids (mostly sugar) in the juice is determined with an Abbe refractometer using one or two drops of juice from a composite sample expressed from a selected number of cherries (usually 25) from each tree or replicate.

Measurements of skin toughness and ability to withstand brusing and cracking can be made from (i) counts of the number of cherries that crack when soaked in water, (ii) the amount of bruising and discoloration that occurs in the harvesting and canning process, and (iii) the force required to pit the fruit or to shear through a given mass of fruit. The L. E. E. Kramer shear press, when equipped with a special pin, can be used to measure the force required to pit 20 cherries (3,5).

If fruit rots appear, 200 fruits per tree are collected in a uniform manner to determine the percentage of fruits affected by each type of rot.

If a new fungicide appears promising, 25–50 small cans or jars of fruit are processed according to commercial procedures. Four to six cans are opened at 60–90-day intervals, and the fruit is rated for color, odor, taste, and acceptability.

**Statistical procedure**—The data should be summarized and analyzed with a suitable statistical procedure, usually Duncan's multiple range test. Conversion of percentage data to equivalent angles, the **square root of X plus 0.5, or other suitable** transformation should be considered before statistical analysis. Mathematical analysis of color differences is not considered necessary.

## Discussion

The method described in this chapter can be as accurate as objectives, work, and expense justify. In a typical experiment, the mean number of leaves remaining on the tree had to differ by 2.6% in June, 6.1% in August, and 10.1% in September to be significantly different. Differences of 9–11 cherries per pound and of 1% fruit solids content can be measured with satisfactory accuracy.

The most desirable fungicide for tart cherries is (i) available for use on two or more deciduous fruits, preferably apples, peaches, and cherries, (ii) has been approved by the Cooperative Extension Service and state and federal agencies, (iii) is effective in leaf spot control, (iv) has no undesirable effects on fruit yield or quality, (v) leaves no visible residue on the fruit 10–14 days after the last spray application, (vi) causes few or no allergic effects among fruit pickers, and (vii) does not cause fruit injury at dosages up to about three times the amount necessary to control leaf spot. It is helpful if the fungicide controls powdery mildew and the fruit rots. Other characteristics might sometimes be important.

A good fungicide trial can be conducted if the worker (i) chooses an experimental plan that is adapted to available trees and equipment, (ii) varies only one factor in each treatment, (iii) applies the treatments uniformly with minimum spray drift, (iv) collects and reports the data to show the effects of the fungicides, and (v) statistically analyzes the data so that both the worker and readers of the reports can judge the variability in the trial.

## Literature Cited

1. EHLERS, C. G., and J. D. MOORE. 1960. Cherry leaf spot, brown rot and blossom blight of cherry. Fungicide Nematicide Tests. 16:34-35.
2. EHLERS, C. G., and J. D. MOORE. 1961. Brown rot of cherry, cherry leaf spot. Fungicide Nematicide Tests 17:34-35.
3. HARTZ, R. E., and K. E. LAWVER. 1965. Effect of sprays on quality factors of canned red tart cherries. Food Technol. 19(3):103-105.
4. KRAMER, A., and B. A. TWIGG. 1970. Quality control for the food industry. The Avi Publishing Company: Westport, CT.
5. LAWVER, K. E., and R. E. HARTZ. 1965. Effect of sprays on quality factors of raw red tart cherries. Food Technol. 19(3):100-103.

# 18. Laboratory and Field Test Procedures for Evaluation of Fungicides for Control of Brown Rot Diseases of Stone Fruits

**J. M. Ogawa, B. T. Manji, H. English, M. A. Sall**

Department of Plant Pathology, University of California, Davis, CA 95616

**W. E. Yates**

Department of Agricultural Engineering, University of California, Davis

**L. Chiarappa**

Food and Agriculture Organization of the United Nations, Rome, Italy

**D. Rough**

Farm advisor, San Joaquin County, Stockton, CA 95207

The methods described in this chapter are used in screening fungicides for the control of brown rot disease of stone fruits, which is caused by *Monilinia fructicola* (Wint.) Honey and *M. laxa* (Aderh. & Ruhl.) Honey. The methods have been developed over a long period of testing numerous fungicides for control of brown rot in California. The methodology combines laboratory and field investigations to determine potential for disease control, inhibition of *Monilinia,* and presence of *Monilinia* strains resistant or tolerant to specific fungicides. Critical stages of the disease cycle and the need for integrating brown rot control measures with existing orchard practices have been considered in developing the methodology. This has proved useful in securing maximum grower cooperation and producing useful experimental results even under conditions of low disease incidence. The resulting procedures are believed to be applicable to semiarid growing areas of the world when proper consideration is given to differences in environment and cultural practices.

## Procedures

**Preliminary laboratory screening**—Preliminary testing in vitro and in vivo eliminates compounds that show little or no inhibition of spore germination or mycelial growth of *Monilinia* spp.

In tests for inhibition of mycelial growth and spore germination, fungicides are incorporated as suspensions or solutions in solvents into hot or warm potato-dextrose (PDA) or synthetic agar media. Fungicides such as captan, which decompose when exposed to high temperature, must be placed in warm media. Some fungicides such as nabam and chlorothalonil are difficult to assess by this method. A 5-mm mycelial plug from an actively growing *Monilinia* culture is transferred to the petri dish or flattened dam test tube with fungicide-containing medium. Linear mycelial growth rate is measured. For *M. laxa* mycelial plugs, two- or three-day-old cultures are best. This growth inhibition test can be used to detect strains of *Monilinia* that are resistant or tolerant to specific

fungicides (14). Spore germination tests can be used to determine $ED_{50}$, $ED_{95}$, or $ED_{100}$ for each fungicide. Some chemicals (eg, benomyl), however, do not inhibit the germination of *Monilinia* conidia. At least three replications are required for each test.

Screening for blossom blight control indicates gross activity of chemicals, but does not resolve small differences in effectiveness between chemicals. This technique is best used to test fungicides with systemic activity. Twigs with unopened blossoms may be surface-sterilized with hypochlorous acid, opened in the laboratory, and sprayed with fungicide. Alternatively, blossoms already sprayed with fungicide in the orchard may be cut and placed in moist sand in moist chambers. The blossoms are inoculated with spore suspensions containing 100,000 spores per ml and incubated at room temperature. Infections are determined by sporulation of the *Monilinia* (11). Laboratory screening using harvested fruit is an excellent tool to evaluate eradication of contaminating spores or suppression of fruit decay (5,12).

**Field tests**—Suitable experimental or commercial orchards must have a history of the disease, as indicated by signs or symptoms (15). They should be in an area where environmental conditions favoring disease development, such as fog or rain showers, are common. Experimental chemicals are initially tested in experimental orchards. Fungicide screening is done occasionally in orchards of cooperative growers who are willing to leave unsprayed control trees, make available their spray equipment and water source, and most importantly, give full assurance they will use no other pesticides without prior consent of the researcher. The grower must be willing to provide twigs, blossoms, and fruit for disease evaluations and residue analyses, and must divert the crop from commercial channels if experimental chemicals are used.

For postharvest treatments, experimental chemicals are usually tested in noncommercial pilot treaters.

**Plot layout**—Experimental design for experimental fungicides consists of randomized, single tree replicated six to ten times. An unsprayed control and a reference chemical in commercial use are included as treatments in each test. Tests are usually made on experimental farms reserved for such studies. Portions of commercial orchards may be used, in which case the crop is destroyed after disease data and residue samples are collected.

**Registered fungicides**—Test plots for registered fungicides or chemicals with temporary tolerance are designed to yield data from which recommendations for use can be made. Randomized, single-tree blocks are set up in at least three locations differing in climatic or cultural conditions. Large-scale applications using airblast ground sprayers may also be made. These use five full-row blocks sprayed to give at least three randomized replications in a Latin-square design. In tests using helicopter or fixed-wing aircraft, blocks must be seven rows wide (140 ft, or 43 m) and replicated at least three times. Long rows can be divided into sections at least 20–30 trees in length. Helicopter plots could be shorter. The trees not sprayed with the test chemical are sprayed with a standard fungicide using the same type of sprayer to leave only a limited number of unsprayed (control) trees (16).

For postharvest treatments, fungicides that are registered with temporary tolerance may be tested in commercial packing houses and the treated fruit handled under normal conditions of storage, transit, and marketing.

**Test plants**—*Blossom and twig blight*—Information should be available on the relative susceptibilities, infection sites, disease symptoms, and pollination requirements of individual crops and varieties. Some of the pertinent information on spring blossom and twig blight of stone fruit crops in California is given in Table 18.1.

*Preharvest fruit decay*—Preharvest fruit decay and cracking are common on varieties that ripen during damp periods of rain, fog, or dew. Decay is also increased by insect injuries that provide a site for the dried fruit beetle to contaminate fruit with *Monilinia* conidia (2,13).

*Postharvest fruit decay*—Susceptibility to postharvest decay appears to differ little or not at all between cultivars. The most common infection site is the stem end where the skin is peeled or loosened during harvest. Quiescent infection appears to be a minor problem in California (4). The brown rot organisms *M. laxa* and *M. fructicola* can be grown for identification on 1% PDA at room temperature under a variable light source (3). Continuous light induces sporulation of *M. laxa* and makes differentiation of the two species more difficult. The brown rot medium that Phillips (9) developed may be used to detect *Monilinia* spores on blossoms or fruit surfaces when contamination from other fungi and bacteria is limited. Variation in host susceptibility to *Monilinia* species should be recognized (3,6,13).

**Inoculation**—Methods for successful large-scale spore inoculation of blossoms in the field have yet to be developed. Fruit inoculations can be made with spore suspensions that are sprayed onto trees with airblast sprayers. Repeated sprays of water from airblast sprayers or overhead sprinklers immediately follow inoculations to keep the fruit wet for 8–12 hr.

Field inoculum levels can be measured with funnels (1), spore traps such as the Hirst Automatic Volumetric spore trap (Brinkman Instruments, Menlo Park, CA), or the Burkard seven-day spore trap (Burkard Mfg. Co., Ltd., Rickmansworth-Hertfordshire, England). The importance of ascospores has not been determined in California. Apothecia of *M. fructicola*, however, can be found in peach and prune orchards with histories of brown rot blossom blight.

Experimental test plots should have high levels of natural inoculum to reduce the number of trees required for each treatment. On almonds and apricots, high inoculum levels can be obtained by injecting spore suspensions into blossom buds before they open. This should be done at least one year before experimentation. In California, *M. laxa* is used on almonds, apricots, cherries, plums, and prunes; *M. fructicola* is used on peaches and nectarines (3,6). Increasing inoculum levels on peaches may be made easier by obtaining *M. fructicola* mummies from commercial orchards, soaking

TABLE 18.1. Information on crops, pathogens, and disease essential to proper testing of fungicides for blossom blight control

| Crop | Test cultivars used[a] | Pollination | Pathogen | Common infection site(s) | Symptoms |
|------|------------------------|-------------|----------|--------------------------|----------|
| Almond (*Prunus amygdalus*) | Drake, Jordanolo, Marriott, Merced, NePlus Ultra, Mission (some orchards) | Self-infertile | *Monilinia laxa* | Stigma, anthers, petals | Blossom and twig blighting |
| Apricot (*P. armeniaca*) | Derby Royal, Blenheim | Self-fertile | *M. laxa* | All floral parts | Same as almond but with more twig blighting Young shoots also killed |
| Cherry (*P. avium*) | All susceptible | Self-infertile | *M. laxa* | Stigma, anthers | Blossom blight (tendency to drop before entering fruit spur) |
| Plum (*P. salicina*) | All susceptible | Mostly self-fertile | *M. laxa* | Stigma, anthers | Blossom blight (tendency to drop before entering fruit spur) |
| Prune (*P. domestica*) | Imperial, French | Self-fertile | *M. laxa*, *M. fructicola* | All floral parts | Same as almonds |
| Peaches and nectarines (*P. persica*, *P. persica* var. *nectarina*) | Dixon, Loadel, Carolyn, Fay Elberta peaches Le Grand nectarine | Self-fertile | *M. fructicola* *M. laxa* (rare) | Anthers only | Blossom and twig blight |

[a]These are considered more susceptible than are other cultivars.

them for 24 hr, and hanging them on the trees.

**Equipment and application rates**—In all tests, specifying the name of the equipment manufacturer, number and placement of nozzles, orifice size, core number if used, pressure at nozzle, forward speed of sprayer, tree spacing, tree height, and total gallons per acre is important.

Fungicide sprays are applied by handgun at the rate of 4–8 gal (15–30 L) per full-size tree to the drip stage. Chemicals are usually mixed in water before placing in the sprayer, and are allowed to recirculate through the spray hose before spraying trees. At least three concentrations of a new chemical must be tested to determine effective dosage. Concentration and volume are calculated to reflect the amount of active chemical used per acre.

In airblast applications, special consideration must be given to differences in air volume discharged by commercial sprayers. This is necessary to insure complete coverage. Wind at time of application is most critical when low-volume sprayers are used. Sprayers must be calibrated in liters or gallons discharged per minute to determine the tractor speed necessary to apply the desired amount of active ingredient per acre.

Where overhead sprinklers are used to apply fungicides, calibration is done by measuring the time required for a dye to move from the input to the furthest point of the discharge. Fungicide applications are made to the drip stage, allowed to dry, and repeated as necessary (8). Increasing the number of applications will increase coverage and deposit. A wind of 5–10 mph appears to increase the rate of drying without affecting application.

In aircraft application, the basic data stipulated above should be recorded and information included on swath spacing, angle of nozzle with respect to airstream, calibration of total flow rate in gallons per minute or liters per hectare, and height of aircraft. Aerial photos of spray blocks are sometimes made available to the pilot and aerial observer to insure more effective application (15).

Specific types of postharvest treatments need to be mentioned in addition to the name of the manufacturer. These are hydrocooling, prerinse, method of chemical application, dryers, and type of final pack. Data on chemical treatments must include the concentration used, the time of fruit exposure, and whether applications were applied as mixtures or with carriers such as waxes or oils. Preharvest chemical treatments should be noted as they could affect postharvest decay.

**Meteorological observations**—Temperature and moisture directly affect disease development. Blossom infection is closely related to temperatures above 55°F (13°C) and to duration of wetness. Warm, dry winds following infection of peach anthers can completely stop progress of the fungus in the filaments. Wetness is also required for preharvest fruit infection. Information is needed on rainfall, dew periods, temperature, relative humidity, and wind speed and direction during critical stages of flowering and ripening. Because local weather station reports usually do not agree with orchard microclimates, meteorologic equipment in orchards is needed to provide such information.

**Data collection**—*Sporodochial counts*—Ten twigs of apricots or almonds that were blighted the preceding year are randomly collected from each quadrant of the tree just before or during bloom. Sporodochia are counted under the dissecting microscope and the total number of spores determined. Spores are placed on agar media to determine germination rates (5,7,10).

*Blossom and twig blights*—Counts are made on each tree quadrant, and the percentage of blighted blossoms or twig blight calculated. For almond and apricot, the incidence of *M. laxa* twig blight is recorded, whereas for peaches, the incidence of blossom blight is observed. The number counted varies according to the degree of disease incidence but should be sufficient to provide a typical picture of disease with a minimum of variation between replications. Blossom blight is evaluated shortly after petal fall. Other causes of blossom blight must be differentiated. Apricot and almond flowers affected by *Pseudomonas syringae* van Hall appear dark brown to black, with a large proportion of consecutive blossoms on a limb being affected. This differs from brown rot in which blighted blossoms are light tan and usually show some sporulation. Twig blight in apricots, almonds, and prunes can be evaluated after petal fall, but should be assessed before newly developed leaves start masking blighted twigs. An acceptable method is not available for evaluating blossom blight in plums with which infected blossoms generally drop before the fungus invades twigs. On sweet cherries, dead flowers are collected before they drop and are examined for sporulation under a dissecting microscope. Isolations must be made from flowers in which sporulation cannot be observed.

*Preharvest fruit infection*—Quantifying infection incidence is difficult in most stone fruits because either disease is limited (apricot and almonds), or decay occurs in fruit clusters (plums and cherries) or is masked by interfering factors (eg, cracks, insects). Fruit rot data can be obtained during harvest by counting infected fruits, or just prior to harvest by counting mummified green fruits on the trees. The number of fruit inspected varies with disease incidence.

*Postharvest decay*—Decay data on apricots, peaches, nectarines, plums, and cherries are obtained by recording cause of decay after typical harvest, packing, storage, and transit conditions. Decay caused by other organisms must be differentiated. Three ten-box replications of these counts are needed. Observations also should be made on toxicity of chemicals to the fruit.

*Data processing*—Where counts are expressed as percentages, angular transformation is used. Duncan's multiple range test is used to test significant differences among treatments at the 5 or 1% level or both.

**Other required information**—Other information collected in these tests include data on phytotoxicity, residues, compatibility with other chemicals, and other possible side effects.

## Discussion

The methodology for screening fungicides for brown rot control in California is aimed at studying interruption of four important stages of the disease cycle: overwintering inoculum, spring infection of blossoms and twigs, preharvest fruit infection, and postharvest fruit infection. The economic value of chemical control of brown rot and the relative efficacy of different fungicides can be assessed better when related to the reduction in crop losses that these chemicals can achieve or to the economic gains that they can produce. These (decreased) losses or (increased) gains need to be assessed over long as well as short periods. For example, the loss of a few blossoms in any one year does not significantly affect yield. Continuous increase in inoculum, however, may well cause severe fruit losses if environmental conditions become favorable to the disease. Likewise, the killing of fruiting wood by twig blight must be evaluated as a potential loss of crop the following year.

## Literature Cited

1. CORBIN, J. B., and J. M. OGAWA. 1974. Springtime dispersal patterns of Monilinia laxa conidia in apricot, peach, prune and almond trees. Can. J. Bot. 52:167-176.
2. EL-BEHADLI, A. H. 1975. Mold contamination and infection of prunes and their control, PhD thesis, University of California.
3. HEWITT, W. B., and L. D. LEACH. 1939. Brown-rot Sclerotinias occurring in California and their distribution on stone fruits. Phytopathology 29:337-351.
4. JENKINS, P. T., and C. REINGANUM. 1965. The occurrence of quiescent infection of stone fruits caused by Sclerotinia fructicola (Wint.) Rehm. Aust. J. Agr. Res. 16:131-140.
5. MANJI, B. T., J. M. OGAWA, and G. A CHASTAGNER. 1974. Suppression of established infections in nectarine and peach with postharvest fungicide treatment. Proc. Am. Phytopathol. Soc. 1:43.
6. OGAWA, J. M., W. H. ENGLISH, and E. E. WILSON. 1954. Survey for brown rot of stone fruits in California. Plant Dis. Rep. 38:254-257.
7. OGAWA, J. M., D. H. HALL, and P. A. KOEPSELL. 1963. Spread of pathogens within crops as affected by life cycle and environment: Air-borne microbes. Symposium of the Society for General Microbiology: London, pp. 247-267.
8. OGAWA, J. M., B. T. MANJI, and W. R. SCHREADER. 1975. Monilinia life cycle on sweet cherries and its control by orchard sprinkler fungicide applications. Plant Dis. Rep. 59:876-880.
9. PHILLIPS, D. J., and J. M. HARVEY. 1975. Selective medium for detection of inoculum of Monilinia spp. on stone fruits. Phytopathology 65:1233-1235.
10. RAMSDELL, D. C., and J. M. OGAWA. 1973. Reduction of Monilinia laxa inoculum potential in almond orchards resulting from dormant benomyl sprays. Phytopathology 63:830-836.
11. RAMSDELL, D. C., and J. M. OGAWA. 1973. Systemic activity of methyl 2-benzimidazolecarbamate (MBC) in almond blossoms following prebloom sprays of benomyl and MBC. Phytopathology 63:959-964.
12. RAVETTO, D. J., and J. M. OGAWA. 1972. Penetration of peach fruit by benomyl and 2,6-dichloro-4-nitroaniline fungicides. Phytopathology 62:784 (abstr.).
13. TATE, K. G., and J. M. OGAWA. 1975. Nitidulid beetles as vectors of Monilinia fructicola in California stone fruits. Phytopathology 65:977-983.
14. TATE, K. G., J. M. OGAWA, B. T. MANJI, and E. BOSE. 1974. Survey for benomyl tolerant isolates of Monilinia fructicola and M. laxa in stone fruit orchards of California. Plant Dis. Rep. 58:663-665.
15. WILSON, E. E., and J. M. OGAWA. 1976. Fungal, bacterial, and certain nonparasitic diseases of fruit and nut trees in California. Department of Plant Pathology, University of California, Davis (class outlines).
16. YATES, W. E., J. M. OGAWA, and N. B. AKESSON. 1974. Spray distribution in peach orchards from helicopter and ground applications. Trans. Amer. Soc. Agr. Eng. 19(4):633-639, 644.

# 19. Field Evaluation of Fungicides for Control of Peach Leaf Curl

## H. L. Dooley

Supervisory plant pathologist, United States Environmental Protection Agency, Northwest Biological Investigations Station, 3320 Orchard Avenue, Corvallis, OR 97330

## I. C. MacSwan

Extension plant pathologist, Department of Botany and Plant Pathology, Oregon State University, Corvallis, OR 97331

Peach leaf curl is caused by the fungus *Taphrina deformans* (Berk.) Tul. This organism attacks most if not all cultivars of peach (*Prunus persica* [L.] Batsch). (The literature on this disease will not be reviewed here, but the reader is referred to material basic to the understanding of the disease [1–17].) In our tests we have used the cultivars Rochester, Elberta, and Early Improved Elberta, because they are susceptible under conditions in Willamette Valley, OR.

The exact relationship between climatic conditions and infection and disease development is still in doubt. Most researchers (1–5, 12–14, 17), however, agree that infection and disease development progress best under wet, cool weather conditions when the buds begin to swell. Most areas of the country that have 20 in. (50.8 cm) or more of rain have more disease than do those

areas with less rainfall and warmer temperatures.

Peach leaf curl is a short-cycle disease. The spores infect the leaves in the swelling buds. Mycelium grows within the leaf and produces ascospores that are discharged before the infected leaf drops (1–5, 7, 12, 14). The ascospores are lodged on the bark and twigs where they may germinate to form bud conidia several times before new bud swelling and infection occur.

Weather conditions play an important role. Research indicates that rain and dew may wash ascospores and bud conidia onto the swollen leaf buds to cause infection and complete the life cycle (1,3,4,7,12,14). Cool weather slows leaf development and allows further time for infection (1–4, 6, 11, 14, 15). During warm weather, the infection pegs may die or the leaf matures so rapidly that the fungus is outgrown and no disease occurs.

A single fungicide application can control peach leaf curl (1,8,12,14). The fungicide must be present throughout the period of bud swelling, opening, and leaf elongation. Although a single spray application has been proved to control the disease (1,3,12,14), two applications are recommended to assure acceptable control in the Willamette Valley. Applications in the Willamette Valley should be made in mid-December and early January.

The objective of this method is to evaluate fungicides used to control peach leaf curl caused by *T. deformans*. The test should be conducted at a site suitable for peach growth, where the disease is endemic.

## Procedure

A randomized block or other valid statistical design should be used, with a minimum of four single-tree replications per treatment. Treatments should be arranged according to Snedecor's randomly assorted digits table (16), or its equivalent.

Spray applications may be made with a handgun sprayer, using 300–350 psi (2,068–2,413 KPa) of pressure. A backpack duster for single-tree applications has been used successfully. If speed sprayers or dusters are used, however, we recommend that blocks of trees be sprayed rather than single trees to reduce spray drift to adjacent plots. To minimize drift, spray or dust applications should be made when there is little or no wind.

Treatments, including an unsprayed check and a standard fungicide, should consist of a minimum of four single-tree replicates. In areas where the disease is severe, it may be advisable to use a less effective fungicide (eg, Bordeaux mixture 8–8–100) on the check trees rather than to leave them untreated. Untreated trees are used to determine severity of the disease. A standard fungicide treatment, such as Bordeaux mixture 12–12–100 (5.44 kg–5.44 kg–379 L), is used to demonstrate that the timing of application was adequate for control. The performance of test fungicides may be compared with that of the standard fungicide or disease incidence in the untreated control trees or both.

## Data Collection

The number of diseased and healthy leaf terminals should be recorded after the leaves have emerged and unfolded in early spring but before defoliation. Leaf terminals consist of the rosette of leaves located at branch tips. The mean percentage of diseased leaf terminals and the total leaf terminals evaluated should be calculated and reported. A minimum of 50 leaf terminals per tree should be observed for a dwarf and 100 leaf terminals for a full-sized tree. The severity of any phytotoxicity as well as a description of injury should be reported. Phytotoxicity may appear as wood or twig burning.

## Results

Although 95–98% of disease control is possible with an effective fungicide and proper timing, a product being evaluated is considered effective if at least 70% of disease control is obtained.

## Reporting

The following information is helpful if reported: (i) name of product, EPA registration number, or experimental permit number, (ii) active ingredient or ingredients and the percentage of each, (iii) the application rate used, (iv) how the product was applied, (v) number of trees used per treatment, (vi) number of replications used, (vii) application date or dates, (viii) total leaf terminals evaluated per treatment, (ix) percentage of diseased leaf terminals, (x) date of bud swelling, and (xi) percentage and type of phytotoxicity if it occurs.

## Literature Cited

1. ANDERSON, H. W. 1956. Diseases of Fruit Crops. McGraw-Hill Book Co., Inc.: New York, pp. 215-222.
2. DUGGER, B. M. 1909. Fungus Diseases of Plants. Ginn and Company, Boston, pp. 176-182.
3. FITZPATRICK, R. E. 1934. The life history and parasitism of Taphrina deformans. Sci. Agr. 14:305-306.
4. FITZPATRICK, R. E. 1935. Further studies on the parasitism of Taphrina deformans. Sci. Agr. 15:341-344.
5. GROVES, A. B. 1938. The relation of concentration of fungicides and bud development to control of peach leaf curl. Phytopathology 28:170-179.
6. KNOWLES, E. 1887. The curl of peach leaves: A study of the abnormal structure induced by Exoascus deformans. Bot. Gaz. 12:216-218.
7. KOCK, L. W. 1934. Studies on the overwintering of certain fungi parasitic and saprophytic on fruit trees. Can. J. Res. 11:190-206.
8. MacSWAN, I. C. 1969. Peach leaf curl (*Taphrina deformans*). Fungicide Nematicide Tests 25:48.
9. MARTIN, E. M. 1925. Cultural and morphologic studies of some species of Taphrina. Phytopathology 15:67-76.
10. MARTIN, E. M. 1940. The morphology and cytology of Taphrina deformans. Amer. J. Bot. 27:743-751.
11. MIX, A. J. 1924. Biological and cultural studies of Exoascus deformans. Phytopathology 14:217-233.
12. MIX, A. J. 1935. The life history of Taphrina deformans. Phytopathology 25:41-66.
13. MIX, A. J. 1936. The genus Taphrina I: An annotated bibliography. Kans. Agr. Exp. Sta. Bull. 34:113-149.
14. PIERCE, N. E. 1900. Peach leaf curl: Its nature and treatment. USDA Division Vegetable Pathology Bulletin 20.
15. REDDICK, D., and L. A. TOAN. 1915. Fall spraying for peach leaf curl. NY Agr. Exp. Sta. Ithaca Bull. 31: 65-73.
16. SNEDECOR, G. W., and W. G. COCHRAN. 1967. Statistical Methods. Ed. 6. Iowa State University Press: Ames.
17. WILSON, E. E. 1937. Control of peach leaf curl by autumn applications of various fungicides. Phytopathology 27:110-112.

# 20. Evaluation of Cytospora Canker Severity in French Prunes

### P. F. Bertrand

Formerly research assistant, Department of Plant Pathology, University of California, Davis. Presently extension plant pathologist, University of Georgia Rural Development Center, PO Box 1209, Tifton, GA 31794

### Harley English

Professor, Department of Plant Pathology, University of California, Davis, CA 95616.

Evaluation of disease severity in plants can be a vexing problem, especially when perennial plants are gradually decimated by a slow-moving disease. Cytospora canker as it occurs on French prunes in California is one such disease.

French prune (*Prunus domestica* L. 'French') accounts for approximately 84,000 acres, or 94%, of California's total prune acreage (3). Cytospora canker caused by *Cytospora leucostoma* Sacc. is one of the most common, easily recognized diseases of French prunes in California. The perfect stage of this fungus has been described as *Valsa leucostoma* Fr. (*Leucostoma persoonii* [Nits.] Hohn.).

In California's major prune-growing region, which is centered around Yuba City and Marysville, orchards in all stages of Cytospora canker may be found. *C. leucostoma* is most virulent in weakened or predisposed trees (1). Work was undertaken in 1972–1974 to determine the various factors that contributed to Cytospora canker. A prerequisite of this work was to make an objective evaluation of disease severity in a given orchard so that comparisons of disease severity and other correlative factors could be made for a wide range of orchards of various ages.

This chapter describes a disease index (DI) for evaluating severity of Cytospora canker of French prune.

Fig. 20.1. Dormant French prune tree showing how cankers were weighted in value according to location. Cankers on small branches (A) received weighted value of 1; on large branches (B), 3; on main scaffolds (C), 9.

## Methods

The severity of Cytospora canker in an orchard was estimated by counting the number of separate fruiting cankers on a tree and giving each canker a weighted value based on its location, but not size, within the tree. Cankers on small branches and terminals were given a value of 1; those on large branches and secondary scaffolds, 3; and cankers on the major, or primary, scaffolds, 9 (Fig. 20.1). Each canker was placed in one of these three categories. Canker location rather than size was used, because commercial surgical removal of diseased tissues nearly always involves branch removal and not detailed scraping out of cankers. Large continuous cankers extending from minor to major branches were given a single value according to the most important limb on which they occurred. Thus, a single continuous fruiting canker from a terminal to a major scaffold would get a value of 9.

The sum of the weighted canker values for all trees evaluated divided by the number of trees evaluated gives the average canker value per tree. The amount of Cytospora canker in a susceptible orchard tends to increase yearly because of canker enlargement and new infection. Dividing the average canker values by an age factor minimizes the effect of age on the DI. Six years was subtracted from the age of the orchard, because Cytospora canker generally is not a problem in California until the trees begin to bear crops, which is generally after about six years' growth. Dividing by the age factor (orchard age minus six years) results in only an approximation of total disease incidence, because it may falsely assume equal yearly increases in the amount of disease. Disease counts were made from December to February, because cankers were most easily seen in the absence of leaves. The DI was calculated by the following formula:

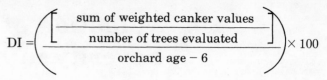

$$DI = \left( \frac{\left[ \dfrac{\text{sum of weighted canker values}}{\text{number of trees evaluated}} \right]}{\text{orchard age} - 6} \right) \times 100$$

Multiplying by 100 adjusts the DI ratings to values greater than 1.

## Discussion

Two considerations must be made before using this DI in other situations: First, bacterial canker caused by *Pseudomonas syringae* van Hall can cause severe damage to young prune and other stone fruit trees. These trees are often found to be severely infected with Cytospora canker. No attempt was made to evaluate Cytospora canker under these conditions by the DI previously described. When bacterial canker is a predisposing condition, an evaluation of both diseases may be necessary.

Secondly, winter injury can be a factor contributing to disease incidence. Although winter injury usually does not occur in the Central Valley of California, it is important in relation to development of Cytospora canker in other fruit areas of the nation (4–6). In these cases, the calculations might have to be modified to take winter injury into account.

In our studies, initial sampling was done on the Eckels' ranch near Yuba City, CA. This orchard is a 20-year-old, 40-acre planting on nonuniform soil. Management practices do not vary within this orchard. A second set of samples was taken from single plots in 13 separate French prune orchards of various ages and management regimes located throughout the northern California prune-growing regions. Comparison of DIs and other data gathered along with them (Tables 20.1 and 20.2) demonstrated that DI was significantly

**TABLE 20.1. Soil-plant data and disease indexes for French prune plots in 20-year-old Eckels' orchard near Yuba City, CA**

| Plot | Clay in root zone[a] (%) | Leaf K (%) | Observed disease index | Predicted disease index[b] |
|------|------|------|------|------|
| A | 53.4 | 1.08 | 80 | 80.4 |
| B | 42.4 | 1.30 | 30 | 44.0 |
| C | 51.5 | 1.01 | 90 | 77.8 |
| D | 47.4 | 0.55 | 96 | 82.3 |
| E | 37.4 | 1.74 | 21 | 16.0 |
| F | 28.0 | 1.60 | 8 | – 4.1 |
| G | 50.0 | 0.81 | 80 | 80.5 |
| H | 36.0 | 1.33 | 17 | 26.1 |
| I | 40.0 | 1.34 | 17 | 36.3 |
| J | 28.0 | 1.40 | 13 | 2.6 |
| K | 36.0 | 0.79 | 39 | 44.1 |
| L | 33.0 | 0.76 | 26 | 37.2 |
| M | 32.0 | 1.14 | 28 | 21.8 |

[a]Root zone is taken as that zone 30–76 cm below soil surface.
[b]Prediction of disease index (DI) based on regression equation developed from percentage of soil clay and leaf potassium content:

$$DI = -24.66 - 33.45 \text{ (leaf K)} + 2.64 \text{ (\% clay)}$$

$$R^2 = 0.8824 \ (P = 0.01)$$

**TABLE 20.2. Soil-plant data and disease indexes for 13 French prune orchards of various ages in northern California**

| Plot | Age | Clay in root zone[a] (%) | Leaf K (%) | Observed disease index | Predicted disease index[b] |
|------|------|------|------|------|------|
| Matthews West | 11 | 29.8 | 3.83 | 0 | 8.8 |
| Reynolds South | 16 | 24.8 | 2.75 | 25 | 10.6 |
| Masera | 11 | 29.8 | 1.38 | 22 | 23.4 |
| La Mantia | 15 | 34.8 | 0.83 | 46 | 31.4 |
| Barbaccia | 8 | 45.4 | 0.47 | 38 | 43.5 |
| Winfrey | 16 | 33.4 | 0.70 | 28 | 30.8 |
| Ruzich | 12 | 16.2 | 1.99 | 8 | 7.0 |
| Libby | 9 | 12.4 | 1.24 | 6 | 8.0 |
| Hall North | 11 | 12.9 | 2.28 | 1 | 2.2 |
| Hall East | 11 | 16.8 | 2.44 | 2 | 4.9 |
| Lindaur | 11 | 17.4 | 2.87 | 9 | 2.9 |
| Pacific | 15 | 21.4 | 1.72 | 1 | 13.5 |
| Butler | 16 | 16.0 | 2.66 | 4 | 2.9 |

[a]Root zone is taken as that zone 30–76 cm below soil surface.
[b]Prediction of disease index (DI) based on regression equation developed from percentage of soil clay and leaf potassium content:

$$DI = 3.77 - 5.97 \text{ (leaf K)} + 0.94 \text{ (\% clay)}$$

$$R^2 = 0.7397 \ (P = 0.01)$$

correlated with soil clay content and the potassium content of prune leaves (2). Potassium content of prune leaves is generally related to total soil potassium. Knowledge of soil factors in relation to subsequent tree performance can be useful in selecting planting sites or making long-range management decisions.

Although we have not used the DI described herein for evaluating pesticide or other treatments used for the control of Cytospora canker, we believe it could be used satisfactorily for this purpose. The DI would appear to be especially useful in an experiment designed to evaluate the use of spray treatments (fungicides or sun-protection paints) or soil treatments (fungicides, nematicides, or fertilizers) extending over several years in orchards of different ages.

## Literature Cited

1. BERTRAND, P. F., and H. ENGLISH. 1976. Virulence and seasonal activity of Cytospora leucostoma and C. cincta in French prune trees in California. Plant Dis. Rep. 60:106-110.
2. BERTRAND, P. F., H. ENGLISH, and R. M. CARLSON. 1976. Relationship of soil physical and fertility properties to the occurrence of Cytospora canker in French prune orchards. Phytopathology 66:1321-1324.
3. CALIFORNIA CROP AND LIVESTOCK REPORTING SERVICE. 1974. California fruit and nut crop acreage estimates, June 1974. Sacramento, CA.
4. HELTON, A. W. 1961. Effect of simulated freeze-cracking on invasion of dry-ice–injured stems of Stanley prune trees by naturally disseminated Cytospora inoculum. Plant Dis. Rep. 46:45-47.
5. HILDEBRAND, E. M. 1947. Perennial peach canker and the canker complex in New York, with methods of control. Cornell University Agricultural Experiment Station Memorandum No. 276.
6. ROLFS, F. M. 1910. Winter killing of twigs, canker and sunscald of peach trees. Missouri State Experiment Station Bulletin No. 17.

# SECTION III.

# B. Vegetable Crops

# 21. Evaluation of Fungicides in the Field for Control of White Rot of Onion

## P. B. Adams

Research plant pathologist, Soilborne Diseases Laboratory, Plant Protection Institute, Agricultural Research Service, USDA, Beltsville, MD 20705

White rot, which is caused by *Sclerotium cepivorum* Berk., is a disease that can occur on all *Allium* spp. In the United States it is an important disease on onion, shallot, leek, and garlic, all of which are planted in the fall and harvested in spring or early summer. The disease is particularly destructive in California, New Jersey, Oregon, and Washington. Since no resistant lines or cultivars of *Allium* spp. have been found, disease control is dependent on use of effective fungicides.

## Methods

The most important consideration in any field test is to determine its purpose: Is the field test designed to screen numerous chemicals to find one or more treatments that will provide a high level of disease control, or to compare several treatments for a cost-benefit ratio? Another important consideration is whether data from the field test will be used to support registration of one or more treatments. Since the answers to these questions may affect the design of the field test, they deserve careful consideration.

If the results from the field test are to be used to support registration of a fungicide, one must determine what kind of data and samples are to be taken. If soil treatments are involved, soil samples may be required at the time of application and at various intervals thereafter. Plant samples are required at harvest for residue analysis and may even be required at various intervals prior to harvest. With some treatments, soil and crop samples for residue analysis may be required on succeeding crops. These factors influence location of the field test, plot size, and number of replications. Investigation and careful consideration of these factors could reduce the number of field tests required to obtain the data necessary to support registration of a fungicide.

**Test plot**—The area of a field chosen for the field test should have a history of severe losses due to white rot. It should be level to prevent interplot interference caused by sclerotia of the pathogen being washed from one plot to another during heavy rains (2). Forming a ditch around each plot to drain the runoff water prevents interplot interference.

The size of the plot is determined by the type and size of the equipment to be used on it. In most of my work, the available equipment formed plant beds 5 ft (1.52 m) wide, with two rows of bunching onions per bed. Plants within each row were spaced less than 1 in. apart. Under these conditions, I used $5 \times 25$-ft ($1.52 \times 7.62$-m) plots. In other fields, with 5-ft (1.52-m) beds and four rows of bunching onions, I used $5 \times 12.5$-ft ($1.52 \times 3.81$-m) plots.

All field tests were set up in a randomized block design with at least four replications (3). Such a design permits statistical evaluation of replications as well as treatments. This is often useful, especially if a part of the test area was used for a field test during the previous growing season.

To compare treatments adequately, at least 50% of the plants in the control plots should become infected with the pathogen. If such a field cannot be found, the plot area can be infested with the pathogen before setting up the field test. For this procedure, *S. cepivorum* was grown on sterilized oat seed (2,000 g of oat seed plus 2,500 ml of water) for four weeks. At least one month before the field was to be used, the inoculum and oat seed was broadcast over the plot area and worked into the soil to a depth of about 6 in. (15 cm).

In all field tests, untreated control plots have been included. Sometimes the standard recommended treatment, 2,6-dichloro-4-nitroaniline (DCNA), also is included. These two sets of controls enable one to determine whether an experimental treatment reduced disease severity and whether it was better than the standard treatment.

The onion cultivar used in the field test is usually that which the local growers use. Cultural practices are the same as those of the local growers, or those of the grower whose field is being used for the test.

**Fungicide application**—Four methods of applying fungicides for the control of soilborne pathogens are available: preplant soil incorporation, in-furrow spray or dust, seed treatment, and foliar sprays. In my field tests, the preplant soil-incorporated chemicals are either sprayed on the soil surface with a hand sprayer or applied with a lawn-type fertilizer spreader, depending on their formulation. Once applied, these materials are worked into the plots with a tractor-mounted Rototiller or Rotovator. Seeds are planted with tractor-mounted equipment. The in-furrow sprays are applied to an open

seed furrow with a 1-gal hand sprayer and the plots planted with a hand seeder (Planet Jr. No. 300 A). For seed treatments, the seeds are treated in a laboratory and planted with a hand seeder. I have not evaluated foliar spray materials for the control of onion white rot, and thus this method of control will not be considered further.

**Data collection**—As a minimum, one should collect data that provide an estimate of treatment phytotoxicity, disease control, and crop yield from each plot. With preplant soil incorporation treatments, in-furrow spray treatments, or seed treatments, evidence of phytotoxicity can be obtained from plant stand counts. These generally were taken about 30 days after planting. Disease control data were obtained at time of harvest and were recorded as percentage of diseased plants. Yield was calculated as the weight of healthy plants for bunching onions and the weight of healthy bulbs for bulb onions. Bulb onions should be graded for size and yield data taken by grade. All data collected from the field test were subjected to an analysis of variance (3) and to Duncan's multiple range test (1).

## Results and Discussion

Table 21.1 shows the results of an in-furrow spray field test for the control of white rot of green bunching onions. Plant stands were counted 30 days after planting to get an estimate of treatment phytotoxicity. A similar count, as well as a field appearance rating (a subjective rating based on plant stand, vigor, and color), was made at harvest to estimate phytotoxicity further. These data indicated no phytotoxicity due to the treatments. Although differences of 1.5% in plant infection were statistically significant, yield data were not significantly different among chemicals and rates tested.

Because of the several methods of applying materials and planting seeds, I have frequently divided the field test into two or three smaller tests comprising preplant soil incorporation treatments, in-furrow spray treatments, and seed treatments. Small tests can be established in fields of different growers. For example, during the 1975–1976 growing season, I had a seed treatment test at Beltsville, MD, and Centerton; in-furrow tests at Centerton and Vineland; and a preplant soil incorporation test at Centerton. One disadvantage

**TABLE 21.1. Control of white rot of green bunching onions in in-furrow spray field test in New Jersey (1973–1974)**

| Treatment[a] | Rate (lb active ingredient per acre) | Plant stand At 30 days | Plant stand At harvest | Field appearance of harvest[b] (0-10) | Yield (lb per acre) | Infection (%) |
|---|---|---|---|---|---|---|
| Thiophanate methyl | 5.3 | 35 A[a] | 52 A[c] | 8.50 AB[c] | 31,098 A[c] | 0.00 A[c] |
| Benomyl | 20.0 | 47 A | 48 A | 8.75 AB | 28,896 A | 0.00 A |
| Thiophanate methyl | 21.0 | 48 A | 56 A | 9.25 A | 30,530 A | 0.00 A |
| Benomyl | 10.0 | 53 A | 58 A | 8.50 AB | 32,388 A | 0.25 AB |
| Thiophanate methyl | 10.5 | 49 A | 52 A | 9.25 A | 32,560 A | 0.75 AB |
| Benomyl | 5.0 | 44 A | 45 A | 7.25 B | 23,960 A | 1.50 B |
| Control (untreated) | ... | 57 A | 29 B | 3.50 C | 86 B | 99.25 C |

[a]Treatments were applied and seeds of evergreen bunching onions planted on September 11, 1973. Plots were harvested on May 13, 1974.

[b]Subjective rating of plants in plots in which 0 was assigned to plants with poor stand, vigor, and color and 10 to plants with excellent stand, vigor, and color.

[c]Values not followed by same letter are statistically different (P = 0.05), as determined by Duncan's multiple range test.

was that comparing treatments in different tests such as a seed treatment was difficult with an in-furrow treatment.

The procedures described in this chapter have been developed during five years of field testing treatments for the control of white rot. Results from some of these field tests are being used in an attempt to obtain registration of several fungicides for white rot control.

## Literature Cited

1. DUNCAN, D. B. 1955. Multiple range and multiple F tests. Biometrics 11:1-42.
2. SHOEMAKER, P. B. 1974. Fungicide testing: some epidemiology and statistical considerations. Fungicide Nematicide Tests 29:1-3.
3. SNEDECOR, G. W. 1956. Statistical Methods Applied to Experiments in Agriculture and Biology. The Iowa State College Press: Ames, IA.

# 22. Procedures for Field Testing Foliar Fungicides for Potato Late Blight Control

### R. C. Cetas

Professor, Department of Plant Pathology, Cornell University, Long Island Horticultural Research Laboratory, Riverhead, NY 11901

### F. E. Manzer

Professor, Department of Botany and Plant Pathology, University of Maine, Orono, ME 04473

### W. E. Fry

Associate professor, Department of Plant Pathology, Cornell University, Ithaca, NY 14853

### O. E. Schultz

Professor, Department of Plant Pathology, Cornell University, Ithaca, NY 14853

### E. K. Wade

Extension plant pathologist, Department of Plant Pathology, University of Wisconsin, Madison, WI 53706

### D. P. Weingartner

Associate professor, University of Florida, Agricultural Center, PO Box 728, Hastings, FL 32045

### H. L. Dooley

Supervisory plant pathologist, United States Environmental Protection Agency, Northwest Biological Investigations Station, 3320 Orchard Avenue, Corvallis, OR 97330

### S. S. Leach

Plant pathologist, United States Department of Agriculture, Science and Education Administration, Federal Research, Greenhouse, University of Maine, Orono, ME 04473

### L. E. Heidrick

Plant pathologist, Chevron Chemical Company, Ortho Division, PO Box 118, Moorestown, NJ 08057

### B. L. Richards, Jr.

Plant pathologist (retired), E. I. du Pont de Nemours & Company, 268 Building, DuPont Experiment Station, Wilmington, DE 19898

Late blight caused by *Phytophthora infestans* (Mont.) DeBary is a major disease of potato (*Solanum tuberosum* L.). It appeared in Europe and the United States about 1830 and was the cause of the 1845–1846 potato famine in Ireland. Late blight has appeared in most potato production areas of the world (5,13).

Many studies have been conducted to determine the environmental conditions that are favorable for *P. infestans* and the disease that it incites. Crosier (6) reviewed much of the early work, and his studies provide good information on the environmental conditions that the fungus requires. Excellent summaries can be found in many textbooks (2,3,25) and other publications (5). In general, the conditions favorable for late blight are 90–100% relative humidity (RH) for 10 hr or more and air temperatures of 7.2–29.4°C (45–85°F). The optimum conditions for the various stages in the life cycle of *P. infestans* are: sporangiospore formation, 18–22°C (64.4–71.6°F) and 100% RH; direct germination of sporangiospores, 24°C (75.2°F) and free moisture on the foliage; zoospore formation (indirect germination of sporangiospores), 9–15°C (48.2–59.0°F) and free moisture; zoospore germination, 12–15°C (53.6–59.0°F) and free moisture; ingress, 15–21°C (59.0–69.8°F) and free moisture; and incubation, 18–23°C (64.4–73.4°F). Under optimum conditions, sporangiospores form in 8–14 hr, indirect germination and infection occur in 2.5–7 hr, and new lesions appear in three to five days

(5,6). Under field conditions, the incubation period usually is five to ten days.

Although many control measures have been developed since 1846, spraying the plants with an effective fungicide has remained the chief control measure since about 1900 when Bordeaux mixture became widely accepted in Europe and the United States (25). The control of late blight constitutes one of the major uses of fungicides. Consequently, many candidate fungicides are screened for control of late blight before being tested for control of other vegetable diseases.

Most agricultural chemical companies have developed their own routine laboratory and greenhouse methods for determining the potential activity of a candidate fungicide. The methods and procedures described in this chapter are intended as a guide for conducting field tests to determine the efficacy of protective foliar fungicides. These procedures can also be used in field tests to determine the proper dosage, interval between applications, and other factors necessary to obtain satisfactory control of late blight with protective foliar fungicides.

## Methods

**Test site**—The tests should be conducted in an area adaptable for potato culture, where climatic conditions favor development of late blight. The soil type and fertility should be as uniform as possible. Irrigation should be available to supplement rainfall during periods of drought. Overhead sprinkler systems are preferred. If possible, the test site should be isolated from commercial or other experimental potato and tomato plantings or both.

**Test plant**—Certified or foundation quality seed of a susceptible cultivar adapted to the area should be planted in the test plots. The cultivar used should have little or no known horizontal or vertical resistance to late blight. Acceptable cultivars include Chippewa, Green Mountain, Hudson, Katahdin, Netted Gem, Norgold Russet, Red LaSoda, and Russet Burbank. All plots within an experiment should be planted with seed of the same cultivar from the same source. Cut seed pieces can be treated with a recommended fungicidal dust (eg, 8% mancozeb or 7% captan) or dip treatment to prevent seed piece decay. The seed pieces should average between 42.5 and 56.7 g (1.5–2.0 oz) and should be spaced 20.3–30.5 cm (8–12 in.) apart in rows 0.76–1.02 m (30–40 in.) apart. The plants should be maintained in good growing condition by use of fertilizer; weed, insect, and nematode control measures; irrigation; and other cultural practices recommended for the area. If nonregistered pesticides or those labeled exclusively for experimental purposes are used, the tubers or foliage must not be used for human consumption or animal feed.

**Plot design**—The experimental design depends on the number of compounds to be tested and on the land and type of equipment that is available. Acceptable designs include paired plots, randomized complete blocks, Latin square, and randomized split plots (4,15,21,22). Treatments should be randomly allocated by using such tables as Snedecor's random digits (4,21) or other appropriate means of randomization. Each treatment should be replicated at least three times. Plot size depends on objectives of the test. For screening and ranking fungicides, small plots one to three rows wide by 3.0–7.6 m (10–25 ft) long are adequate. For determining fungicidal rates, application intervals, or specific materials for recommendation, larger plots three to ten rows wide by 7.6–15.2 m (25–50 ft) long may be used. Ends of plots should be separated by a buffer zone of sufficient length, at least 1.5 m (5 ft), to prevent drift or cross-contamination or both with equipment being used. The buffer zone can be planted to potatoes. The plots should be arranged so that the test row or rows are not damaged by the application equipment.

Recently, interplot interference has been discussed considerably (11,19,24). Inclusion of a no-fungicide or ineffective fungicidal treatment in the design may result in underestimating the efficacy of more effective treatments. Conversely, placing a highly effective fungicidal treatment next to an inferior one can result in overestimating the efficacy of the inferior fungicide. Consequently, many researchers prefer that the no-fungicide control plots be adjacent to but separated from the experimental area by three or more rows that are sprayed with an effective fungicide on a seven- to ten-day schedule. Such plots should be located down wind from the experimental area to reduce the inoculum pressure and interplot interference on the experimental plots, and should be of sufficient width and length to insure that drift does not contaminate the center rows. If a no-fungicide control is included in the experimental design, it should be located in a manner that exposes all other treatments equally to the interplot interference it may cause. Uniformly high inoculum or disease pressures can be achieved by having one or more no-fungicide guard (buffer) rows adjacent to each treatment plot. This is desirable when screening and ranking fungicides for effectiveness, but may be undesirable when comparing application intervals.

A standard fungicide should be included in each experiment. The standard selected should be a fungicide that commercial growers commonly use in the area, eg, mancozeb or maneb. These are usually applied at the rate of 1.34–1.79 kg of active ingredient per hectare (1.2–1.6 lb of active ingredient per acre) per application. Other acceptable standards include chlorothalonil, captafol, and metiram. These should be used at locally recommended dosages.

**Fungicide application**—The sprayer should be tractor-mounted or self-propelled and should provide a constant speed that can be adjusted readily. The sprayer should be calibrated to apply between 468 and 936 L/ha (50–100 gal/acre) at sufficient pressure for good coverage. High-volume, high-pressure boom sprayers usually are operated at $1.7 \times 10^6$ to $2.8 \times 10^6$ Pa (250–400 psi). The maximum speed should not exceed 6.4 km/hr (4 mph). The sprayer should have either a brush or modified-brush boom or booms, with at least three dripless hollow-cone nozzles per row. The nozzles should be angled forward about 0.26 rad (15 degrees) from the vertical. The sprayer must be equipped with either paddle or an overflow agitator or agitators that keep wettable powders in suspension.

The first spray should be applied when the plants are 20.3–25.4 cm (8–10 in.) tall. Subsequent sprays should be applied at weekly (seven-day) intervals for routine evaluations. When application intervals are being studied, the standard should be compared at the same

64

intervals as the candidate fungicide or fungicides.

**Organism and maintenance of cultures**—Race O (field race) of *P. infestans* is preferred. Cultures of the fungus have been maintained by serial transfer on frozen lima bean agar (20), dry lima bean agar (23), rye agar (7), potato tuber slices (13), foliage of suscepltible cultivars of potato and tomato, and in potato tubers (17,28) and liquid rye medium (7). Cultures grown on agar are incubated at 4–20°C (39.2–68.0°F). The cultures must be transferred at weekly to biweekly intervals when incubated at 20°C (68°F) and monthly when at 4°C (39.2°F). Stock cultures under mineral oil have remained viable for at least two years (7). Isolates maintained on agar frequently become attenuated and should be tested for pathogenicity on plants of susceptible potato cultivars before being used to produce inoculum for field plots. Some workers have found autoclaving the rye agar for 30 min at 116°C (240.8°F) necessary to eliminate bacterial contaminants from the rye seeds.

Pathogenic cultures of *P. infestans* have been successfully maintained on tuber slices for several years. With this method, firm, rot-free tubers of a susceptible cultivar are surface sterilized in 0.525% sodium hypochlorite solution for 10 min or longer. The tubers are blotted dry with clean paper towels and cut into slices about 5 mm thick with a flame-sterilized knife. The slices are placed in suitable moist chambers, eg, 250 × 80 mm glass dishes with covers and metal screen (hardware cloth) false bottoms, and inoculated with *P. infestans*. Natural polypropylene-plastic sterilizing pans (31.8 × 25.4 × 13.3 cm [12.5 × 10 × 5.25 in.]) covered with two layers of cheesecloth and single-strength window glass make excellent moist chambers in which to incubate tuber slices. The capacity of the pans can be doubled by fitting them with a two-tiered tray (about 30.5 × 25.4 cm [12 × 10 in.]) made of 6.3-mm (0.25-in.) mesh metal screen (hardware cloth). The tiers are held about 7.0 cm (2.75 in.) apart by five stove bolts, one in each corner and one in the center. The bottom tier is held between the bolt head and a single nut, while the top tier is affixed between two nuts.

Several methods have been used to inoculate the tuber slices. One is to transfer a few sporangiophores and sporangiospores from the margin of a tuber slice culture to the damp surface of freshly cut slices. This can be done with a flame-sterilized dissecting needle. A second method is to place sporulating tuber slice cultures on a hardware cloth tray above the freshly cut tuber slices in a moist chamber. Thus, the sporangiospores drop onto the freshly cut surfaces of the tuber slices below. A third but more laborious method is to make a suspension (sterile distilled water) of sporangiophores and sporangiospores harvested from the margins of tuber slice cultures. One drop of suspension is placed on each tuber slice.

When incubated at 12.8°C (55°F), the tuber slice cultures begin sporulating in about five days and should be transferred about every seven days. Bacterial contamination has been one of the main drawbacks to maintaining cultures of *P. infestans* on potato tuber slices. Bacterial contamination, however, can be minimized by weekly transfers and by using the direct transfer method rather than the spore suspension method. The interval between transfers probably could be lengthened by decreasing the incubation temperature.

Cultures of *P. infestans* have been maintained by collecting infected tubers from field plots and storing them at 4.5°C (40°F) and 85–95% RH. Another method is to inoculate tubers in the laboratory with a sporangiospore or zoospore suspension (17,28) and store them as above.

Some workers prefer to maintain their cultures on plants of susceptible cultivars of potatoes. With this method, the plants are sprayed to runoff with an aqueous suspension of sporangiospores or zoospores (500–1,000/cm$^3$) and incubated at 20°C (68°F) and 100% RH. Twenty to 25 hr after inoculation, the plants are transferred to a greenhouse bench. Two to three days later, the leaves with very young lesions are harvested and stored at 4°C (39.2°F) and high RH for seven to ten days. The infected leaves are then transferred to a moist chamber maintained at 18–22°C (64.4–71.6°F) for 24 hr. Spores washed off these leaves are used to prepare inoculum for inoculating additional potato plants.

Other workers prefer to maintain their cultures on susceptible cultivars of tomatoes. The tomato plants are grown under a low fertilizer regimen. They are sprayed to runoff with an aqueous suspension of sporangiospores or zoospores (500–10,000/cm$^3$) and incubated for 20–24 hr at 20°C (68°F) and 100% RH. The plants are then cut into sections and packed loosely in plastic or glass containers (plastic shoe boxes are ideal). These are incubated at 20°C (68°F) for three to four days, or until the foliage is covered with sporangial growth. This material can be stored for seven to ten days at 4–5°C (39.2–41.0°F) without serious deterioration. The spores are washed off by dipping the foliage in chilled (5–10°C [41–50°F]) distilled water. The suspension is adjusted to the desired spore concentration and used to inoculate plants either in the greenhouse or field plots. The cultures must be transferred every 7–14 days to maintain viability.

**Inoculum for field plots**—Sporangiospores of *P. infestans* may be produced by any one of the methods described above. They may also be produced by planting infected tubers in a special "blight garden" where environmental conditions favorable for late blight development are maintained.

The sporangiospores are washed off the culture medium (lima bean agar, potato tuber slices) with distilled water or with nonchlorinated or nonfluorinated tap water or both. Tap water, especially if copper pipe is involved, should be tested for its effect on spore germination before being used to prepare spore suspensions. If it seriously inhibits sporangiospore or zoospore germination or both, it should not be used. The suspension is filtered through four or more layers of cheesecloth to remove debris and adjusted to a concentration of at least five to ten sporangiospores per ×100 microscope field (5,000–10,000/cm$^3$) or to an equivalent measure of zoospores.

**Inoculation of field plots**—One or more foliar fungicidal sprays should be applied before the plots are inoculated when protective foliar fungicides are being evaluated. At least 12 hr should elapse between application of the fungicide and inoculation of the plants. The plants should be wet with dew, rain, or irrigation water when inoculations are made. The foliage should remain wet for 10 hr or more following inoculation. This requirement is often met when the plants are inoculated in the evening. If the natural environment does not provide these conditions, overhead sprinkler irrigation (0.25 cm/hr [0.10 in./hr]) can be used to provide moisture for spore germination and ingress.

The plots may be inoculated by spraying the spore suspension onto the plants in the no-fungicide guard or buffer rows adjacent to each plot, onto all plants in the test (middle) row or rows of each plot, or onto plants in a swath about 0.91 m (3 ft) wide across all rows in the center of each plot. Many researchers prefer the first method.

The inoculum can be applied with a small hand sprayer or with a one-nozzle-per-row compressed-air or power sprayer. The sprayer must be free of all chemicals and preferably be one used solely for this purpose. Repeated inoculations may be required or desired or both to establish an epiphytotic severe enough to distinguish the fungicides that are less effective than the standard.

Environmental conditions favorable for development and spread of late blight must be maintained following completion of the primary cycle. If not provided naturally, supplemental overhead sprinkler irrigation (about 0.25 cm/hr [0.10 in./hr]) can be employed, in the morning or evening, to extend natural wet periods attributed to dew on the potato foliage.

## Data

Foliar blight, phototoxicity, and yield data are collected from most routine efficacy tests. Some researchers examine the tubers for late blight on a routine basis. When candidate fungicides reach a given stage in development, manufacturers request tuber samples for residue analyses. Occasionally, information may be desired concerning the effect of the fungicidal treatments on specific gravity, taste, and cooking and processing qualities of the tubers.

**Foliar blight**—The Horsfall-Barratt grading system for measuring plant disease (8,18) is commonly used to estimate disease severity visually. Some researchers, especially in Canada and Europe, prefer the British Mycological Society key (1,5). Disease severity should be determined at least twice following inoculation. One reading should be made, if possible, before the plants in the no-fungicide controls are 50% defoliated, and the second when they are 90–100% defoliated. The final reading, however, should be immediately before the plants are killed by frost or by chemicals in preparation for harvest. If the plants are allowed to die naturally, the final reading should be made before natural senescence prevents proper estimation of disease severity. If disease progress curves are desired, visual estimates of disease severity should be made at intervals of five to seven days beginning with the completion of the primary cycle.

Disease severity is commonly expressed as percentage of defoliation. Some researchers, however, prefer to express their data as percentage of control. When percentage of disease control is reported, it should be in addition to percentage of defoliation. Defoliation data provide information on the severity and reliability of the test. Percentage of disease control can be calculated according to the following formula:

$$PDC = \frac{DIC - DIT}{DIC} \times 100$$

where PDC is percentage of disease control; DIC, disease incidence in control (no-fungicide); and DIT, disease incidence in treatment.

**Phytotoxicity**—Symptoms of phytotoxicity seldom are the same for any two compounds. Some symptoms that may be observed are necrotic flecking or spotting of the leaflets, marginal chlorosis or necrosis or both, stunting due to shortening of the internodes, smaller or darker green leaflets or both, thicker and crisper leaflets, upward rolling of the margins of the leaflets, and deformed leaflets. In addition to describing the phytotoxic effects, indicating their severity in relation to dosage, number of sprays, and stage of plant growth is important.

**Yield**—In the absence of late blight and other foliar diseases, nonphytotoxic foliar fungicides seldom have a significant effect on yield of tubers. In the presence of late blight or other foliar diseases, the difference in yield between a fungicidal treatment and the no-fungicide control depends on when the epidemic begins in relationship to the stage of tuber development, rate at which the epidemic develops, the severity of the epidemic, and the effectiveness of the fungicidal treatment (9–12, 16). Consequently, many researchers think that yield data are relatively unimportant in evaluating protective foliar fungicides. Yield data, at least from some tests at most locations, are required to demonstrate that the candidate fungicide has no adverse effect on yield and are required by the governmental agencies responsible for registering and labeling pesticides.

Yield data are obtained by harvesting the tubers from the center row or rows of each plot. They are usually reported as total and U.S. No. 1 size A (4.8 cm or 1.875 in. minimum diameter) and are expressed as either kilograms or pounds per plot, quintals or metric tons per hectare, or hundredweight per acre.

**Tuber rot**—The amount of late blight tuber rot that develops depends on the amount of fungicidal residue on the soil surface, the persistence of the residue, the soil type, depth at which the tubers develop, amount of rainfall or supplemental irrigation or both applied during periods of sporulation of the fungus, and the amount of fungicide used. Consequently, the amount of tuber rot that develops often is not what might be expected based on the amount of foliar blight. When these data are taken, all size A tubers are washed and examined for tuber rot within two weeks of harvest. A second reading taken one or two months later would be beneficial. These data can be expressed as percentage of rot based on either weight or number of tubers.

**Residues**—Residues of candidate fungicides in the tubers must be determined to comply with the rules and regulations of governmental agencies responsible for registering and labeling pesticides. This normally is the responsibility of the manufacturer, who usually prescribes how and when the tuber samples should be taken. The samples normally consist of a composite of 2.3–4.5 kg (5–10 lb) of tubers from the desired treatment or treatments. An equal number of tubers of similar size are selected randomly from each of three or more replications of each treatment. To prevent contamination, the no-fungicide control sample or samples should be collected first. The tubers normally are not washed and are placed in heavy (1.5–2 mil) plastic bags or in special sample bags that the manufacturer provides. Unless otherwise indicated, the samples are frozen at −17.8°C (0°F) immediately after harvest and packed with dry ice in insulated boxes when shipped to the residue laboratory for assay.

**Specific gravity**—Specific gravity is a measure of the internal quality of the tubers and is correlated with total solids. Nonphytotoxic protective fungicides usually have no significant effect on specific gravity of the tubers (16). If desired, the specific gravity of the tubers can be determined by a 3.6-kg (8-lb) capacity potato hydrometer (available from Potato Chip Institute International, 946 Hanna Building, Cleveland, OH 44115) or by the weight-in-air versus weight-in-water method. These readings should be based on a random sample of size A tubers from each plot.

**Off-flavor**—Taste tests normally are not considered in evaluating protective foliar fungicides. As we move into the use of systemic compounds, this may become a more important aspect of fungicidal evaluation programs. An effective compound is useless if it imparts an off-flavor to the tubers. Therefore, sufficient taste tests should be conducted to demonstrate that the candidate fungicide does not adversely affect the flavor of the tubers. Samples for taste tests should be collected in a manner similar to those described for residue analyses. Preliminary taste tests can be conducted in the researcher's own laboratory. Peeled tubers should be boiled until tender, and a panel of three or more people should taste the cooked potatoes. More refined tests often can be conducted in cooperation with home economics schools or colleges in many universities.

**Cooking and processing**—Tests to determine the effect of foliar fungicides on the cooking and processing qualities of the tubers normally are not part of the routine evaluation of fungicides for late blight control. Situations may arise, however, when one may wish to consider these factors. Tuber samples for such tests can be collected from routine fungicidal experiments by a method similar to those described for residue analyses.

**Additional tests**—The manufacturer or cooperating agency should conduct additional tests to determine the following, which are beyond the scope and intent of this publication: (i) efficacy of candidate fungicides when applied by ultralow- or low-volume ground sprayer, row-crop airblast sprayer, aircraft, overhead sprinkler irrigation, or any other application method that may be developed, (ii) effect of various adjuvants on efficacy of the candidate, including studies to determine the minimum and maximum amount of the adjuvant required to improve efficacy or prevent drift or both, and (iii) compatibility of the candidate fungicides with other pesticides that commercial growers may include in the spray mixture.

**Statistical analysis**—All data should be subjected to the analysis of variance and the proper test for significance applied whenever possible. Duncan's multiple range test at the 5% level (15) is acceptable. Recently, many researchers have been using Duncan's new multiple range test and the Student-Newman-Keuls multiple range test. Percentages, especially when they cover a wide range, should be converted to equivalent angles (angle equals arc sine times square root of percentage) before analysis (21,22).

**Effectiveness**—Effective fungicides should be nonphytotoxic, provide a degree of control equal to or better than the standard, and have no detrimental effect on yield or quality of the potato tuber.

## Reports

Each investigator has the responsibility of reporting the results of his or her tests to the manufacturer of the candidate fungicide and, where applicable, to the public. In reporting to the public, the format that the publishing agency recommends should be followed. The reports to the manufacturer or manufacturers are used to make decisions that concern the future development of the candidate fungicide or fungicides. These reports often are used in preparing applications for registering and labeling the candidate or candidates with governmental agencies for commercial growers' use. These reports should be as brief and complete as possible. The following is a check list of information that most manufacturers require:

1. Product name or code number, including trade name, chemical name, or common name, or all three if established and not confidential.

2. Formulation and percentage of each active ingredient unless confidential.

3. Formulation of spray mixture. This includes the amount of water and formulated product put into the spray tank.

4. Rate of application. This includes the amount of spray mixture and the amount of pesticide, expressed as amount of active ingredient or formulated product, applied per hectare or acre.

5. Actual dates of application, not the intended interval between applications.

6. Method of application, including a brief description of the application equipment.

7. Experimental design. This includes such specifications as the number of replications, distance between rows, spacing of plants within the row, number of treated rows per plot, length of treated plots, number of nontreated rows between plots, length of buffer areas between blocks or ranges of plots, and location of no-fungicide control plots.

8. Cultivar used.

9. Date or dates and method or methods of inoculation.

10. Number and dates of irrigation, including the approximate amount of water in centimeters or inches applied per irrigation and the time required to apply the water for each irrigation.

11. Brief summary of the environmental conditions between planting the potatoes and termination of the test. This should include the amount of rainfall in centimeters or inches per month, number of rainy periods each month, the monthly average maximum and minimum temperatures, and the dates and severity of late blight infection periods as indicated by either Blitecast (14), the Wallin system (26,27), or other method or methods of forecasting late blight that may be developed.

12. Other pesticides used, including rates, dates of application, and whether they were applied separately or in combination with the fungicides.

13. Percentage of defoliation on each date that data were taken.

14. Percentage of disease control if calculated.

15. Type and severity of phytotoxicity.

16. All other data taken from experimental plots.

17. A brief written evaluation and interpretation of the data to indicate the confidence that the investigator has in the results.

## Literature Cited

1. BRITISH MYCOLOGICAL SOCIETY. 1947. The measurement of potato blight. Trans. Brit. Mycol. Soc. 31:140-141.

2. BUTLER, E. J., and S. G. JONES. 1949. Plant Pathology. MacMillan & Co., Ltd.: London.

3. CHUPP, C., and A. F. SHERF. 1960. Vegetable Diseases and Their Control. The Ronald Press Company: New York.

4. COCHRAN, W. G., and G. M. COX. 1950. Experimental Designs. John Wiley and Sons, Inc.: New York.

5. COX, A. E., and E. C. LARGE. 1960. Potato blight epidemics throughout the world. U.S. Department of Agriculture, Agriculture Handbook No. 174. U.S. Government Printing Office: Washington, D.C.

6. CROSIER, W. 1934. Studies in the biology of Phytophthora infestans (Mont.) DeBary. Cornell University Agricultural Experiment Station Mem. No. 155.

7. HODGSON, W. A., and P. N. GRAINGER. 1964. Culture of Phytophthora infestans on artificial media prepared from rye seeds. Can. J. Plant Sci. 44:583.

8. HORSFALL, J. G., and R. W. BARRATT. 1945. An improved grading system for measuring plant diseases. Phytopathology 35:655 (abstr.).

9. JAMES, W. C., L. C. CALLBECK, W. A. HODGSON, and C. S. SHIH. 1971. Evaluation of a method used to estimate loss in yield of potatoes caused by late blight. Phytopathology 61:1471-1476.

10. JAMES, W. C., C. S. SHIH, L. C. CALLBECK, and W. A. HODGSON. 1971. A method for estimating the loss in tuber yield caused by late blight of potato. Am. Potato J. 48:457-463.

11. JAMES, W. C., C. S. SHIH, L. C. CALLBECK, and W. A. HODGSON. 1973. Interplot interference in field experiments with late blight of potato (Phytophthora infestans). Phytopathology 63:1269-1275.

12. JAMES, W. C., C. S. SHIH, W. A. HODGSON, and L. C. CALLBECK. 1972. The quantitative relationship between late blight of potato and loss in tuber yield. Phytopathology 62:92-96.

13. JONES, L. R., N. J. GIDDINGS, and B. F. LUTMAN. 1912. Investigations of the potato fungus Phytophthora infestans, U.S. Department of Agriculture, Bureau Plant Industry Bulletin No. 245. U.S. Government Printing Office: Washington, DC.

14. KRAUSE, R. A., L. B. MASSIE, R. A. HYRE. 1975. Blitecast: A computerized forecast of potato late blight. Plant Dis. Rep. 59:95-98.

15. LE CLERG, E. L., W. H. LEONARD, and A. G. CLARK. 1962. Field Plot Technique. Ed. 2. Burgess Publishing Company: Minneapolis.

16. MANZER, F. E., R. C. CETAS, R. C. PARTYKA, S. S. LEACH, and D. MERRIAM. 1965. Influence of late blight and foliar fungicides on yield and specific gravity of potatoes. Am. Potato J. 42:247-252.

17. MILLS, W. R., and R. B. STEVENS. 1967. Zoospore production by the late blight fungus and techniques in inoculation. In KELMAN, A. Sourcebook of Laboratory Exercises in Plant Pathology. W. H. Freeman and Company: San Francisco, pp. 28-29.

18. REDMAN, C. E., E. P. KING, and I. F. BROWN, JR. 1962. Tables for converting Barratt and Horsfall rating scores to estimated mean percentages. Eli Lilly and Company: Indianapolis.

19. SHOEMAKER, P. B. 1974. Fungicide testing: Some epidemiological and statistical considerations. Fungicide Nematicide Tests 29:1-3.

20. SMOOT, J. J., F. J. GOUGH, H. A. LAMEY, J. J. EICHENMULLER, and M. E. GALLEGLY. 1958. Production and germination of oospores of Phytophthora infestans. Phytopathology 48:165-171.

21. SNEDECOR, G. W., and W. G. COCHRAN. 1967. Statistical methods, Ed. 6. Iowa State University Press: Ames, IA.

22. STEEL, R. G. D., and J. H. TORRIE. 1960. Principles and Procedures of Statistics. McGraw-Hill Book Co.: New York.

23. THURSTON, H. D. 1957. The culture of Phytophthora infestans. Phytopathology 47:186.

24. VAN DER PLANK, J. E. 1963. Plant Disease: Epidemics and Control. Academic Press: New York.

25. WALKER, J. C. 1952. Diseases of Vegetable Crops. McGraw-Hill Book Co.: New York.

26. WALLIN, J. R. 1951. Forecasting tomato and potato late blight in the north-central region. Phytopathology 41:37-38 (abstr.).

27. WALLIN, J. R. 1962. Summary of recent progress in predicting late blight epidemics in United States and Canada. Am. Potato J. 39:306-312.

28. ZALEWSKI, J. C., J. P. HELGESON, and A. KELMAN. 1974. A method for large scale laboratory inoculation of potato tubers with late blight fungus. Am. Potato J. 51:403-407.

# 23. Field Evaluation of Protectant Fungicides for Controlling Celery Late Blight

H. L. Dooley

United States Environmental Protection Agency, Northwest Biological Investigations Station, 3320 Orchard Avenue, Corvallis, OR 97330

*Septoria apiicola* Speg. is the causal organism of late blight of celery, *Apium graveolens* var. *dulce* (2). Celery late blight develops best under cool weather conditions. Temperatures favorable for spore germination and infection are 16–21°C. Spattering and windblown water are the major means of spore dissemination (1,4).

This chapter describes a method to evaluate protectant foliar-applied fungicides for control of celery late blight in the field.

## Method

A test site is selected where the soil is suitable for celery growth and the climatic conditions are favorable for disease development.

**Test plants**—Giant Pascal celery is highly susceptible to *S. apiicola* and is suitable as a test plant. Other highly susceptible varieties can also be used.

Disease-free seeds were planted in 2-in. (51-mm) bands

or in peat pots in the greenhouse three months before transplanting in the field.

**Procedure**—Disease-free celery seedlings were transplanted into a field that had been tilled to seedbed condition. Rows were made on 30-in. (762-mm) centers, and seedlings were planted 6–8 in. (152–203 mm) apart within the row.

Treatments were arranged in a randomized block design according to Snedecor's randomly assorted digits table (6) or the equivalent. Other statistical plot designs are also acceptable. Each replication consisted of at least 15 linear ft (4.572 m) of row and at least four replications per treatment. Larger plot sizes may be desirable under varying inoculum pressure or with use of larger application and tillage equipment. A 2–4-ft (609–1,219-mm) buffer zone planted to celery or left fallow separated each treatment within rows. The first of every three rows was left unsprayed to serve as a source of inoculum, while the second and third rows were used for experimental treatments.

Treatments began soon after the celery seedlings were transplanted in the field. To begin after the first signs or symptoms of disease appear may be desirable. Treatments may be applied with (i) a row-crop sprayer adjusted to deliver the desired rate in gallons per acre or liters per hectare, (ii) a hand sprayer, using 30 psi (207 KPa) of pressure to apply the desired rate in gallons per acre or liters per hectare, and (iii) a hand duster, applying measured amounts of dust per foot, acre, or hectare.

After the first treatments were applied to the second and third of every three rows, the first row was inoculated with a freshly prepared fungus spore suspension prepared from air-dried leaves that were heavily infected with viable spores. Two- or three-year-old leaves that have been held at room temperature in the dark have been used successfully. One pound (454 g) of dried infected leaves in 2.5 gal (9.463 L) of water was used to prepare the spore suspension. It was mixed thoroughly, allowed to set for ten minutes, and then filtered through cheesecloth to remove plant debris from the spore suspension. Inoculum was applied with a clean sprinkling can. Inoculation should be done late in the afternoon or evening, or during cloudy, humid periods to keep the spores moist.

The entire plot was sprinkle-irrigated for 3 hr three times each week unless rain occurred with similar frequency to spread the fungus spores and produce heavy infection on young celery plants.

A standard fungicide such as Bordeaux mixture 4–4–100 (1.8 kg–1.8 kg–379 L) was included in each test. Other fungicides known to be effective may also be used as standards. The standard fungicide treatment helps to determine if control can be obtained under the test conditions and should be replicated as the other treatments.

An untreated control was used as a basis for comparison of disease severity and was included and replicated as other treatments.

**Data collection**—The test may be terminated when the disease incidence on the foliage of the untreated controls reaches 50% or more. The test should be continued to within two weeks of harvest, however, if crop yield and grade are to be considered. Data are recorded one week after the last treatment by visually estimating the percentage of disease incidence for each treatment and replication. The mean percentage of disease incidence is determined for each treatment and the mean percentage of disease control calculated according to the following formula (3,5):

$$ MPDC = \frac{MPDIC - MPDIT}{MPDIC} \times 100 $$

where MPDC is mean percentage of disease control; MPDIC, mean percentage of disease incidence in control; and MPDIT, mean percentage of disease incidence in treatment.

The severity of phytotoxicity, if any, as well as a description of injury was recorded. Phytotoxicity may occur as, for example, stunting, leaf burning, or chlorosis.

Visible fungicide residue is determined and recorded as none, light, moderate, or heavy. Residue color should also be reported.

**Results**—Effective products provide a mean percentage of disease control of 70 or more. Phytotoxicity should be at or near zero.

**Reports**—The following information would be helpful to others if included when reporting the results: (i) product name, EPA registration number, or experimental permit number, (ii) active ingredient or ingredients and percentage of each, (iii) celery variety used, (iv) rate of application, (v) how the product was applied, (vi) number of plants per treatment and spacing, (vii) the statistical design and number of replications, (viii) spraying or dusting dates, (ix) mean percentage of disease incidence, (x) mean percentage of disease control, (xi) percentage of phytotoxicity, (xii) types of phytotoxicity, (xiii) visible residue present in marketable portions of plants, (xiv) statistical analysis of data, and (xv) marketable yield and grade.

### Literature Cited

1. COCHRAN, L. C. 1932. A study of two Septoria leaf spots of celery. Phytopathology 22:791-812.
2. GABRIELSON, R. L., and R. G. GROGAN. 1964. The celery late blight organism Septoria apiicola. Phytopathology 54:1251-1257.
3. HORSFALL, J. G., and J. W. BARRATT. 1945. An improved grading system for measuring plant diseases. Phytopathology 35:655.
4. OWENS, C. E. 1925. Principles of plant pathology. John Wiley and Sons, Inc., New York, pp. 469-474.
5. REDMAN, C. E., E. P. KING, and I. F. BROWN, JR., 1962. Tables for converting Barratt and Horsfall rating scores to estimated mean percentages. Eli Lilly and Co., Indianapolis, IN.
6. SNEDECOR, G. W., and W. G. COCHRAN. 1967. Statistical Methods. Ed. 6. Iowa State University Press: Ames, IA.

# C. Diseases of Tropical Plants

# 24. Laboratory and Field Evaluation of Fungicides for Control of Coffee Berry Disease

## D. M. Okioga

Senior research officer and head, Pathology Section of the Coffee Research Foundation, Coffee Research Station, P.O. Box 4, Ruiru, Kenya

Coffee berry disease (CBD), which is caused by *Colletotrichum coffeanum* Noack (sensu Hindorf), is a devastating disease of *Coffea arabica* L. Under favorable conditions for CBD infection and in the absence of control measures, loss of coffee beans due to this disease may be up to 100%.

*C. coffeanum* habitually colonizes the maturing coffee bark of 7–14 internodes of the branches. The fungus invades the cortex external to the developing phellogens. Inoculum from the maturing bark initiates primary outbreaks of CBD (2–4). Control of this disease is effected through use of fungicides that inhibit sporulation of the pathogen. Sporulation capacity of the pathogen has been defined as the number of conidia produced per square centimeter of the coffee bark per hour (1,5). Use of fungicides is at present the only means of controlling the disease, although research now in progress is expected to provide resistant varieties requiring little fungicidal protection.

The laboratory and field methods reported in this chapter were used to determine the efficiency of foliar-applied fungicides in controlling CBD.

## Materials and Methods

**Laboratory evaluation of fungicides that inhibit CBD**—For laboratory evaluation of new fungicides that inhibit CBD, naturally colonized mature coffee branches selected from the unsprayed plots of *Coffea arabica* cultivar SL28 or SL34 or other susceptible cultivars were used. The coffee twigs, about 3.5 mm in diameter, were cut into pieces about 45 mm long. These were washed in a solution of Teepol to remove any initial spore load. Twelve sections were laid on a moist paper towel in plastic (sandwich) boxes measuring $17.5 \times 11.5$ cm$^2$ and 3.5 cm deep.

Solutions of the candidate fungicide were prepared, ranging from 0.1 to 1.0% based on the active ingredients of the fungicide. The coffee twigs in each box were then sprayed with 10 ml of each concentration of fungicide. Treatments were replicated three times, with standard and control treatments included in the experiments. In the standard treatment, twigs were sprayed with captafol (N-[1,1,2,2-tetrachloroethylthio]-Δ⁴-tetrahydrophthalimide) (Difolatan, Chevron Chemical Co., Ortho Division, San Francisco, CA 94104) 80% WP at the rate

of 0.32% of active ingredient. In the control treatment, twigs were sprayed with water. After incubation (around 22°C for three to five days), the spores were removed by vigorously shaking the 12 twigs from each box into 10 ml of water. The number of spores per milliliter in the suspension was determined by hemocytometer count; twig sections were measured and their total volume obtained by displacement. The rate of spore production (spores per square centimeter of bark per hour) was calculated using the following formula:

$$\frac{\text{spores/ml}}{\text{surface area of twigs}} \times t$$

where t is the length of incubation period. The surface area of the twigs is:

$$\sqrt{LV \times 3.545}$$

where L is total length of the twigs (in centimeters) and V is total volume of the twigs (in cubic centimeters). Since the time (t) and 3.545 were constant throughout the experiment, these were eliminated in the calculation and the rate of spore production was calculated thus:

$$\frac{\text{spore count}}{\sqrt{LV}}$$

Reduction, or stimulation, of spore production in the treatments was expressed as the percentage of sporulation of the control.

Fungicides that inhibited sporulation of *C. coffeanum* as efficiently as or better than the standard treatment (captafol) were selected for field evaluation.

**Field evaluation of fungicides for CBD control**—Field trials to determine the efficacy of new fungicides against CBD were arranged on the standard randomized block design, with four replications. Each replicate consisted of 25-tree plots that were five trees square. Each chosen concentration of fungicide was applied with a motorized sprayer at the rate of 63.6 L/100 trees, following the CBD spray program. Captafol was included at the rate of 0.32% of active ingredients as the standard treatment. Control plots received no fungicidal sprays. Shortly after flowering, one branch of five to ten nodes from each of the nine central trees was selected randomly. A label was affixed to this branch, and a piece of plastic wire twisted around it proximally and distally to the five nodes under observation. Monthly counts were made of the healthy and infected pinheads and the green and ripe berries on marked nodes. The percentage of CBD infection was then obtained.

The total yield of ripe berries from all 25 trees per plot was recorded in kilograms, and the average yield per plot calculated. Both records of CBD infection and yield data were used as criteria for determining the efficacy of a fungicide. A new fungicide that was as effective as or better than the standard treatment of captafol was selected for further field trials. If the fungicide proved effective for at least two consecutive years under field trials, it was recommended as an effective fungicide against CBD.

## Discussion

During laboratory screening of a new fungicide, certain fundamental assumptions inherent in the method may affect the usefulness of the test. Certain *Colletotrichum* spp., namely, *C. coffeanum* Noack (sensu Hindorf), *C. acutatum* Simmonds, and *C. gloeosporoides* Penz. (both the acervulus and mycelial forms), live on the coffee bark. Of these species, only *C. coffeanum* causes CBD. The rest of the *Colletotrichum* spp. or types are mainly saprophytic cohabitants. Some groups of fungicides, particularly those belonging to the 2-methyl benzimidazole carbamic acid (MBC) group, stimulate the sporulation of these saprophytic colletotricha. On the other hand, MBC fungicides completely inhibit the sporulation of pathogenic *C. coffeanum*. In practice, therefore, these MBC fungicides are effective against CBD. In contrast, during the laboratory screening tests, these MBC fungicides may perform poorly, since the production of large numbers of spores belonging to a saprophytic *Colletrotrichum* sp. may interfere with the interpretation of the results in vitro. This is solely because the laboratory test does not determine whether the recorded number of spores belong to pathogenic or saprophytic *Colletotrichum* spp.

In the context of pollution and the declining quality of the environment, insuring through agricultural disease control that chemicals do not contribute to pollution by leaving residues on edible parts of the crop is an ever-increasing demand. Although significant residues in coffee beans resulting from use of fungicides are virtually absent at present, researchers should not remain complacent concerning the issue. Therefore, before a recommendation is given for a new chemical for control of CBD, stringent field tests should be done to find out whether the fungicide leaves residue on the coffee beans. Should the residue be found in quantities that are unacceptable from the viewpoint of public health, the fungicide should be disqualified for use in controlling CBD.

Fungicides in which active ingredients are found to be water soluble are of little value and need not be included in field trials. This is because the fungicidal sprays for control of CBD begin just before the rains and continue, normally at three- to four-week intervals, throughout the rainy season. Rain easily washes off (erodes) water-soluble candidate materials, and thus their efficacy is reduced or even nullified.

Fungicides that are highly toxic to man; inconvenient to handle or store; or decompose easily by heat, sunlight, or ultraviolet light or under storage are also unsuitable for control of CBD in the tropics.

## Literature Cited

1. GIBBS, J. N. 1969. Inoculum sources for coffee berry disease. Ann. Appl. Biol. 64:515-522.
2. NUTMAN, F. J., and F. M. ROBERTS. 1960a. Investigations on a disease of *Coffea arabica* caused by a form of *Colletotrichum coffeanum* Noack. I. Some factors affecting infection by the pathogen. Trans. Brit. Mycol. Soc. 43:489-505.
3. NUTMAN, F. J., and F. M. ROBERTS. 1960b. Investigations on a disease of *Coffea arabica* caused by a form of *Colletotrichum coffeanum* Noack. II. Some factors affecting germination and infection, and their relation to disease distribution. Trans. Brit. Mycol. Soc. 43:643-659.
4. NUTMAN, F. J., and F. M. ROBERTS. 1961. Investigations on a disease of *Coffea arabica* caused by a form of *Colletotrichum coffeanum* Noack. III. The relation between infection of bearing wood and disease incidence. Trans. Brit. Mycol. Soc. 44:511-521.
5. VINE, B. H., P. A. VINE, and E. GRIFFITHS. 1973. Evaluation of fungicides for control of coffee berry disease in Kenya. Ann. Appl. Biol. 75:359-375.

# 25. Laboratory and Field Evaluation of Fungicides for Control of Coffee Leaf Rust

## D. M. Okioga

Senior research officer and head, Pathology Section of the Coffee Research Foundation, Coffee Research Station, P.O. Box 4, Ruiru, Kenya

In Kenya, *Hemileia vastatrix* Berk. & Br. causes coffee leaf rust. Infection from this fungus occurs only on the underside of the leaves. The infection site becomes encrusted with spore masses that form characteristic orange pustules. Sometimes the whole leaf may become infected, but only one or two pustules will cause premature leaf fall. In most cases, therefore, the infected trees become heavily defoliated. Because defoliation results in inadequate starch formation, the tree begins to use starch reserves in the roots and stem, leading to undernourishment and dieback of small, woody rootlets. Similar dieback is common on branches. Since coffee demands a high carbohydrate supply for successful maturation of the crop, leaf rust has a distinctly adverse effect on yields.

Use of fungicides is presently one of the practical means of controlling the disease. Breeding work is also in progress to help to develop disease-resistant plant varieties.

The laboratory and field methods reported in this chapter were developed to determine the efficacy of foliar-applied fungicides in the control of leaf rust epidemics.

## Materials and Methods

**Laboratory evaluation of new fungicides against leaf rust**—Detached leaves from the naturally infected coffee plants were used in the laboratory to evaluate new fungicides for leaf rust control. Naturally infected leaves were obtained from *Coffea arabica* cultivars SL28 and SL34, which are highly susceptible to infection from this pathogen.

Leaves with uniformly spaced lesions were selected. Two leaves were laid with the lower side face up on a moistened paper towel in plastic (sandwich) boxes measuring $17.5 \times 11.5$ cm$^2$ and 3.5 cm deep. In one box, the spores were removed by gently mopping the infected leaves with a wet paper towel. The uredospores were not removed in the other box. Solutions of the candidate fungicide ranging in concentrations from 0.1 to 1.0% based on the formulation of the fungicide were prepared. The leaves in each box were sprayed with 10 ml of each concentration; treatment was replicated three times. Two standard treatments, cuprous oxide (Perenox) and pyracarbolid (Sicarol), and a control treatment sprayed with water were included in the experiment. In standard treatments, the leaves were sprayed with a 0.7% solution of Perenox (containing 50% copper) or with Sicarol at the rate of 0.4% of formulated material. The sprayed leaves were incubated for about five days around 22°C. After incubation, each leaf was rated visually for new uredospore formation, lesion development, and phytotoxicity (blackening of leaves characterized phytotoxicity). The fungicides that were selected for field trials were those that inhibited new uredospore formation (on leaves with spores removed) or that effectively killed or eradicated or both the uredospores (on infected leaves with spores) as effectively as or better than Perenox and Sicarol, provided that they did not stimulate sporulation of *Colletotrichum coffeanum*

Noack, the organism that causes coffee berry disease to develop on naturally colonized coffee twigs if sprayed with a new fungicide.

**Field evaluation of fungicides for leaf rust control**—Field trials to determine the efficacy of new fungicides against leaf rust were evaluated in a randomized block design, with four replications. Each replicate consisted of 25-tree plots that were five trees square. Each concentration of fungicide was applied with a motorized knapsack sprayer at the rate of 63.6 L/100 trees, following the leaf rust spray program (1). Perenox, which was applied at the rate of 0.7%, was included as standard treatment. Control (unsprayed) plots received no fungicide sprays.

Records of leaf rust infection were made at three- to four-week intervals throughout the experimental period. Seventy infected leaves were taken from each of nine central trees using modified point quadrant samplers. These consisted of wooden staves about 200 cm long with sampling pins, or 15-cm lengths of doweling inserted at 15-cm intervals along the length of the staff. The unit was inserted into the tree canopy, and the leaves touched by the ends of the ten pins were picked. The process was repeated seven times for each tree.

The total number of leaves infected in relation to the total number of leaves picked was used to calculate the percentage of leaf rust infection. A new fungicide that was as effective as or better than the standard treatment of Perenox was selected for further field trials. If the fungicide proved effective for at least two consecutive years under field trials, it was recommended as an effective fungicide for leaf rust control.

## Discussion

Anti-leaf rust fungicides should not be water soluble, because rainwater easily erodes them after application. Similarly, they should not be highly toxic to man and should be stable when exposed to sunlight, ultraviolet light, and high temperatures. They also should not leave persistent residues in the coffee beans. These factors are discussed in more detail in chapter 24.

## Literature Cited

1. 1977. Control of coffee berry disease and leaf rust in 1977. Kenya Coffee 42:7-11.

# 26. Screening Fungicides for Control of Blister Blight of Tea in Northeast India

G. Satyanarayana, G. C. S. Barua, and K. C. Barua

Mycology Department, Tocklai Experimental Station, Tea Research Association, JORHAT 785008, Assam, India

Blister blight of tea (*Camellia sinensis* [L.] O. Kuntze) is caused by *Exobasidium vexans* Massee, an obligate parasite that attacks young succulent leaves and stems. It is prevalent throughout northeast India. Relatively

cool, cloudy weather with light rain favors the disease (1). Young nursery plants may die as their stems break off at the point of infection. Crop losses may be considerable if the disease becomes severe during the period when the plants are recovering from pruning or cutting back. Injury is especially undesirable when it occurs on the shoots carrying two leaves and a bud, because these are harvested for the manufacture of black or green tea (2).

## Materials and Methods

**Location of the experiment**—Darjeeling is the preferred location for fungicide trials, because the disease occurs almost regularly from June to September. Sites are selected in commercial tea estates where blister blight has been a problem. Tea areas pruned in the preceding cold season and situated on the northern or northeastern slopes of the hills or adjoining forests are usually considered ideal. Cool, misty weather, which is favorable to disease development, prevails in such areas where abundant new growth following pruning further accentuates the disease.

**Plot design and experimental procedure**—A randomized complete block design is used, with four or five replicates per treatment. Each replicate consists of 30 to 60 bushes planted in one or two rows (distance usually is from 90 to 120 cm within rows and 120 cm between rows). Two guard rows are left to protect against spray drift. The treatments include a standard fungicide and an untreated control. Depending on disease severity, four or five sprays are applied after a plucking round, the interval for which may vary from seven to ten days. A seven-day interval is usually used.

**Collection and treatment of data**—In all experiments, an assessment of disease incidence is made prior to the first fungicide application. In this and in the final evaluation of the treatments, shoots with three leaves and a bud are plucked. Pluckings from each plot are combined, and 100 shoots are picked at random.

Only blisters on the third leaf are counted, because lesions on the second leaf frequently do not advance to the stage at which they are easily visible. Results of the disease counts are expressed as "total number of blisters on 100 third leaves." Results are statistically analyzed, and the efficacy of the test fungicides is compared with the standard and the untreated control.

## Discussion

The standard fungicides are copper formulations containing oxides and oxychlorides of 50% metallic copper. With mist-blower power sprayers, 625 g/ha in sufficient water to allow distribution of the fungicide onto the top hamper of the plants is recommended. Equal results can be obtained with a 1:800 dilution of fungicide applied with a hand-operated, pressure-retaining knapsack sprayer (2). Dithiocarbamate fungicides are effective (4). Sicarol 15% dispersion (2 methyl-5, 6-dihydro-4H pyran-3 carboxylic acid anilide) (Hoechst-Roussel Pharmaceuticals Inc., Somerville, NJ) has recently been found effective at 500 ml/ha in weekly sprays, or 700 ml/ha at biweekly intervals (3).

## Literature Cited

1. SARMAH, K. C. 1960. Diseases of tea and associated crops in northeast India. Memo No. 26, Indian Tea Association.
2. SATYANARAYANA, G., and G. C. S. BARUA. 1975. Trends in disease control with special reference to tea in northeast India. Pestic. Annu. pp. 106-114.
3. SATYANARAYANA, G., G. C. S. BARUA, K. C. BARUA, and P. K. TAMANG. 1975. A new systemic fungicide against blister blight in northeast India. Two Bud 22(2):87-88.
4. SATYANARAYANA, G., S. K. SARKAR, G. C. S. BARUA, K. C. BARUA, and P. C. CHAKRAVORTY. 1974. Blister blight of tea and its control. Two Bud 21(1):1-2.

# 27. Procedures for Screening Fungicides for Control of Greasy Spot, Melanose, and Scab on Citrus Trees in Florida

### J. O. Whiteside

Plant pathologist, University of Florida, Institute of Food and Agricultural Sciences, Agricultural Research and Education Center, PO Box 1088, Lake Alfred, FL 33850

The number of fungicide applications that can be used to control greasy spot (*Mycosphaerella citri* Whiteside), melanose (*Diaporthe citri* [Fawc.] Wolf), and scab (*Elsinoe fawcetti* Bitanc. and Jenk.) in Florida citrus groves is economically limited. Any candidate material that would have to be applied more than twice annually to control any one of these diseases probably could not compete with materials that are presently recommended (1). Satisfactory control of greasy spot is usually provided by one spray of benomyl or a copper fungicide

in early summer, or if the disease pressure is not too heavy, by a 1% oil-water emulsion alone (6). Adequate control of melanose can usually be achieved with one application of copper fungicide in late April or early May (10,11). Scab can often be controlled satisfactorily by one spray of captafol in late winter before shoot growth commences (3,8), or one spray of benomyl at bloom (2).

The number of times that citrus trees can be sprayed each year is limited by the expense of the relatively high volumes of spray required to cover the characteristically dense canopy. Because the capital outlay is high, each spraying unit is expected to handle a large acreage, and treatments usually must be four to six weeks apart. Another consideration is that only part of the acreage is likely to be sprayed at the most appropriate time epidemiologically. Therefore, only those materials that have long residual protectant action or inoculum-reducing ability or both are satisfactory. The screening procedures for materials for control of greasy spot, melanose, and scab are necessarily severe. Initially, only one application of a test material is made at a time when it is likely to have maximum effect. Because the major infection periods for each of the diseases are different, materials are tested against each one in separate experiments.

## Methods for Greasy Spot Control

**Test sites and cultivars used**—Tests are made on bearing grapefruit (*Citrus paradisi* Macf.) and sweet orange (*C. sinensis* Osbeck) trees that have shown moderate to severe greasy spot in previous years. Grapefruit trees are used to test the effectiveness of spray materials against greasy spot rind blemish, because this facet of the disease is much more conspicuous on grapefruit than on sweet orange.

The most serious aspect of leaf infection is the defoliation it causes during the winter after infection. Since more premature defoliation usually occurs from the spring growth flush than from summer flushes, disease ratings usually are obtained from the previous spring flush. The length and vigor of shoots and the number of leaves per shoot on the spring flush tends to become more uniform as trees grow older. Therefore, trees less than six years old are not used for fungicide screening tests against greasy spot.

**Organism**—Ascospores released from perithecia produced on partially decomposed fallen citrus leaves are the major source of inoculum (4). Conidia of the imperfect (*Stenella*) stage of this fungus come only from transient mycelial growth that develops on the surface of leaves and fruit following ascospore or conidia germination. This growth requires a combination of high relative humidity and temperature and thus flourishes mostly during the humid summer months (7). In Florida citrus groves, conidia are produced in much smaller numbers than are ascospores, and they are of minor epidemiologic importance (4). Host penetration occurs only through stomata. Development of the external mycelial growth considerably increases the chances of hyphae reaching the stomata.

The supply of infected fallen leaves reaches a peak in May following heavy leaf drop through winter and spring (4). With more frequent rainstorms in May and June, perithecial development and maturation is hastened. Ascospore release (mostly by rain) reaches a peak in June or July (4). During August or September, ascospore numbers decline rapidly as a consequence of (i) advancing decomposition of the leaf litter, (ii) perithecial exhaustion, and (iii) fewer leaves dropping from citrus trees during the summer to replenish the inoculum supply (4). Combined high temperatures and prolonged high humidity that occur almost nightly in Florida from June through September favor infection. Because ascospores are fewer after July, however, more infection occurs in June and July than in August or September.

Normal cultural practices are followed in the test groves. Routine disc harrowing in late spring buries much of the leaf litter, but enough of it stays under the tree to provide inoculum for fungicide screening purposes.

**Procedure**—The first tests with a new material are made either on small six- to ten-year-old trees or on 1.8 m (6 ft-wide sectors of the canopy of large trees. On very large trees, several such sectors can be plotted on each tree. Confining all of the plots of one replication to either the north or the south side of the tree is important, however, because defoliation from greasy spot occurs earlier on the shaded north side of the canopy (Whiteside, unpublished data). The single-tree or canopy-sector plots are replicated at least four times.

After determining that a material is effective against *M. citri* in small-tree or tree-sector plots, it is tested on large trees using one- or two-tree plots replicated at least six times in a randomized block design.

During testing, the risk of interference from drifting spray droplets is slight on calm days, because few such droplets are likely to settle on the underside of the leaves. Fungicides must be deposited on the lower leaf surface (to which the stomata are confined) to prevent infection with *M. citri* (5). Therefore, unsprayed trees are not left as buffers between plots unless the space between the canopy of neighboring trees is less than 1.8 m (6 ft). As a precaution, however, sprays are not applied to nonbuffered trees if the wind speed exceeds 8 kmph (5 mph).

Materials are applied as dilute sprays to dripoff, and special care is taken to wet the underside of the leaves thoroughly. Application of sprays to canopy-sector plots is made with pressure-retaining hand sprayers. Whole trees are sprayed with single-nozzle handguns at 28 kg/cm$^2$ (400 psi). Volume of spray applied amounts to 76 L (20 gal) per tree for 6-m (20-ft) trees and to 30 L (8 gal) per tree for 3-m (10-ft) trees.

The test materials are applied only once any time between mid-June and late July. Unsprayed check trees and standard treatments of 1% oil-water emulsion alone or copper or benomyl or both are included in every test. No other fungicides or oil sprays are applied to the trees during the year of test. Promising materials are later tested in combination with spray oil (meeting FC 435-66 specifications of Florida Pesticide Law, Rules and Regulations), which is commonly applied in the summer spray to control several pests as well as greasy spot.

**Data collection**—Greasy spot has a long incubation period and symptoms do not normally appear before November or December. By this time, identifying shoots of the spring growth can be difficult. Spring growth can be confused with late growth of the previous year and even with subsequent summer growth of the current year. Therefore, labeling shoots of the spring flush in May soon after they have fully expanded is necessary so that they can be identified the following winter. For this purpose, a copper wired white 1.5 × 9.0-cm plastic tag is tied around an internode near the middle of each shoot to be sampled. The number of shoots thus labeled on each canopy sector or tree plot varies from 20 to 50 per plot. To aid retrieval, the tags are placed in groups at the same locations on each plot. On a canopy-sector plot, they are all placed in a group near the center of the sector. When whole trees are used, tags are placed at the same two or four compass points on each tree.

In some tests the procedures used to assess severity of greasy spot involved picking all leaves from each labeled shoot and examining them closely in the laboratory for symptoms. The leaves were judged merely as diseased or healthy, or severity of the disease on each leaf was rated quantitatively according to the percentage of the leaf-blade area affected. In the latter case, the chlorotic areas as well as the greasy spots were included as representing the total area diseased (6). These methods require that leaves be examined before substantial defoliation commences in unsprayed checks and poorer treatments. Thus, picking the leaf samples long before winter ends may be necessary. More recent procedures (Whiteside, unpublished data) allow for the fact that greasy spot does more harm by causing premature leaf abscission than by partially impairing photosynthesis. Data now include the amount of defoliation that occurs before the new shoots appear the following spring. All leaves on each shoot are counted the previous summer before greasy spot appears, or the stems are examined for leaf scars when the final count is made. From these data the mean percentage of defoliation is calculated. Leaves that do not abscise are examined for greasy spot to provide an additional rating based on the percentage of remaining leaves that are diseased.

The effectiveness of materials for controlling greasy spot rind blemish is determined mostly from tests in older grapefruit groves, because for unknown reasons, greasy spot rind blemish seldom appears on the fruit of young grapefruit trees. Sufficient fruit is picked at random from the lower part of the canopy on each one- or two-tree plots to fill 88-L (2 × 1.25-bu) boxes. This provides a fruit count of 80 to 100 per plot. The sample is washed mechanically to remove sooty mold and other obscuring deposits and then is graded for presence of greasy spot rind blemish. Sometimes the disease is also rated as a percentage of fruit surface showing symptoms.

Examinations are also made on washed fruit for any evidence of phytotoxicity caused by the spray materials.

All greasy spot percentage data receive arc sine transformation to degrees of an angle before being subjected to an analysis of variance and Duncan's multiple range test.

## Methods for Melanose Control

**Test sites and cultivars used**—All citrus cultivars are susceptible to melanose, but tests usually are made on grapefruit because melanose has a greater commercial impact on this host. Severity of melanose is determined partly by the amount of recent twig dieback in the tree canopy. Old trees usually carry more dead twigs than do young trees. Trees over 40 years old therefore have been used for most fungicide tests against melanose. Since considerable differences in vigor, and hence dead twig content, can occur between trees, weaker trees are excluded to achieve a more uniform disease pressure.

**Organism**—In Florida, *D. citri* produces airborne ascospores and water-liberated and splash-dispersed pycnidiospores. Only the latter, however, are responsible for serious melanose development on fruit

rind. Rind blemish occurs when rain washes large numbers of pycnidiospores onto the fruit from dead twigs above. No pycnidia occur on the melanose pustules as long as the substrate remains alive.

Melanose tends to be unevenly distributed. Fruit on the lower part and inside of the canopy carry more melanose than those located at the top and outside.

Severity of melanose on fruit rind varies considerably from year to year. The amount of infection depends on the occurrence of sufficiently long fruit-wetting periods before the rind becomes immune to infection 10–12 weeks after petal fall (11). In some years, melanose is insufficient to permit thorough testing of fungicides.

Usually, more infection occurs in May or June than in March or April (10,11). This is partly because of differences in the number of days of rain in these months. Also, precipitation after April occurs mostly as afternoon and evening thunderstorms that are often conducive to overnight wetting periods. Any rain that does fall in March and April is usually associated with fast-moving cold fronts and rapid drying. The probability of infection after cold-front precipitation is reduced further by the rapid drop in temperature that generally follows it. At 25°C, a minimum of 10–12 hr of continuous wetting is required for infection, but at 15°C, the minimum time increases to 18–24 hr (10,11).

For fungicide screening tests, a record of probable infection days is maintained. This record is based on data provided by a continuously recording rain gauge and hygrothermograph and is done by inspection of fruit for raindrops in early morning and by periodic examinations of fruit for early symptoms of melanose.

**Procedure**—Test materials are applied to one- to four-tree plots replicated at least six times in a randomized block design. A single line of unsprayed buffer trees must separate the large old test trees to avoid spray drift effects. Dilute sprays are applied to dripoff at 28 kg/cm² (400 psi) with a single-nozzle handgun using a maximum of 76 L (20 gal) per tree. No other fungicides are applied to the trees during the year of test. Rust mites are controlled by timely spraying with chlorobenzilate or ethion before their fruit russetting can obscure symptoms of melanose.

Basic copper sulfate, 1 g of metallic copper/L (0.8 lb/100 gal) has been included as a standard treatment in all tests. Sometimes it is applied also at half this rate, particularly when different copper fungicides are to be evaluated (10). Unsprayed check trees are included in every test, because disease pressure can vary greatly from year to year.

Fungicides are tested first as postbloom fruit-protectant sprays. For this purpose, the test materials are applied in late April or early May—about four weeks after petal fall and before the six- to eight-week period when infection is more likely to occur. Spraying at this time also means that the fruit will be much larger and will hold more fungicidal material than it would have shortly after petal fall.

Materials that are effective protectants against *D. citri* must be sprayed before fruit set to determine if they can reduce inoculum at the source and if they can be redistributed enough to protect fruit that sets after treatment, as in the case of captafol (3,12).

**Data collection**—Because melanose pustules are more obvious after the rind has colored, collection of fruit for disease assessment is usually delayed until

December. The sample consists of fruit picked only from the lower part of the canopy where melanose is likely to be more severe. Sufficient fruit is picked at random to fill 88-L ($2 \times 1.25$-bu) boxes per tree, thus providing a sample of at least 80 fruit per tree.

Fruit is washed to remove superficial deposits and graded as (i) essentially melanose-free (up to 100 small pustules per fruit are allowed if they are widely scattered and none exceeds 1 mm in diameter) or (ii) pustules sufficiently conspicuous and numerous to detract from fruit appearance. Data are presented as a percentage of melanose-free fruit. An analysis of variance and Duncan's multiple range test are done after transforming the percentage values to degrees of an angle.

## Methods for Scab Control

Field screening of fungicides for scab control is conducted in two stages (8). A test material is first included in small-scale experiments to determine if it has sufficient activity against the scab fungus to justify further interest. These tests are made on small trees and are based on the ability of a material to reduce the amount of inoculum at source or to become redistributed or both, thereby protecting shoots that emerge after treatment. Materials that survive primary screening tests are then applied to bearing trees to determine their effectiveness in preventing fruit infection.

**Test site and cultivars**—*Primary screening*—These tests are made on trees of a rough lemon (*Citrus jambhiri* Lush.) variant (author's accession No. 27) that is highly susceptible to scab but resistant to the Alternaria leaf spot that renders most rough lemons less suitable for work on scab. Although other scab-susceptible citrus species such as sour orange (*C. aurantium* L.) could be used, this rough lemon variant is preferred because it is so vigorous that the trees recover quickly when scab pressure is reduced. Thus, the trees can be used again for further tests after allowing only a short period for recovery. The site is irrigated with overhead sprinklers to encourage scab attack when rainfall is inadequate. After completion of a test, the canopy is pruned periodically to keep tree size within bounds and to promote abundant new growth that can become infected with scab. The trees are not used again for further tests until disease uniformity between trees has been restored.

*Grove screening*—Bearing trees of the commercially important Temple (*C. temple* Hort. ex Y. Tan.) cultivar are used for grove testing of promising candidate materials. Temple fruit react to infection by producing conspicuous blister-type scab lesions. Temple trees are, therefore, more suitable for testing fungicides for scab control than are citrus cultivars such as grapefruit, which produce a flat, scurfy type of lesion that can be confused with wind scar. Temple trees less than 3.6 m (12 ft) high are used for fungicide screening tests, making buffer trees against spray drift unnecessary.

**Organism**—The scab fungus overwinters on infected living leaves, twigs, and fruit. Successive crops of conidia are produced from old scab lesions when wetted by rain or dew. The perfect stage of the fungus has not been observed in Florida. Conidia are mostly water liberated and splash dispersed. The disease therefore tends to be unevenly distributed throughout the tree canopy, with the severity of fruit infection depending on proximity to the inoculum source. Some dry dispersal of conidia by wind also occurs. Such conidia can withstand desiccation and then germinate in dew the following night (9). While this type of dispersal may be important in long-distance dissemination, it probably does not cause heavy infection by itself. Primary infections started by such windborne conidia, however, could give rise to abundant splash-induced secondary infections if unusually frequent rain showers occurred during the fruit-susceptible period.

Only young tissue is susceptible to attack. Shoot and leaf infection is not of major economic importance except on susceptible rootstock varieties such as rough lemon and sour orange in nurseries. Shoot infection is, however, of epidemiologic importance, because it builds up inoculum to infect fruit. Fungicides are required to inhibit conidia production on scab lesions when new shoots and young fruit are susceptible to infection. Critical spraying times for scab control are just before shoot emergence in early spring and before major fruit set.

Conidia of *E. fawcetti* require only a short period of wetting to germinate and penetrate the host; even at 15°C, 3–4 hr will suffice (9). In this respect, scab is different from melanose. It can be promoted by cold-front precipitation and relatively short periods of overhead sprinkler irrigation. Furthermore, only a short period of wetting is required for conidia development. Operation of overhead sprinklers for 4–6 hr has promoted conidia formation and infection, even when the scab lesions previously had none (9).

**Procedure**—*Primary screening*—Sprays are applied to dripoff from a 57-L (15-gal) motorized sprayer at 18 kg/cm$^2$ (250 psi). Each material is applied to at least five trees arranged in a randomized block design. The materials are sprayed either in late winter just before spring shoot growth is expected or during the summer at times when no shoot extension is present. If one sees in summer tests that the next flush of growth may be long delayed or uneven, the trees are pruned lightly to promote new shoot growth. If rainfall is inadequate, scab infection is assured by running the overhead sprinkler irrigation periodically for 4–6 hr during the early phases of shoot expansion. Standard treatments of captafol or benomyl or both are included in every test.

*Grove screening*—Treatments are applied to two- or three-tree plots of Temple trees replicated at least six times in a randomized block design. Sprays are applied in February before growth or shortly before major petal fall or both. Standard treatments of captafol and benomyl at the late dormant stage or at bloom or both are included in all tests. No other chemicals need to be added to the spray mixes at these times. The volume of dilute spray amounts to about 37 L (10 gal)/3.6-m (12-ft) Temple tree. Unsprayed check treatments are included in each test, but no buffer trees are left between plots.

**Data collection**—In the primary screening tests, the degree of control is based on the number of new shoots that show scab symptoms. No attempt is made to index disease severity on each shoot. Even if only one visible lesion is present, the shoot is considered diseased. After the new shoots have fully expanded, 50 are selected at random from the lower half of the canopy of each tree. Examinations have to be made on turgid shoots, because wilting obscures symptoms. Data are expressed as the percentage of shoots diseased.

In the grove tests, scab assessments are made only on the fruit. Because scab symptoms are conspicuous on immature Temple fruit, data can be obtained before normal picking time. Usually, however, the crop is allowed to mature before sampling. Washing fruit samples before examination is not necessary, because scab lesions show clearly through any surface deposits. Grading of fruit as healthy or scab infected therefore can be done in the grove at harvest.

The size of the fruit sample has varied considerably. In some tests, the whole crop has been picked and graded. Usually, however, the disease assessment is based on a sample of mature fruit that has been picked randomly from the middle of the canopy. Enough fruit is picked from each tree to fill one 44-L (1.25-bu) box. This provides a count of between 100 and 140 fruit per tree. Percentage of fruit and shoot infection data are transformed to degrees of an angle and then subjected to analysis of variance and Duncan's multiple range test.

## Results and Discussion

**Greasy spot**—Coefficients of variation in screening tests for greasy spot control have fallen mostly between 15 and 25%. While greater precision might be achieved by increasing the sample size and numbers of replications, the current procedures are sufficiently precise for screening and ranking. Standard treatments of copper fungicides or benomyl have always been effective (6) and have provided a satisfactory reference for interpreting test results. Oil sprays, however, have been less reliable as standards due to their variable results (6).

Defoliation data provide the best means of expressing greasy spot severity. Nevertheless, sometimes greasy spot develops too late or is insufficiently severe to cause substantial defoliation before the end of winter. In these cases, useful comparative data can still be obtained by rating the disease according to the percentage of leaves or of leaf area affected.

Spray oil is applied almost routinely to Florida citrus groves in the summer to control various pests as well as greasy spot. It is most effective in June or July when fungicides give optimum control of greasy spot. Fungicides for greasy spot are used as a supplement to oil to help insure control. To avoid a separate spraying operation, any fungicide for greasy spot control must be compatible with oil. Testing of candidate materials in combination with oil is, therefore, an essential part of the screening procedure.

A candidate material must have certain qualities to be effective against greasy spot. Long residual action is of primary importance. A fungicide spray that is applied in June must last into July and perhaps even longer to control the disease throughout the major infection period. It must also kill on contact any mycelial growth of *M. citri* already present on the leaf and fruit surface. The screening procedures for testing fungicides applied after some ascospores have reached the leaf and germinated allow for the need for an eradicant as well as residual protectant action.

**Melanose**—Coefficients of variation in fungicide screening tests against melanose in recent years have ranged from 18 to 28% (10). While greater precision is preferred, practical considerations impose limitations.

The amount of plot spraying that can be accomplished in one day restricts the number of trees per plot and the number of replications that can be covered. While the level of precision is considered adequate for screening purposes, more replication and larger fruit samples would be required to detect any differences of lower magnitude statistically.

Relatively few materials have equalled copper fungicides in their ability to control melanose, presumably because few other materials are sufficiently stable to provide long-term residual protection. The practice of applying test materials only once during the period of fruit susceptibility highlights the residual qualities of chemical treatments.

**Scab**—In Temple groves, scab severity varies from tree to tree, resulting in coefficients of variation approaching 50% in some tests. This problem has been partly solved by using the parts of Temple groves that showed the heaviest, most uniform infection the previous year, and by increasing the number of replications.

Inclusion of highly effective standards (captafol or benomyl or both) in each test has assisted considerably in interpreting results from candidate materials. Copper fungicides and ferbam are no longer used as standards, because their performance is inconsistent and often poor (2,3,8).

The possibility that wind-blown conidia may cause interplot interference must be considered in tests for scab control. Results could be misinterpreted if new shoots or young fruit become infected with conidia from external sources, particularly if substantial secondary infection follows. In sprayed plots, conidia from external sources could obscure eradicant effects on conidia from old scab lesions. For this reason, unsprayed buffer trees are omitted, and for grove tests against scab, the trees have to be correspondingly small to avoid spray drift problems. Including in the grove tests only those materials that seemed highly promising in the primary screening tests also minimizes interplot interference. Unsprayed check plots are necessary, however, to measure the variability in scab pressure that can occur over the experimental area.

## Literature Cited

1. Florida citrus spray and dust schedule. 1977. Fla. Coop. Ext. Serv. Circ. 393-C. University of Florida, Gainesville.
2. HEARN, C. J., J. F. L. CHILDS, and R. FENTON. 1971. Comparison of benomyl and copper sprays for control of sour orange scab of citrus. Plant Dis. Rep. 55:241-243.
3. MOHEREK, E. A. 1970. Disease control in Florida citrus with Difolatan fungicide. Proc. Fla. State Hortic. Soc. 83:59-65.
4. WHITESIDE, J. O. 1970. Etiology and epidemiology of citrus greasy spot. Phytopathology 60:1409-1414.
5. WHITESIDE, J. O. 1972. Spray coverage requirements with benomyl for effective control of citrus greasy spot. Proc. Fla. State Hortic. Soc. 85:24-29.
6. WHITESIDE, J. O. 1973. Evaluation of spray materials for the control of greasy spot of citrus. Plant Dis. Rep. 57:691-694.
7. WHITESIDE, J. O. 1974. Environmental factors affecting infection of citrus leaves by Mycosphaerella citri. Phytopathology 64:115-120.
8. WHITESIDE, J. O. 1974. Evaluation of fungicides for citrus scab control. Proc. Fla. State Hortic. Soc. 87:9-14.
9. WHITESIDE, J. O. 1975. Biological characteristics of Elsinoe fawcetti pertaining to the epidemiology of sour orange scab. Phytopathology 65:1170-1177.
10. WHITESIDE, J. O. 1975. Evaluation of fungicides for the

control of melanose on grapefruit in Florida. Plant Dis. Rep. 59:656-660.

11. WINSTON, J. R., J. J. BOWMAN, and W. J. BACH. 1927. Citrus melanose and its control. U.S. Dep. Agric. Bull. No. 1474.

12. YAMADA, S., H. TANAKA, M. KOIZUMI, and S. YAMAMOTO. 1966. Studies on rationalization and labor saving of fungicide application for citrus disease control (English summary). Bull. Hortic. Res. Stn. Ser. B. Okitsu, Japan. No. 5, pp. 75-87.

# 28. Methods for Fungicidal Control of Fruit Diseases of Avocado

## H. T. Brodrick

Atomic Energy Board, Private Bag X256, Pretoria, 0001, South Africa

Anthracnose disease, caused by *Colletotrichum gloeosporioides* Penz on avocado (*Persea americana* Miller), has not been regarded as so serious a problem in South Africa as to warrant fungicidal sprays. Within a few seasons, however, a fruit-spotting disease has threatened the avocado industry in this country. It has now been identified as Cercospora spot or blotch caused by *Cercospora purpurea* Cke. (1,3,7). Application of fungicides for control of this disease has become standard practice in several areas, and the economic importance and control of anthracnose disease have been investigated.

Greenhouse studies are considered impractical, since a significant amount of space would be needed to cultivate bearing avocado trees. Therefore, this chapter is concerned with field testing various methods for control of these two fungal diseases.

## Materials and Methods

**Test plants**—To reduce possible variation that may occur in field trials, young, vigorously growing avocado trees should be selected. They should be uniform in growth and free from obvious root disorders (eg, root rot caused by *Phytophthora cinnamommi* Rands). The cultivar chosen for the experiment (eg, Fuerte) should be susceptible to anthracnose.

**Test site**—The site chosen should consist of trees with a known history of infection over several seasons. Several sites should be selected in avocado areas with different climatic conditions, because severity of infection varies from season to season.

**Pathogen**—With the recently reported fruit-spotting disease in South Africa, the first task is to determine the periods when the avocados are most susceptible to infection. Thus, fruit-bagging experiments and staggered fungicidal sprays were employed.

**Procedure**—For this type of experiment, the suggested experimental layout is a randomized block design with eight to ten replications of single-tree plots. Where only a few fungicides are tested, a factorial design is also recommended in which each fungicide is used in all possible combinations, with all others used in the test (2,4).

In each experiment, untreated check trees should always be included. A standard fungicide with known effectiveness should also be included in each test.

To test for compatibility, various insecticides and trace elements used in avocado orchards should be combined with the experimental fungicide.

When beginning an experiment, recording certain facts about each test fungicide is important, including the common name (if established), chemical name, formulation, percentage of active ingredient, rate or rates recommended for use, and the name of the manufacturer or supplier. Three dosage rates should be tested: (i) the supplier's recommended rate, (ii) one half of the recommended rate, and (iii) twice the recommended rate, to test for possible phytotoxicity.

When possible, the infection period should be determined over several seasons and in different areas by one of the following two procedures.

(i) Bagging—The fruits are covered with a reinforced paper bag and tied tightly onto a cotton wool pad around the fruit stem, 10 to 15 cm below the fruit at times shown in Table 28.1. Kotzé (6) used this technique successfully to establish infection periods for other diseases.

(ii) Staggered fungicidal sprays—A further experiment using staggered fungicidal sprays should be conducted at the same time. The procedure illustrated in Table 28.1 could be followed except instead of bagging the fruit, the trees are sprayed with a fungicide at different times (shown by x in Table 28.1).

In the screening trial, the candidate fungicides should be applied to give protection from just after fruit-set until five months later, ie, during the susceptible period for infection.

When systemic fungicides are applied as curative sprays, a spray adjuvant must be added (eg, benomyl plus a light summer oil) to allow for greater efficacy (5).

When comparing high- and low-volume applications, the same quantity of product material must be applied to each tree. This is calculated with a standard fungicide of known concentration using high-volume equipment and spraying until runoff.

**Data collection**—Weather conditions during spraying, eg, temperature (sun and shade), humidity, and wind (direction and velocity), should be recorded. These data are essential in the event of phytotoxicity or reduced efficacy of the product.

The effects of the fungicidal sprays on other avocado diseases and insect problems should always be observed

**TABLE 28.1. Periods during which avocado fruits are protected from infection by bagging (x) during the season**

| Treatment number | Fruit-set | Weeks after fruit-set | | | | | | | |
|---|---|---|---|---|---|---|---|---|---|
| | | 3 | 6 | 9 | 12 | 15 | 18 | 21 | 24 |
| 1 | 0[a] | 0 | 0 | 0 | 0 | 0 | 0 | 0 | 0 |
| 2 | x[b] | 0 | 0 | 0 | 0 | 0 | 0 | 0 | 0 |
| 3 | x | x | 0 | 0 | 0 | 0 | 0 | 0 | 0 |
| 4 | x | x | x | 0 | 0 | 0 | 0 | 0 | 0 |
| 5 | x | x | x | x | 0 | 0 | 0 | 0 | 0 |
| 6 | x | x | x | x | x | 0 | 0 | 0 | 0 |
| 7 | x | x | x | x | x | x | 0 | 0 | 0 |
| 8 | x | x | x | x | x | x | x | 0 | 0 |
| 9 | x | x | x | x | x | x | x | x | 0 |
| 10 | x | x | x | x | x | x | x | x | x |

[a]0 = Bag removed.
[b]x = Bag in position.

**TABLE 28.2. Presentation of data for disease and phytotoxicity of avocado**

| Treatments | Fruit spot | Phytotoxicity | Percentage of nonmarketable fruits due to Anthracnose | |
|---|---|---|---|---|
| | | | Local storage[a] | Shipping[b] |
| 1. | | | | |
| 2. | | | | |
| 3. | | | | |
| 4. | | | | |
| Etc. | | | | |
| LSD .05 | | | | |
| .01 | | | | |

[a]Local storage conditions at ambient room temperature (20–25°C) for approximately ten days.
[b]Simulated shipping (export) conditions at 5.5°C for three weeks, followed by five days at 15–20°C.

carefully. Some common diseases, pathogens, and arthropods are sooty mold (*Stomiopeltis citri* Bitanc), algal spot (*Cephalearos cirescens* Kze), scab (*Sphaceloma perseae* Jenkins), scale insects, and mites.

Blotch or spotting disease becomes noticeable on the fruit during the season. Several preliminary assessments for disease incidence and phytotoxicity may be made by examining 50 fruits at random between waist and shoulder height around the circumference of each tree. In the event of a possible mishap at picking (eg, experimental or total wipeout by hail or wind), these preliminary results could be useful.

During the season, observations should also be made for all possible signs of leaf scorch and defoliation.

Finally, results are taken at picking on the total crop from each tree. Results on fruit spot and anthracnose are based on an evaluation scale from 1 to 5, where 1 is no disease; 2, slight disease; 3, moderate disease; 4, severe disease; and 5, very severe disease (ie, 0, 1–5, 6–9, 10–49, and 50–100% of the fruit surface is affected, respectively).

Fruits in categories 1 to 3 (having less than 10% of their surface affected) are regarded as marketable, whereas fruits in categories 4 to 5 are nonmarketable.

Normally, anthracnose symptoms appear only when the fruit is mature and then develop rapidly as the fruit softens. Assessments should be made on each fruit as it reaches the edible ripe stage. This usually requires daily examination until the last fruit in the experimental batch has ripened.

Since avocados are either marketed locally or refrigerated in transit to distant markets at 5.5°C, experimental fruit should be kept under like conditions. In South Africa, locally marketed avocados are kept at ~20–25°C for approximately ten days, while export avocados are shipped at ~5.5°C for three to four weeks before being exposed in European markets to temperatures of 15–20°C for up to five days.

Experimental results should include data obtained with avocados kept at ambient temperature (local market) and with fruit kept under simulated export conditions. From observations of experiments thus far, anthracnose symptoms apparently are more evident on cold-stored fruits.

Determination of fungicidal residues on the fruit may be required. This entails randomly selecting approximately 20 fruits at various heights and positions on each tree. The samples usually are frozen prior to analyses of the skin and pulp.

The preferred method for data presentation is given in Table 28.2.

## Discussion

Normally with these fruit experiments, analysis of variance of the data shows a coefficient of variation ranging from 10 to 30% or more. Emphasis must be placed on careful selection of all experimental trees. Trees showing any signs of root disorders should be rejected. Where possible, the trees should be selected during the winter (ie, when they are at the full-bearing stage). Careful checks should be made regularly until the final selections of the experimental trees are made just before starting the experiment.

## Literature Cited

1. BRODRICK, H. T., W. J. PRETORIUS, and R. T. FREAN. 1974. Avocado Diseases. Farm. S. Afr. Avocado Ser. No. H 1/1974.
2. COCHRAN, W. G., and G. M. COX. 1957. Experimental Designs. Ed. 2. John Wiley & Sons, Inc., New York.
3. DARVAS, J. M. 1977. Cercospora spot. Proc. Tech. Com. S. Afr. Avocado Growers Association. pp. 3-5.
4. HOLTON, C. S., G. W. FISCHER, R. W. FULTON, H. HART, and S. E. A. McCALLAN (eds). 1959. Plant Pathology, Problems and Progress, 1908-1958. University Wisconsin Press, Madison, WI.
5. KELLERMAN, C. 1975. Recommendations for the control of

pre-harvest citrus diseases. III. Citrus Subtrop. Fruit J. 501:19-23.
6. KOTZE, J. M. 1963. Studies on the black spot disease of citrus by *Guignardia citricarpa* Kiely, with particular reference

to its epiphytology and control at Letaba, DSc thesis, University of Pretoria.
7. RUEHLE, G. D. 1963. The Florida Avocado Industry. University of Fla. Bulletin No. 602.

# 29. Methods for Control of Anthracnose and Other Diseases of Mango

## H. T. Brodrick

Atomic Energy Board, Private Bag X256, Pretoria 0001, South Africa

Anthracnose, caused by the fungus *Colletotrichum gloeosporioides* Penz, is the most common, widespread disease of the mango *Mangifera indica* L. In addition to attacking twigs, leaves, and inflorescences, this disease results in a marked lowering of the grade and quality of mango fruit (4). A program of spraying is essential if the fruit is to be marketed successfully (1). According to Ruehle and Ledin (4), two other fungus diseases—scab (*Elsinoe mangiferae* Bitanc. & Jenkins) and powdery mildew (*Oidium mangiferae* Berthal)—have to be considered in the spray program in Florida. In India, Pakistan, and South Africa, a bacterial spot disease (previously described as *Erwinia mangiferae* Doidge; now reported to be caused by *Pseudomonas mangiferaeindicae* Patal et al.) has resulted in heavy fruit losses (5,6,7).

In this chapter, field experiments to control anthracnose (and related problems) will be discussed. Procedures may be modified when other diseases present serious problems.

## Materials and Methods

**Test plant**—Young, vigorously growing, bearing mango trees not more than 3 m in height (to facilitate both spraying and picking) should be selected.

**Test site**—Trees in the test site should have a known history of disease over several seasons. Several sites should be selected in the mango-producing areas with different climatic conditions.

**Procedure**—A randomized block design with eight to ten replications of single-tree plots is normally recommended for this type of experiment. A factorial design may also be used, especially when various combinations of fungicides and bactericides have to be considered (eg, for anthracnose and bacterial spot control) (2,4).

Each experiment should have a check treatment and a standard treatment of known efficacy. Tests should also be conducted where various insecticides and trace elements (known to be used in mango orchards), are combined with the experimental material.

With each experiment, recording certain information about each test chemical is important, including the common name (if established), chemical name, formulation, percentage of active ingredient, rate or rates recommended for use, and the name of the manufacturer or supplier. Three rates for each test material should be used—the supplier's recommended rate, one half of the recommended rate, and twice the recommended rate, to test for possible phytotoxicity.

In cases in which the control of mango bacterial spot disease also must be considered, it is essential that (i) trees selected for the experiment have 80–95% of the crop affected and (ii) the infection times be determined over several seasons and in different areas by covering the

**TABLE 29.1. Mango fruits protected from infection (by bagging) at different periods during season**

| Treatment number | Fruit-set | Weeks after fruit-set | | | | | | | |
|---|---|---|---|---|---|---|---|---|---|
| | | 3 | 6 | 9 | 12 | 15 | 18 | 21 | 24 |
| 1 | 0[a] | 0 | 0 | 0 | 0 | 0 | 0 | 0 | 0 |
| 2 | x[b] | 0 | 0 | 0 | 0 | 0 | 0 | 0 | 0 |
| 3 | x | x | 0 | 0 | 0 | 0 | 0 | 0 | 0 |
| 4 | x | x | x | 0 | 0 | 0 | 0 | 0 | 0 |
| 5 | x | x | x | x | 0 | 0 | 0 | 0 | 0 |
| 6 | x | x | x | x | x | 0 | 0 | 0 | 0 |
| 7 | x | x | x | x | x | x | 0 | 0 | 0 |
| 8 | x | x | x | x | x | x | x | 0 | 0 |
| 9 | x | x | x | x | x | x | x | x | 0 |
| 10 | x | x | x | x | x | x | x | x | x |

[a]0 = Bag removed.
[b]x = Bag in position.

fruits for different periods. This involves covering the fruit with an open-bottom reinforced paper bag or a cardboard cap and tying tightly onto a cotton wool pad around the fruit stem just below the fruit. This method effectively protects the fruit, since the infective bacteria originate primarily from shoot cankers. The periods during which fruits should be protected are shown in Table 29.1.

Once the infection period for bacterial black spot has been determined, experiments can be planned to determine the most effective timing of sprays. When systemic fungicides are applied as curative sprays for the control of fungal disease, adding a spray adjuvant (eg, benomyl plus a light summer oil) is essential for greater efficacy (3).

When comparing high- and low-volume applications, the same quantity of product material must be applied to each tree. This is calculated with a standard chemical of known concentration using high-volume equipment and spraying to runoff.

**Data collection**—Weather conditions during spraying should be recorded, eg, temperature (sun and shade), humidity, and wind (direction and velocity). These data are essential in the event of phytotoxicity or reduced efficacy of the product.

The effect of the various chemical sprays on other mango pests and diseases should also be noted carefully. Some common diseases and pests are powdery mildew, caused by *Oidium mangiferae* Berthal; blossom blight, caused by *Physalospora* sp.; mango scab, caused by *Elsinoe mangiferae* Bit. & Jenkins; red rust, caused by *Cephaleuros virescens* Kunze; scale insects, eg, *Coccus mangiferae* Green, and *Pinnaspis strachani* Cooley; thrips, eg, *Selenothrips rubocinctus* (Giard.); and mites, eg, *Paratetranychus yothersi* (McG.).

Since all of these are preharvest problems (except for anthracnose in which symptoms appear on the fruit in both the preharvest and postharvest stage), several preliminary assessments for pests and diseases as well as phytotoxicity should be taken during the course of the season. This may be done by examining 50 fruits, leaves, or shoots selected at random between waist and shoulder height around the circumference of each test tree.

During the season, observations should also be made for all possible signs of leaf scorch and defoliation.

Finally, results should be taken at picking on the total crop from each tree. Results are based on an assessment scale from 1 to 5, where 1 is no disease; 2, slight disease; 3, moderate disease; 4, severe disease; and 5, very severe disease (ie, 0, 1–5, 6–9, 10–49, and 50–100% of the fruit area affected, respectively).

Fruits in categories 1–3 (having less than 10% of their surface affected) are regarded as marketable, whereas fruits in categories 4 and 5 are nonmarketable.

Normally, anthracnose symptoms appear only when the fruit is mature and then develop rapidly as it softens. Assessments should be made on each fruit as it reaches the edible ripe stage. This usually requires daily examination until the last fruit in the experimental batch has ripened.

Since mangoes are either marketed locally or refrigerated at 11°C in transit to distant markets, experimental fruit should be kept under like conditions. In South Africa, locally marketed mangoes are kept at ~20–25°C for approximately 14 days, while exported mangoes are shipped at ~11°C for three to four weeks before being exposed in European markets to temperatures of 15–20°C for up to seven days.

Experimental results should include data obtained with mangoes kept at ambient temperatures (local market) and with fruit kept under simulated export conditions.

A determination of chemical residue on the fruit may be required. This entails randomly selecting approximately 20 fruits from various heights and positions on each tree. The samples are usually frozen prior to analyses of the skin and pulp.

The preferred method for data presentation is given in Table 29.2.

## Discussion

Variation in field trials depends on careful selection of experimental trees. The trees should be selected during the winter preceding the experiment. Regular inspections should be made from then until the final selections are made just before beginning the experiment.

In the case of mango bacterial spot disease, control experiments should be repeated on the same site over two to three seasons, applying the same treatments to the original candidate trees. The effect of the spray treatments on the formation of cankers on branches, twigs, and leaf petioles can only be determined properly over this time span. These cankers are important, because they serve as the main source of infection of the fruit during wet periods.

**TABLE 29.2.** Presentation of data for disease and phytotoxicity of mango

| Treatments | Percentage of nonmarketable fruits due to | | | |
| | Bacterial spot | Phytotoxicity | Anthracnose | |
| | | | Local storage[a] | Shipping[b] |
| --- | --- | --- | --- | --- |
| 1. | | | | |
| 2. | | | | |
| 3. | | | | |
| 4. | | | | |
| Etc. | | | | |
| LSD .05 | | | | |
| .01 | | | | |

[a]Local storage conditions at ambient room temperature (20–25°C) for approximately 14 days.
[b]Simulated shipping (export) conditions at 11°C for three weeks, followed by seven days at 15–20°C.

## Literature Cited

1. BRODRICK, H. T. 1971. Mango diseases. Leaflet No. 71. Subtrop. Fruit Ser. No. 12, Department of Agriculture Technical Services, Pretoria, South Africa.
2. COCHRAN, W. G., and G. M. COX. 1957. Experimental Designs, Ed. 2. John Wiley & Sons Inc., New York.
3. KELLERMAN, C. 1975. Recommendations for the control of pre-harvest citrus diseases. III. Citrus Subtrop. Fruit J. 501:19-23.
4. RUEHLE, G. D., and R. B. LEDIN. 1955. Mango growing in Florida. Fla. Univ. Agric. Exp. Stn. Tech. Bull. 574.
5. STEYN, P. L., N. M. VILJOEN, and J. M. KOTZE. 1974. The causal organism of bacterial spot of mangoes. Phytopathology 64:1400-1404.
6. VILJOEN, N. M., and J. M. KOTZE. 1972. Bacterial spot of mango. Citrus Grow. Subtrop. Fruit J. 462:5-8.
7. VILJOEN, N. M., P. L. STEYN, and J. M. KOTZE. 1972. Bakteriese swartvlek van mango. Phytophylactica 4:93-94.

# SECTION III.

# D. Ornamentals and Turf

# 30. Methods for Evaluating Fungicides for Controlling Powdery Mildew, Black Spot, and Rust of Roses

H. L. Dooley

Environmental Protection Agency, Northwest Biological Investigations Station, 3320 Orchard Avenue, Corvallis, OR 97330

Powdery mildew, black spot, and rust are the three most serious diseases of roses (*Rosa* spp.). Powdery mildew is serious under both field and greenhouse conditions. Black spot is a major problem under field conditions throughout the United States. Rust is particularly destructive among field roses in the western United States. The methods described in this chapter were designed to evaluate the efficacy of foliar-applied protectant and eradicant fungicides.

Rose powdery mildew is caused by *Sphaerotheca pannosa* (Wallr.) Lev. var. *rosae* Wor. Spore germination and infection are favored by temperatures of 18–24°C and relative humidity of 95–99% (7,8,13,15). This fungus is an obligate parasite and must be maintained on susceptible plant cultivars.

Pathogenic races of *Diplocarpon rosae* Wolf., the causal agent of black spot, differ in their ability to cause disease in host cultivars (2,24). Temperatures of 16–21°C favor spore germination and infection, and a 6- to 7-hr period of continuous leaf wetness is required for infection (2,4,14,16).

Rose rust is caused by *Phragmidium* spp., nine of which have been found on roses in the United States (1). The causal fungus in Oregon is *P. mucronatum* (Fr.) Schlecht., and in California, *P. disciflorum* (Tode) James. Ramsbottom (21), however, has pointed out that the two species are synonomous, and *P. mucronatum* (Fr.) Schlecht. is the valid binomial. *P. mucronatum* is the only species that commonly infects cultivated roses. Temperatures of 18–21°C favor spore germination and infection (5,6), which requires a wet leaf surface for a period of 4 hr (5).

## Methods

**Environmental conditions**—The soil at the test site should be suitable for rose growth, and the climatic conditions should favor disease development. Diseases should be endemic and inoculum potential high in the test area.

**Inoculum**—If powdery mildew, black spot, and rust inoculum potentials are high in the field area, no additional inoculum should be necessary to provide adequate disease pressure. Under greenhouse conditions, plants heavily infected with powdery mildew are the source of inoculum. After the first treatments are applied, heavily infected plants are dispersed down the center of the test bench 25.4–30.5 cm (10–12 in.) apart. One heavily infected plant is used for every four test plants. Conidia are discharged daily with a 413–552-kPa (60–80-psi) air jet. This is done best by blowing the spores, starting from a different corner of the bench each day. Strong air currents from fans will also discharge conidia and can be used instead of the pressurized air, provided the conidia are equally dispersed.

Successful artificial inoculation of plants with *D. rosae* can be accomplished in two ways. First, leaves severely infected with black spot are collected from the vicinity of the test area to prevent introduction of new fungus strains into the area. These leaves may be used immediately, or stored for later use. If stored, the leaves should either be frozen or air-dried and then stored dry not longer than one year. The leaves are placed in distilled water or water that has neither been chlorinated nor fluorinated to prepare a spore suspension. The leaves are soaked for 10 min and then filtered into another container through two layers of cheesecloth to remove debris. The spore count is determined and the plants are spray inoculated. With the second method, the fungus may be grown on nutrient agar for 10–12 weeks and a spore suspension prepared from these cultures. Further information may be obtained from the literature (3,9,12,17,18,20).

Artificial inoculation with rose rust can be achieved with dispersal of heavily infected leaves among the test plants and subsequent irrigation with overhead sprinklers for 4–6 hr. The collected leaves should contain urediospores for best results. We have not been successful in inoculating with a rust spore suspension.

**Test plants**—Dwarf Crimson Rambler, Command Performance, or White Knight rose cultivars are especially useful in field tests for powdery mildew, because they are susceptible to this disease but not to rust or black spot. The peduncles and flowers of Dwarf Crimson Rambler become infected with powdery mildew, and under heavy infection pressure, the leaves and stems may also become infected. Command Performance and White Knight are both hybrid tea roses, and foliar infection is more serious than is floral infection. Fungicides react differently on rambler roses than on hybrid tea roses in the field. This may be due to different races of the fungus as well as to the difference in the physiology of the two types of roses. We have not worked with miniature or tree roses, but would also expect them to react differently.

In the greenhouse, Dwarf Crimson Rambler roses are used as test plants and to maintain powdery mildew inoculum, because they require less space than do hybrid teas. Under greenhouse conditions, the stems, leaves, peduncles, and flowers of Dwarf Crimson Rambler are all susceptible to infection. Diseased plants for inoculum should be maintained in a greenhouse separate from the test plants until after the first protectant treatments are made. All plants should be grown in 152-mm (6-in.) pots containing one part sand, one part loam, and two parts peat moss. Alternate cool-white and warm-white fluorescent tubes provide adequate light during the winter months. A 14-hr day has proved successful when supplementary light is necessary for good plant growth. The greenhouse is maintained at 18–24°C during the day and 13–18°C at night for good growth and disease development. Plants are maintained in a vigorous growing condition by fertilizing as needed (approximately every 21 days) with an appropriate fertilizer (3).

Test plants for black spot should be susceptible only to that disease and highly resistant to other foliar diseases. If the test plant is susceptible to both black spot and rust, one cannot distinguish which disease caused defoliation, nor can a true reading of infection be recognized. Juliett hybrid tea rose is a cultivar susceptible only to black spot and is commercially available from United Rose Growers Inc., Woodland, WA. Other cultivars that are specifically susceptible to black spot but not to rust are acceptable and should be selected by observations from the test locality. Races of the black spot fungus may vary from one locality to another.

Rust test plants should be from a variety susceptible only to rose rust and highly resistant to other foliar diseases for the reason given above. McGredy's Scarlet is such a cultivar. It is an old cultivar and can be purchased commercially from United Rose Growers Inc. Other cultivars susceptible only to rose rust are also acceptable.

**Plot design**—In the field and greenhouse tests, treatments were arranged in a randomized block design according to Snedecor's randomly assorted digits table (23), or its equivalent. Each rose cultivar and treatment tested was replicated at least four times. A standard registered fungicide and an untreated control were always included among the treatments.

The standard registered fungicide should serve as a basis for comparison with new compounds to determine whether test materials provide enhanced protection. The standard material also helps to evaluate whether the test procedure was correct.

The untreated control is always included to determine the inoculum pressure during the test. This allows the researcher a basis for judging comparative efficacy of the various treatments.

**Fungicide application**—Applications of test fungicides should be made at predetermined intervals coinciding with the objective of the test. A minimum of four applications should be made to distinguish phytotoxicity of repeated applications. Treatments should be continued until the untreated controls become heavily infected. Protectant fungicides are applied to disease-free plants on the first day of the test, and usually at weekly intervals thereafter. Eradicant fungicides are applied at the first sign or symptom of disease, and at predetermined intervals thereafter.

Greenhouse test plants are removed from the growing area for treatment to prevent spray drift, which may cause cross-contamination. Sprays are applied with a DeVilbiss paint gun sprayer with a No. 30 head and using 172–207 kPa (25–30 psi) of air pressure. Plants should be sprayed to runoff as they rotate on a compound turntable (19). Pressurized spray cans such as aerosols and low-pressure cans should be used as the manufacturer suggests and plants sprayed while on a compound turntable. If the plants are dusted, a vacuum duster should be used (10) with sufficient dust in the duster cup for good coverage of the plant foliage when using a 20.26-Pa (6-in.) Hg vacuum. If container-duster applicators are used, each plant should be placed in a 0.093-m² (1-ft²) area and dusted as the manufacturer suggests. After fungicide applications are made, the plants are returned to the greenhouse bench.

Field treatments may be applied with the following equipment: (i) a power sprayer, applying fungicide to upper and lower leaf surfaces to runoff, (ii) a hand sprayer, using 207 kPa (30 psi) of pressure applied to runoff, (iii) a hand duster, applying dust for good coverage, (iv) a hose-end sprayer, and (v) aerosol or low-pressure sprayers used according to manufacturer's directions.

**Irrigation and plot maintenance**—Under greenhouse conditions, pots are watered daily, taking care not to wet the foliage. The foliage of greenhouse plants is washed the day before dust treatments are made to remove excess residue from previous treatments.

The field garden is sprinkle irrigated for 3–4 hr in late afternoon 24 hr after treatment to provide a period of high humidity.

**Controls**—Standard and untreated controls help to determine disease severity and provide a basis for comparison. Standard fungicides determine whether disease control is possible under the prevailing test conditions. The standard fungicide for control of powdery mildew is parinol 1.3% emulsifiable concentrate applied at the rate of 14.8 ml/3.785 L (0.5 oz/gal) of water as a protectant, and 29.6 ml/3.785 L (1.0 oz/gal) of water as an eradicant. A standard fungicide of known effectiveness such as benomyl 50W applied at 227 g/379 L (8 oz/100 gal) of water, or maneb 80W applied at 680 g/379 L (1.5 lb/100 gal) of water should be included as a protectant standard for black spot. Maneb 80W applied at the rate of 680 g/279 L (1.5 lb/100 gal) of water is used as a rust control protectant standard.

Other known effective registered chemicals may be used as standards, provided their identity and dosage are specified.

**Data collection**—All field and greenhouse tests are terminated one week after the last fungicide application if the untreated controls are heavily infected. The foliage of each plant is carefully inspected, and the percentage of infection is visually estimated as zero to 100% disease incidence. Rust and black spot data are based on the percentage of infected leaves. Both rust and black spot cause extensive defoliation if not controlled; therefore, evaluating the test before defoliation occurs is important. In the greenhouse, data on powdery mildew are based on the percentage of infection of all foliar plant parts. For hybrid tea roses in the field, powdery mildew infection is determined from the percentage of infected leaves, but infection of Dwarf Crimson Rambler roses is usually determined from the percentage of infection that occurs on the peduncles and flowers. Under severe disease conditions, however, the leaves and stems must also be examined. The mean percentage of disease incidence for each treatment was determined and the mean percentage of disease control calculated using the following formula (11,22):

$$\text{MPDC} = \frac{\text{MDIC} - \text{MDIT}}{\text{MDIC}} \times 100$$

where MPDC is mean percentage of disease control; MDIC, mean percentage of disease incidence in the untreated control; and MDIT, mean percentage of disease incidence in the treatment.

The severity of phytotoxicity as well as a description of injury should be reported. Phytotoxicity may occur in such ways as stunting, marginal leaf burning, vein chlorosis, spot necrosis, or spot chlorosis.

Presence of visible fungicide residue is determined and recorded as none, light, moderate, or heavy. Residue color should also be reported. Such information is important to rose growers producing flowers for exhibition.

## Results and Conclusions

Effective fungicides used for control of powdery mildew and black spot should provide a mean percentage of disease control of 70 or more under either greenhouse or field conditions with weekly spray applications. Powdery mildew control on Dwarf Crimson Rambler roses in the field is an exception, since 70% of control usually cannot be achieved with many presently registered fungicides.

Effective fungicides for control of rust should provide at least 65% of disease control. Presently registered fungicides usually provide 70–80% of disease control.

By using cultivars susceptible to only one disease, interactions of fungi do not complicate evaluation of fungicide effects on that disease. This is most important when conducting tests for rust or black spot control, because infection with either of these organisms may cause severe defoliation.

A cultivar susceptibility index is needed for all parts of the country. Such information is not available in any one publication, and the information presently in the literature may not be accurate for all locations due to the presence of many physiologic races of these fungi.

Registered eradicant fungicides used to control rust and black spot of roses are lacking at present. This area

of research should be pursued.

Since test information is used for either supporting registration or regulatory actions, reports should contain the following to make the findings more meaningful to other researchers, industry, and regulatory agencies: (i) product name, EPA registration number, or experimental permit number, (ii) active ingredient or ingredients and percentage of each, (iii) field or greenhouse test indicated, (iv) how the product was applied, (v) the dosage rate, (vi) number of plants per treatment and the number of replications used, (vii) number of varieties used, (viii) spraying or dusting dates, (ix) mean percentage of disease incidence, (x) mean percentage of disease control, (xi) description and severity of phytotoxicity, (xii) statistical analysis of data, (xiii) residue information as to color, and (xiv) visible fungicide residue on the foliage rated as none, light, moderate, or heavy.

## Literature Cited

1. ARTHUR, J. C. 1934. Manual of the Rusts of the United States and Canada. Purdue Research Foundation, Lafayette, IN. pp. 78-91.
2. BAKER, K. F., and A. W. DIMOCK. 1969. Powdery mildew of roses. In Pennsylvania Flower Growers, New York State Flower Growers Assn. Inc., Roses Inc. Roses. pp. 172-184.
3. BEACH, O. R. 1966. Fertilizers and disease/insect problems. Am. Rose Annu. 51:31-34.
4. CHESTER, K. S. 1950. Nature and prevention of plant diseases. The Blakiston Company, Philadelphia.
5. COCHRANE, V. W. 1946. The common leaf rust of cultivated roses. Am. Rose Annu. 31:131-135.
6. COCHRANE, V. W. 1946. The common leaf rust of cultivated roses, caused by Phragmidium mucronatum (Fr.) Schlecht. N.Y. Agric. Exp. Stn. Ithaca Mem. 268:3-39.
7. COYIER, D. L. 1961. Biology and control of rose powdery mildew, PhD thesis, University of Wisconsin, Madison.
8. DIMOCK, A. W., and J. TAMMEN. 1969. Powdery mildew of roses. In Pennsylvania Flower Growers, New York State Flower Growers Assn. Inc., Roses Inc. Roses. pp. 163-171.
9. DODGE, B. O. 1931. A further study of the morphology and life history of the rose black spot fungus. Mycologia 23(6):446-462.
10. FARRAR, M. D., W. C. L'KANE, and H. W. SMITH. 1948. Vacuum dusting of insects and plants. J. Econ. Entomol. 41:647-648.
11. HORSFALL, J. G., and J. W. BARRATT. 1945. An improved grading system for measuring plant diseases. Phytopathology 35:655.
12. JENKINS, W. R. 1955. Variability of pathogenicity and physiology of Diplocarpon rosae Wolf, the rose blackspot fungus. Am. Rose Annu. 40:92-97.
13. LONGREE, K. 1939. The effect of temperature and relative humidity on the powdery mildew of roses. N.Y. Agric. Exp. Stn. Ithaca Mem. 223:1-43.
14. MASSEY, L. M. 1946. Blackspot can be controlled. Am. Rose Annu. 31:111-118.
15. MASSEY, L. M. 1948. Understanding powdery mildew. Am. Rose Annu. 33:137-145.
16. MASSEY, L. M. 1955. Tests with fungicides for blackspot. Am. Rose Annu. 40:63-91.
17. MASSEY, L. M., and J. A. NAEGELE. 1956. Tests with fungicides for blackspot. Am. Rose Annu. 41:49-60.
18. MASSEY, L. M., J. A. NAEGELE, J. S. MELCHING, and H. ALLER. 1958. Fungicide-insecticide combinations for garden roses. Am. Rose Annu. 43:37-53.
19. McCALLAN, S. E. A., and R. H. WELLMAN. 1943. A greenhouse method of evaluating fungicides by means of tomato foliage diseases. Contrib. Boyce Thompson Inst. 13:93-134.
20. PALMER, J. G., and P. SEMENIUK. 1961. Comparable susceptibilities of 50 species and hybrid roses inoculated with black spot fungus from plants field-grown in

Maryland in 1959. Am. Rose Annu. 46:125-133.
21. RAMSBOTTOM, J. 1914. Notes on the nomenclature of some rusts. Trans. Br. Mycol. Soc. 4:331-340.
22. REDMAN, C. E., E. P. KING, and I. F. BROWN, JR. 1962. Tables for converting Barratt and Horsfall rating scores to estimated mean percentages. Eli Lilly and Company, Indianapolis.
23. SNEDECOR, G. W., and W. G. COCHRAN. 1967. Statistical methods. Ed 6. Iowa State University Press, Ames.
24. STEWART, R. N., and P. SEMENIUK. 1965. Report on rose research in the U.S. Department of Agriculture. Am. Rose Annu. 50:99-105.

# 31. Field Procedures for Evaluating Fungicides for Control of Pythium Blight of Turfgrasses

**Herbert Cole, Jr., C. G. Warren, and Patricia L. Sanders**

Professor, former research aide, and research associate, respectively, Department of Plant Pathology and Pesticide Research Laboratory and Graduate Center, The Pennsylvania State University, University Park, PA 16802

Pythium blight is a devastating disease of various turfgrass species, especially bentgrasses and ryegrasses. Environmental conditions for disease development are critical, however, and relying on natural inoculum and the natural environment for disease occurrence in a research plot often results in failure. When commercial turf areas are used to evaluate fungicides for control of *Pythium,* weather conditions favorable for disease outbreaks often pass before test plots can be established in affected sites. When this occurs, little or no disease develops. Therefore, we have devised a field inoculation procedure that combines movable plastic-covered humidity chambers with artificial inoculation. This allows us to induce Pythium blight whenever summer temperatures are favorable for infection.

**Test site and grass species**—This procedure may be used anywhere that susceptible grass species are available. The only requirement is that ambient air temperatures be high enough to reach 30°C in the humidity chambers for the duration of the incubation period.

Pythium blight is a major problem on bentgrasses *(Agrostis* sp.) and ryegrasses *(Lolium* sp.). These species may be established and maintained according to recommended practices for the turfgrass situation being evaluated. For example, if the fungicides are being studied for use on golf course greens, the grass should be of an appropriate species and the cultivar maintained under greens conditions. The grass population, soil, drainage pattern, and slope of the test site should be uniform throughout. Uniformity of the environment and grass population minimize experimental error and increase the precision with which fungicide treatment differences can be measured.

For most experimentation, the grasses of choice are Penncross creeping bentgrass *(Agrostis palustris* Penncross), which is maintained at 0.6-cm (0.25-in.) cutting height, and Manhattan perennial ryegrass *(Lolium perenne* Manhattan), which is maintained at 2.5-cm (1-in.) cutting height. Test results with these two cultivars represent most golf course and home lawn problem situations, and are suitable for development of fungicide registration data.

**Organisms**—Some controversy exists regarding the *Pythium* spp. that are responsible for inciting the foliar blight of turfgrasses. The two species most commonly associated with Pythium blight of turf are *P. ultimum* Trow. and *P. aphanidermatum* (Edson) Fitzpatrick. The latter is the primary foliar blight pathogen in Pennsylvania; we consequently use *P. aphanidermatum* in all of our fungicide evaluation experiments. Isolates are maintained at room temperature on kernels of rye grain autoclaved in tubes of distilled water. Pathogenicity is determined on bentgrass in pots in the greenhouse before they are incorporated into fungicide field inoculation experiments.

**Humidity chamber design**—The movable humidity chambers (Fig. 31.1) are rectangular and made of single 2.5 × 15.2-cm (1 × 6-in.) boards with a 0.0025-cm (4-mil) transparent polyethylene plastic top. The plastic is attached to the wooden frame with staples. Cross bracing of the wooden structure provides rigidity. To facilitate ease of handling, we restrict the chamber size to 0.9 × 4.6 m (3 × 15 ft).

**Plot design**—The procedure consists of fungicide treatment followed by immediate inoculation with the selected fungal isolates and successive weekly inoculations to evaluate long-term fungicidal suppression of disease. Each successive inoculation is made on a "new" noninoculated portion of the plot. The humidity chamber is moved to the new site each time. The width of individual treatment plots is 0.9 m (3 ft), to

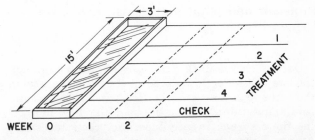

Fig. 31.1. Placement of humidity chamber across treatments after inoculation enhances disease development.

conform to the width of the sprayer boom. The length of the individual treatment plots depends on the number of weekly inoculations to be done. The area of each treatment plot used for individual inoculations is $0.9 \times 0.9$ m ($3 \times 3$ ft). If the test fungicides are to be evaluated at weekly intervals over a six-week period, the individual treatment plots should be 5.5 m (18 ft) long to allow for six successive weekly inoculations. A buffer strip between plots has not been necessary in our experiments.

Our humidity chamber is $0.9 \times 4.6$ m ($3 \times 15$ ft). When placed across the treatment plots, this size accommodates the simultaneous weekly inoculation and incubation of four adjacent treatments plus an unsprayed check. If more fungicide treatments are used, longer or more chambers should be used. Our procedure requires three such chambers to permit three replications of each weekly inoculation. The overall plot design is a replicated randomized block.

**Fungicide application**—Long-range weather forecasts are monitored during midsummer. When a period of prolonged warm weather is anticipated, the entire plot area is treated with the test fungicides. Since the procedure imposes no restriction on the application methods, fungicides may be applied as granules, foliar sprays, or soil drenches. Individual fungicides are applied in 0.9-m (3-ft) strips. The length of these treatment strips depends on the anticipated duration of the experiment. With the advent of long-residue systemic *Pythium* fungicides, a four- to eight-week evaluation period after treatment is sometimes required.

**Inoculum preparation, inoculation, and incubation**—Inoculum is prepared by growing virulent isolates of *P. aphanidermatum* on autoclaved rye grain for approximately one week. Rye flasks are prepared by placing 225 g of rye grain, 4 g of $CaCO_3$, and 275 ml of water in 1,000-ml Erlenmeyer flasks; stopping with cotton plugs; and autoclaving for 30 min at 1 kg/cm² (15 psi). Inoculation is done immediately after fungicide application, and weekly thereafter until the residual effect of the test fungicides is no longer apparent. Inoculation consists of spreading the Pythium-infested rye kernels to achieve an average inoculum density of one kernel per 5.8 cm² (9 in.²) of turfgrass. After inoculation, the contaminated areas are irrigated with approximately 0.32 cm (1/8 in.) of water and covered with humidity chambers. If rainfall occurred in the preceding 4 hr, this irrigation may not be needed.

When the sun is high, air temperatures beneath the canopy must be checked. During periods of bright sun, the canopy must be lifted (5.1–15.2 cm [2–6 in.] above grass surface) to prevent interior air temperatures from exceeding 38°C. Elevation may be accomplished on steel pins, with lateral arms to support the frame. During early evening, night, and early morning hours, the canopy sides should rest directly on the grass surface. The chamber insures the high temperatures and humidities in the foliar microclimate that are essential for Pythium infection and development.

Disease severity usually is evaluated seven days after inoculation, but under favorable conditions, may be rated sooner. The canopy is kept in place until disease ratings are made. It is removed for ratings, and severity is evaluated visually, in terms of percentage of the turfgrass area exhibiting blight symptoms. Phytotoxicity evaluations should also be made. Injury symptoms on turfgrass vary with test materials, but may include tip burn, stunting, yellowing, overall burning, and morphologic malformations. Data obtained are subjected to analysis of variance and Duncan's modified least significant difference test.

This experimental procedure insures infection pressure at regular intervals after fungicide treatment, and helps to offset some of the difficulties of uncertain environment encountered in field testing fungicides for control of Pythium blight of turfgrass. Fungicide efficacy can be evaluated from immediately after treatment to many weeks later if temperatures are suitable and sufficient test area is available. Rainfall and weather records should be maintained for the duration of the experiment.

# 32. Field Evaluation of Fungicides for Control of Ascochyta Blight of Chrysanthemums[1]

## Arthur W. Engelhard

Professor and plant pathologist, IFAS, University of Florida, Agricultural Research and Education Center, Bradenton, FL 33508

Outdoor chrysanthemum culture started in Florida in the late 1940s. For flower production, cuttings are set in the soil starting around the first part of August and continuing at intervals until about March so that the last flowers are cut in early June. This practice does vary among growers, as a few of them produce chrysanthemum flowers during the summer months. Those producing cuttings grow plants throughout the year.

Ascochyta blight, also called ray blight, is the major fungus disease of the flowers, leaves, and stems. It exists throughout the year, but occurs more frequently and seriously during December, January, and February. The disease can cause extensive damage and loss during

[1]Florida Agricultural Experiment Stations Journal Series No. 182.

rainy periods with moderate temperatures. It can attack chrysanthemums at any stage of growth in the field or in the rooting bench.

Ascochyta blight is incited by *Mycosphaerella ligulicola* Baker, Dimock, and Davis, the perfect stage of *Ascochyta chrysanthemi* Stev. Pycnidia and pycnidiospores are produced year round on flowers, stems, and sometimes, leaves. Perithecia and ascospores are found relatively infrequently in Florida. The optimum temperature for perithecium development is 20°C (68°F), and for conidia, 26°C (79°F) (4). The pycnidiospores are spread by water, blowing rain, or infected plant material or by being on hands, tools, or black cloth. Ascospores are spread in a similar manner, and also are wind-borne.

**Objectives**—The method described in this chapter is appropriate for evaluating foliar fungicides for control of Ascochyta blight of chrysanthemum (*Chrysanthemum morifolium* [Ramat.] Hemsl.). This system evolved over a period of ten years with more than 20 experiments conducted on crops grown during the fall (September to December) and spring (February to June) (1,2).

**Inoculum preparation**—Inoculum may be prepared by harvesting diseased flowers and stems from a previous season's unsprayed plants. Diseased stems and flowers are harvested, dried at room temperature in the laboratory, and stored in a dry place until needed. Pycnidia are produced in abundance on diseased stems and flowers and, occasionally, on leaves. The pathogen can be stored for at least one year at room temperature. When a spore suspension is needed, the dried, diseased material is placed in deionized or distilled water for 15 min to allow the spores to eject from the pycnidia. Wetting agents are not needed. The suspension is filtered through a double layer of cheesecloth. Several gallons of spore suspension can be prepared in 30 min with this method.

A second method of preparing inoculum is to grow *M. ligulicola* on potato-dextrose agar. Most isolates produce pycnidia and pycnidiospores when grown under continuous cool-white fluorescent lights in an incubator at 24–27°C (75–80°F). Both spores and mycelia cause infection. A suspension can be prepared for inoculation by adding either distilled or deionized water to cultures in petri plates for 15 min after conidia are produced and filtering through cheesecloth, or the entire contents of the petri plate can be macerated in a Waring Blendor and filtered through cheesecloth. The inoculum should be sufficiently free of particles to be applied with a backpack sprayer. This method is more laborious than the one mentioned above. Hadley and Blakeman (3) have discussed factors affecting conidiospore production in culture in detail.

**Inoculation procedure**—Inoculation of plants has been necessary to insure infection at this location in Bradenton, FL. Test plants should be inoculated during a rain, or at least during a rainy period. A spore suspension containing 150,000 spores/ml yields good infection under rainy conditions. Ninety linear meters (300 linear ft) of bed 0.9 m (3 ft) wide can be inoculated with 11.5 L (3 gal) of spore suspension applied with a backpack sprayer. The plants should be sprayed uniformly, once from each side of the bed and once directly down over the top of the bed. A wet period of 24 hr is adequate for obtaining good field infection when

the temperature is in the 21–27°C (70–80°F) range. Under these conditions, symptoms become apparent within 30 hr on flowers and two to three days on leaves. Test plants may be inoculated after the first fungicide application, in which case the fungicides would be evaluated as protectants, or the plants may be inoculated 24, 48, or 72 hr before the first fungicide application to test the eradicant activity of the fungicides. Plants may be inoculated at any age, but perferably within two to three weeks after transplanting in the field. Early inoculation allows the disease to spread naturally in the test area and spreader rows for the rest of the crop's duration (12–14 weeks). In the absence of adequate natural moisture, disease development and spread can be facilitated with overhead irrigation applied late in the day to keep plants wet overnight.

**Culture of chrysanthemums**—Culture of chrysanthemums varies with location. A successful procedure employed in Florida uses well-prepared, well-drained sandy soil that is fumigated to control fungi, nematodes, and weeds. Since good drainage is necessary for chrysanthemum culture, the plants usually are grown on raised beds that are about 10–15 cm (4–6 in.) high and 0.9 m (3 ft) wide. This is especially important if the area is subject to sudden heavy rains. Bedding wire with $15 \times 20$-cm ($6 \times 8$-in.) or $20 \times 20$-cm ($8 \times 8$-in.) cells is placed on the beds, and two plants are placed in each cell. The wire is raised to support the plants as they grow. Initially the crop is grown vegetatively under long days (more than 14.5 hr of light) for two to four weeks; thereafter the chrysanthemums are induced to flower by exposure to the reproductive cycle (short days of less than 13 hr of light). Light is controlled during natural long days by covering the plants with black cloth. Flowering occurs within nine to ten weeks after the initiation of short days. A nutritional program that allows for vigorous plant growth throughout the test period should be maintained.

**Plot design**—Each experimental treatment, standard fungicide, and a water control should be replicated a minimum of three, preferably four, times. Each replicate should contain approximately 18 plants. This would be nine cells ($3 \times 3$) planted on the bed, each with two plants per cell. On a bed 0.9 m (3 ft) wide, a different cultivar can be planted on each side of the bed so that each treatment can be sprayed onto two cultivars at one time (split plots). The treatments may be placed in a randomized block design. Narrow, unsprayed spreader rows (two-cell wide beds) should be placed between adjacent treatment plots to increase the establishment and spread of disease in the test area. A row across the bed may be left vacant on each side of a treatment plot to facilitate placing thin boards, 0.6 m × 0.9 m × 6.5 mm (2 ft × 3 ft × 1/4 in.), on each side of a plot to control spray drift to adjacent plants. Stakes in appropriate positions are needed to support the boards.

**Fungicide application**—The first fungicide should be applied either before the plants are inoculated (protectant test) or 24–72 hr after the plants are inoculated (eradicant test). Subsequent sprays should be applied weekly for the entire season. Variations can be made to suit individual needs. Fungicides are conveniently applied with a $CO_2$-pressured hand sprayer at 380 kPa (55 psi) of pressure. Extreme care should be taken to apply sprays to the lower and upper surfaces of the leaves (infection occurs through both), especially those near the soil where conditions are best for disease development and spread.

Cultivars suitable for evaluating fungicides are Mrs. Roy or any of the Albatross or Indianapolis cultivars. These are standard types (large blooms) of chrysanthemums that are susceptible to Ascochyta blight. Mrs. Roy is so susceptible that under ideal conditions for disease development, plants may be killed. The other cultivars mentioned are generally less susceptible, and fewer plants are killed under ideal conditions for disease development. Stem and leaf lesions are common with these cultivars. The foliage and stems of Iceberg cultivars are too tolerant to be used satisfactorily.

**Standard fungicide**—A standard treatment is benomyl 50W at 300 mg/L (0.25 lb/100 gal) of water tank-mixed with either zinc ion maneb 80W, captan 50W, or chlorothalonil 75W at 900 mg/L (0.75 lb/100 gal), respectively. Mixtures containing benomyl will control Ascochyta blight when applied as late as 72 hr after inoculation. Tolerance to benomyl is not known to occur when combinations are used consistently. Other standard treatments that may be used are captan 50W, chlorothalonil 75W, or zinc ion maneb 80W (or any equivalent) at 1,800 mg/L (1.5 lb/100 gal) of water.

**Rating system**—Plants are rated at harvest by cutting the plants in the plots and counting individual lesions. The percentage of disease control may be obtained using the following formula:

$$PDC = \frac{\text{Total lesions in control} - \text{Total lesions in treatment}}{\text{Total lesions in control}} \times 100$$

where PDC is percentage of disease control. The weights of the plants or the numbers of diseased plants or both may be used as further indicators of the effect of disease on plant production. The data should be subjected to adequate statistical procedures.

**Phytotoxicity**—The crop should be carefully inspected before and after each fungicide application for any indication of chemical phytotoxicity. Some systemic chemicals may not show phytotoxicity symptoms until eight days after application. Common symptoms include chlorotic or necrotic leaf margins, white or necrotic areas on leaves, and necrotic spots and tipburn on petals.

### Literature Cited

1. ENGELHARD, A. W. 1973. Chrysanthemum, Ascochyta blight (*Mycosphaerella ligulicola*). Fungicide Nematicide Tests 29:102.
2. ENGELHARD, A. W. 1970. Chrysanthemum, Ascochyta blight (*Mycosphaerella ligulicola*). Fungicide Nematicide Tests 26:112.
3. HADLEY, G., and J. P. BLAKEMAN. 1968. Asexual sporulation of *Mycosphaerella ligulicola* in relation to nutrition. Trans. Br. Mycol. Soc. 51:653-662.
4. McCOY, R. E., R. K. HORST, and A. W. DIMOCK. 1972. Environmental factors regulating sexual and asexual reproduction by *Mycosphaerella ligulicola*. Phytopathology 62:1188-1195.

## SECTION III.

# E. Seed Treatments

# 33. Techniques for Evaluating Seed-Treatment Fungicides[1,2]

**Earl D. Hansing**

Professor of Plant Pathology, Kansas State University, Manhattan, KS 66506

Treating seed with fungicides to control crop disease is an important phase of plant pathology (4). Many plant pathologists test fungicides in the field as seed treatments. Since techniques for testing vary, they need to be improved and standardized so that test results can be used more effectively.

Seedborne pathogens are of special concern to growers because of the risk of introducing the causal agent into a new area. Seedborne pathogens become established early, when plants are often most vulnerable. In addition, their presence throughout the growing season greatly increases the probability of an epidemic. Because seed treatments prevent or reduce

[1]Reprinted, with modifications, from Fungicide Nematicide Tests 30: 1-3, 1974.
[2]Contribution No. 662-T, Kansas Agricultural Experiment Station.

expected losses from seed decay, seedling blights, damping-off, and other diseases caused by seed-inhabiting and soil-inhabiting organisms, more acres of cultivated crops are planted each year with fungicide-treated seed than with seed treated with any other pesticide. In the United States, more than 99% of corn (*Zea mays* L.), cotton (*Gossypium hirsutum* L.), and sorghum (*Sorghum vulgare* Pers.) seed and various percentages of seed from other agronomic, horticultural, and forest crops are treated before being planted. The percentage treated depends largely on the expected economic benefits from treatment.

Seed may be infested or infected with a disease-causing organism. For example, pathogenic fungi on cereals may be on or in the lemma and palea, on or in the caryopsis seed coat, or in the meristematic tissue of the embryo. They may also be mixed with seed in the form of sclerotia, smut balls, or infested plant parts. Seed are treated to reduce all such sources of infection. Seed also are treated to protect them from infection with such soilborne fungi as *Fusarium*, *Helminthosporium*, *Pythium*, and *Rhizoctonia* spp., which also may cause seed decay and preemergent and postemergent seedling blights. Fungicides kill or inhibit pathogenic fungi in the soil near treated seeds.

Many fungicides are available as seed treatments for testing and commercial use. They may be organic or inorganic, mercurial or nonmercurial, volatile or nonvolatile, systemic or nonsystemic. The treatments are formulated and applied to seed as powders, wettable powders, liquids, or flowables or by the Evershield® (Cargill Incorporated, Minneapolis, MN 55402) process. Methods and techniques discussed here are restricted to fungicide tests with true seed.

**Selecting a cultivar**—The cultivar selected for an experiment is important. Ideally, it should be so highly susceptible that nearly 100% of the plants from nontreated seed become infected. For example, under favorable environmental conditions when Red Chief wheat (*Triticum aestivum* L.) seed infested with bunt (*Tilletia foetida* [Wallr.] Liro) is planted, 90% of the spikes subsequently produced may be bunted. Using an effective seed treatment fungicide reduces bunted spikes to 1% or less. This gives a researcher a range of 0 to 90% for determining the effectiveness of many seed treatment formulations at different rates. For most diseases, however, plant cultivars are less susceptible than in this example. A cultivar with a maximum of 60% susceptibility is preferred to one with 30%, and one with 30% susceptibility to one with 10%. In any case the researcher must develop procedures to ensure adequate disease controls.

**Seed**—Viability and vigor are highly important, especially in experiments involving seed decay and seedling blight. Experiments with seed of high germination and strong vigor that are planted under highly favorable conditions may not show significant differences between treatments. Thus, experiments should be conducted with seed of average germination and vigor to approximate field conditions more closely.

Using uniform seed from the same source in all seed-treatment experiments is important, particularly in tests involving seed decay, seedling blights, and recording yields of forage or grain. Percentages of germination and moisture near the top of a bushel (35.2 L) of seed that has been sacked for some time may differ significantly from those in the middle or bottom. Similarly, vigor of the seed may vary. Seed for experiments must be thoroughly mixed. One method is to transfer 0.95 L (1 qt) of seed at a time to two large plastic containers, thoroughly mix seed in each container, simultaneously pour the contents of both containers into a third container, thoroughly mix, and repeat the entire process at least five times. Some seed such as peanuts (*Arachis hypogaea* L.) and soybeans (*Glycine max* [L.] Merr.) must be handled carefully to avoid damage during the mixing procedure.

Naturally infested or infected seed is preferred, but the inoculum level must be considered carefully. A low level of infestation will not produce sufficient infection for evaluating treatments, but a high infestation may require impractical or injurious quantities of fungicide or both to effect control.

Large quantities of seed are preferred to small quantities. They may be in units of weight or volume. For example, when seed is treated at a rate equivalent to 28.3 g (1.0 oz) of formulated fungicide per bushel (35.2 L), using 500 ml of seed is more accurate than using 100 ml.

**Fungicide rates for initial evaluation**—Including one well-known standard treatment at the recommended rate is advisable in all seed-treatment tests. Including treatments at one half and twice the normal rates also may be advisable. Rates for entries should differ by a factor of at least two. For example, rates of 14.2, 28.3, and 56.7 g of active ingredient (ai) per 35.2 L (0.5, 1.0, and 2.0 oz ai/bu) may give early information of phytotoxicity and limits of fungicidal activity. Although rates are based on active ingredients, appropriate formulation rates are preferred in reports. The formulation of the fungicide, the amount of product used, and the quantity of seed treated in tests, however, always should be stated clearly.

**Mixing fungicide with seed**—The fungicide should be mixed with the seed as uniformly as possible, either with a mechanical mixer or by hand. Many workers mix by hand, as follows: A clear glass jar is used to observe when most of the fungicide is on the seed. The volume of the jar should be about four times that of the seed, eg, a 2-L jar for 0.5 L (2-qt jar for 1 pt) of seed. When the seed and product are in the jar, it is capped quickly. With hands on the bottom and top of the jar, it is rotated rapidly at a 45-degree angle, first clockwise and then counterclockwise. The jar is inverted and rotation is continued until the glass is relatively clear. The treated seed then may be poured into a double heavy-duty paper sack. Nontreated seed may also be mixed and used as a control to assess mechanical injury to the seed.

For powder formulations, the seed is placed in the jar, the calculated amount of powder scattered over the top of the seed, and all mixed as described.

Wettable powder formulations must not be applied directly to seed, because some will receive more fungicide than others. Wettable powders should be mixed with the correct amount of water in a beaker and agitated with a magnetic stirrer to prevent settling. The jar is held at a 45-degree angle, the correct amount of suspension is added by pipette to one side of the jar, the jar is rotated 180 degrees, and the seed is added while the jar is held at a 45-degree angle. Mixing is done as described above.

For liquid formulations, the jar is held at a 45-degree angle and the correct amount of liquid fungicide is added by pipette on one side of the jar. The jar is rotated 180 degrees and is held at a 45-degree angle while the seed is added. Mixing is done as described above. When less than 28.3 g (1 oz) of product per 35.2 L (1 bu) is to be applied and the fungicide is miscible in water, uniform distribution may be accomplished by diluting one part fungicide to four parts water and applying five times the quantity to compensate for the dilution. In both cases, the jar must be rotated rapidly, because the first seeds to come in contact with the liquid fungicide may absorb more fungicide than may those contacted later.

For flowable formulations, which are viscous suspensions containing 30–50% ai, the correct amount is weighed rather than metered by pipette. With the jar held at a 45-degree angle, the suspension is scraped from the pan to the side of the jar with a scoopula. The jar is rotated 180 degrees while being held at a 45-degree angle; then the seed is added and mixed as described above. If adding all of the formulation from the weighing pan to the seed is difficult, more formulation may be weighed and the remaining weight checked. For example, if 1.25 g is to be applied, one can start with 1.60 g. The fungicide is added to the jar with a scoopula until the balance on the pan is 0.35 g. Material must not be poured from the pan to the seed, because the suspension settles slightly while being weighed.

To compensate for loss by the fungicide's sticking to containers, the theoretical amount of formulation to be used must be increased by 2.8 g/35.2 L (0.1 oz/bu). Flowable fungicides give better coverage of seeds than do wettable powders and liquid formulations at low rates. They create essentially no chemical dust during application, and when dry, create less dust than do wettable powders. A flowable formulation should be mixed continuously and kept agitated with a magnetic stirrer.

Evershield is a patented formulation registered with the Environmental Protection Agency for seed treatment of all crops. It is a stable, cream-colored emulsion containing 48% solids and a binder to inhibit dust formation. The fungicide adheres to the seed, and although the coating breaks down after treated seed is planted, the fungicide remains close to the seeds in the soil, thus providing more protection during early stages of growth. Dye may be added and mixed with the dilution.

For Evershield formulations, the concentrates are kept thoroughly mixed with a magnetic stirrer and diluted with water (1:5); the proper amount is pipetted as with liquid formulations. Seed should be added after the suspension and mixed as described above. To compensate for loss of fungicide in pipetting and by adherance to containers, the theoretical amount of formulation to be used must be increased by 10%.

Most fungicides can be formulated in an Evershield concentrate, resulting in 30% ai by weight. Since Evershield formulations are compatible with one another, combinations of two or more fungicides may be applied to the seed.

Volatile formulations of fungicides are applied by the same methods as for nonvolatile fungicides. The treated seed, however, should be left in the airtight jar 5–10 min and then aerated in a double heavy-duty paper sack for 24–48 hr before being planted. Aeration is especially important when the lemma and palea remain attached to each seed as they do with most cultivars of barley (*Hordeum vulgare* L.), oats (*Avena sativa* L.), and grasses.

**Pelleting seed**—Pelleting is another method used to apply fungicides to seed when the desired amount of seed treatment exceeds the amount that adheres completely to a given amount of seed. Pine (*Pinus* spp.) and other conifer seed are pelleted primarily to protect them against soil fungi; pelleting also has been used to protect onion (*Allium cepa* L.) seed against the soilborne smut fungus *Urocystis cepulae* Frost (4).

Methyl cellulose may be prepared as follows: The weighed quantity of methyl cellulose is placed in a flask and dried at 110°C for several hours. Then 40% of the final volume of water at 80–90°C is added during vigorous stirring. When the particles of methyl cellulose are wetted thoroughly, 50% of the final volume of water is added at room temperature and vigorous stirring is continued until the mixture cools. The mixture is stored at 4–5°C until the methyl cellulose is completely dispersed (about 12 hr). Finally, water is added by weight to being the final solution to the desired concentration.

The amount of sticker (eg, methyl cellulose) required per 0.45 kg (1 lb) of seed depends on size of seed, sticker used, and amount of fungicide to be fixed to the seed. To pellet 0.45 kg (1 lb) of red pine (*P. resinosa* Ait.) seed, a jar is held at a 45-degree angle and 56.7 g (2 oz) of 4% methyl cellulose is added to one side of the jar. The jar is rotated 180 degrees, and seed is added and mixed as described above. The process is repeated using the desired amount of fungicide powder. The fungicide should be uniformly mixed with cellulose as the seeds are coated.

**Mercurial fungicide**—Although methyl and ethyl mercurial fungicides no longer are approved for commercial use as seed treatments for most crops, they are valuable as standard treatment checks, because they are effective and have been used for many years. Most of them have broad-spectrum activity, while nonmercurial organic fungicides are more specific, ie, some are active against the lower fungi and others against the higher fungi. Mercurial fungicides are used at much lower rates than are nonmercurial fungicides. Cyano(methylmercuri)guanidine was used to treat wheat seed at 0.5 g ai/35.2 L (0.016 oz ai/bu). Nonmercurial organic fungicides are generally used at 14.2–28.3 g ai/35.2 L (0.5–1.0 oz ai/bu) of seed.

**Systemic fungicide**—During the first half of this century, the only seed treatment to control loose smut of wheat caused by *Ustilago tritici* (Pers.) Rostr. and loose smut of barley (*U. nuda* [Jens.] Rostr.) was the hot water method. It was not widely used, because temperature was difficult to control and drying facilities were required. Various modifications of the water-soak method were developed after 1950, but quick drying was still essential. Systemic fungicides were first used as seed treatments for wheat about 1965 (1). After seeds germinate, the seedling tissues take up the chemical.

Carboxin inhibits development of several systemic diseases. It now has federal registration for use as a seed treatment for a few of these crops. It not only will control loose smut of wheat and barley but also is an excellent replacement for volatile mercurial fungicides to control *U. avenae* (Pers.) Rostr. and *U. kolleri* Wille, which cause loose smut and covered smut of oats, respectively (2), and *U. hordei* (Pers.) Lagerh. and *Helminthosporium gramineum* Rabh., which cause covered smut and stripe of barley, respectively (3). Systemic fungicides may be applied by methods similar to those used for nonsystemic fungicides.

**Combinations of fungicides and insecticides**—Organic nonmercurial fungicides developed during the last 35 years are more specific than are ethyl and methyl mercurial fungicides. Some nonmercurial fungicides are effective against the lower fungi and others against the higher fungi. Combinations of two or more selected nonmercurial fungicides are often active against a broad range of pathogens. Some combinations are registered for commercial use. They may be applied to seed using methods similar to those used to apply one fungicide. Insecticides as seed treatments should be evaluated in combination with fungicides, because insecticides may predispose seedlings to seed decay and seedling blights. The organic nonmercurial fungicide may be used at 14.2 to 28.3 g ai/35.2 L (0.5–1.0 oz ai/bu and an insecticide (eg, lindane), about 7.1 g ai/35.2 L (0.25 oz ai/bu). This dosage of insecticide controls insects under most storage conditions before planting and during light infestations in the field immediately after planting. For high infestations of insects in the field, additional insecticide may be applied to seed in the drill box.

**Phytotoxicity**—A good seed treatment leaves no residues that are detrimental to seed germination, seedling development, plant growth, or final yield of foliage or seed. Phytotoxicity may be determined by observing the germination of seedlings. Generally, if a product is phytotoxic, about 50% of the seedlings emerge a day to a day and a half late, and stands are reduced. In some cases leaf tips turn yellow. In addition, the plants should grow to maturity to observe possible reductions in height and to obtain yield data. Because some diseases also reduce the percentage of emerged seedlings, relatively noninfested seed and soil should be used to test fungicides for phytotoxicity. In addition to the regular treatments of 14.2 and 28.3 g ai/35.2 L (0.5 and 1.0 oz ai/bu), it is well to have treatments at 56.7, 113.4, and 226.8 g ai/35.2 L (2, 4, and 8 oz ai/bu). Often a phytotoxicity test may involve two replications of 100–200 seeds per plot. The active or inert materials or both may cause phytotoxicity. In general, liquid formulations are more likely to be phytotoxic than are other formulations. Therefore, first evaluating a new fungicide as a powder is advisable.

**Soils**—It is important that sites for seed treatment tests have uniform soil, topography, and cropping histories, because these factors can influence the activity of soilborne pathogens and the performance of seed treatment chemicals. In field tests with seed treatments for control of soilborne diseases, soils are usually not artificially infested, because most of these pathogens are ubiquitous in cultivated soils. Many procedures, however, enhance infection and disease (eg, minimum tillage, incorporating organic matter, fertilization, irrigation, and flooding).

**Planting**—In general, crops used in seed treatment experiments should be planted when environmental conditions favor the disease, not the crop. This procedure accentuates the value of seed treatments compared with controls and permits better evaluation of treatments. The incidence of disease is higher when seed is planted slightly deeper than normal. For systemic diseases such as wheat bunt and barley stripe, deeper planting gives the fungus additional time to become established in the meristematic tissue of the seedlings. Infected seedlings must survive, however, so that experimental data may be taken as the crop matures. For these experiments, researchers should try for a balance between high infection and seedling survival.

In seed decay and seedling blight tests of sorghum, however, the seedlings need not survive. Most seedlings die from either seed decay or preemergent seedling blights. Data are collected when about 50% of the seedlings are in the two-leaf stage. Generally, planting slightly deeper than normal gives the fungi more time to infect and destroy the sorghum seedlings. One should attempt to obtain a 50% decrease in stand for the control plot.

One planting date is usually adequate for seed treatment experiments, but for certain diseases (eg, wheat bunt, damping-off), soil temperature and moisture for the first few days after planting are critical in determining disease level. Therefore, two or more planting dates eliminate unfavorable soil conditions.

**Replications and plots**—The number of replications depends on the experiment conducted, susceptibility of cultivar, range in expected data, and soil variations. For example, when Red Chief wheat seed is infested with bunt teliospores and then planted, 90% of the spikes subsequently produced may be bunted. Treating the seed with an excellent fungicide reduces the infected spikes from 90% to less than 1%. In this case, only three replications are required. Experiments with seed decay and seedling blights of the same crop, however, require at least five replications. When seed is treated before planting, 75% of the emerged seedlings may be observed compared with 60% when the seed is not treated. Replications should not be reduced to test more formulations or rates.

Plot size depends on the crop, disease, and data desired, eg, number of emerged seedlings, infected seedlings, infected heads, and yield. Differences expected between the control and treated plots also affect plot size—the greater the difference, the smaller the plot needed. For control of bunt of wheat, the plot should be a 3-m (10-ft) row with 0.3 m (1 ft) between rows; for covered kernel smut of sorghum (*Sphacelotheca sorghi* [Lk.] Clint.), it should be a 12-m (40-ft) row with 0.6 m (2 ft) between rows. For yield of barley or wheat, it should be four or six rows 3.7 m (12 ft) in length. The middle two or four rows should be harvested to minimize border effects from adjacent plots. Size of plots should not be reduced to test more formulations or rates.

**Cooperative regional or individual tests**—Regional and individual tests have advantages and disadvantages. Regional tests include the same cultivars, seed lots, formulations, and rates at several locations to measure variations in soil type, soil moisture, and temperature during and after planting. The data are obtained from a number of environments on specific treatments, chemical formulations, and rates. Individual experiments permit a researcher to use several cultivars, seed lots naturally infected with various fungi, many chemical formulations with various application rates, and different methods of treating seed. Both cooperative regional tests and individual tests should be continued.

**Publishing seed treatment data**—In general, percentages of emerged seedlings, tillers, final stand of plants, infected seedlings, infected plants, and infected heads should be expressed in whole numbers. Figures rounded off to one or two decimals contribute little additional information compared with whole numbers. Furthermore, comparing whole numbers is much easier for readers. Yields of grain or foliage may be expressed in either whole numbers or with one decimal, whether kilograms or metric tons per hectare (bushels, hundred weights, or tons per acre) are used. Variations are too great to use two decimals.

Soil type and conditions should be stated, eg, organic matter content, pH, moisture, and temperature, as well as recent cropping history and weather conditions during the test. Seed treatment data should be analyzed statistically to show the magnitude of significant differences between treatment means. Duncan's multiple range and least significant difference (LSD) tests are commonly accepted methods (5).

### Conclusion

Extension and research and development personnel involved in seed treatment testing may follow the above suggestions, but they will have to make modifications depending on the crop, cultivar, disease, and type of data desired. Each worker benefits from experience in developing and improving seed treatment tests to meet individual needs and to satisfy professional standards and legal requirements.

### Caution

As with all pesticides and chemicals, seed treatment fungicides must be used with caution. All safety precautions and directions on the label of the container or accompanying fact sheet that the manufacturer provides should be followed. Direct contact with the chemical can be avoided by wearing impermeable gloves, working in a vented chemical hood or a well-ventilated room, and using a bulb with a pipette. All treated seed should be identified (preferably with a dye) and safely stored in a locked room to prevent accidental human or animal consumption.

### Acknowledgment

Suggestions from colleagues are acknowledged and gratefully appreciated.

### Literature Cited

1. HANSING, E. D. 1967. Systemic fungicide for control of loose smut (*Ustilago tritici*) of winter wheat. Phytopathology 57:814.
2. HANSING, E. D. 1970. Control of seed-borne fungi with systemic fungicides. Proceedings of the Fifteenth International Seed Testing Association, Palmerston North, New Zealand, November 1968. 35:655, 660-662, 815-820.
3. HANSING, E. D. 1973. Carboxin, a replacement for methyl mercurial fungicides to control stripe (*Helminthosporium gramineum*) of barley (*Hordeum vulgare*). Second International Congress of Plant Pathology 2:0343.
4. HANSON, E. W., E. D. HANSING, and W. T. SCHRODER. 1961. Seed treatments for control of disease. In Seeds, the Yearbook of Agriculture. U.S. Department of Agriculture, Washington, DC. pp. 272-280.
5. LeCLERG, E. L., W. H. LEONARD, and A. G. CLARK. 1966. Field Plot Technique. Ed. 2. Burgess Publishing Co., Minneapolis.

# 34. Screening Fungicides for Seed and Seedling Disease Control in Plug Mix Plantings[1]

### Ronald M. Sonoda

Associate professor of plant pathology, University of Florida, Institute of Food and Agricultural Sciences, Agricultural Research Center, Fort Pierce, FL 33450

Seedling diseases are an important factor in reducing plant stand, uniformity of growth, and yield in many crops. *Pythium aphanidermatum* (Edson) Fitzp., *P. myriotylum* Drechs., *P. arrhenomanes* Drechs., and *Rhizoctonia solani* Kuehn frequently incite seedling diseases of tomatoes in south Florida (6,7,8). Except for fumigation with expensive wide spectrum biocides, satisfactory control of these pathogens has not been possible in direct field-seeded plantings.

The recent introduction and expanding commercial use of plug mix planting (3), a direct field-seeding technique, has provided an efficient, simple means by which excellent control of these diseases can be obtained inexpensively (8). By the plug mix method, a 1:1 mix of horticultural vermiculite and shredded sphagnum peat

moss containing seed and starter fertilizer (to which fungicide can be added) is deposited at desired intervals with a hand-operated dispenser or a tractor-drawn mechanical planter (3). The method provides quicker-growing, more uniform plants (3) and greater yields (2) than when seed alone are placed in soil. My observations indicate that when fungicides are not used, *Pythium*-incited diseases are more severe in plug mix plantings than in plantings made directly in soil. Perhaps the high moisture-holding ability of the planting mix affects disease development. The horticultural benefits of this method of planting and its usefulness in disease control, however, greatly outweigh this disadvantage.

By incorporating effective chemicals into the planting mix, a fungicide-protected, pathogen-free environment is provided for seeds and young seedlings. A small

[1]Florida Agricultural Experiment Stations Journal Series No. 314.

amount of fungicide, eg, about 10 g active ingredient (ai) of Bay 22555 (sodium p-[dimethylamino] benzenediazosulfonate) (Dexon) or 40 g ai of ETMT (5-ethoxy-3[trichloromethyl]-1,2,4-thiadiazole) (Truban) per hectare of a 7,500-plant per hectare crop (one-plant per hill) effectively controlled Pythium root and stem rot of tomato in previously uncropped soil (8).

The techniques described below, used for screening tomato fungicides for seedling disease control and phytotoxicity, can be used to determine efficacy of candidate fungicides on other plug mix-seeded crops.

## Techniques

**Sites**—Initial screening of candidate fungicides for phytotoxicity to tomato and efficacy against seedling pathogens were conducted in the greenhouse. Oldsmar fine sand from uncropped virgin soil areas was used for the phytotoxicity tests. The virgin soil received dolomitic lime at the equivalent of 1.1 MT/ha and 4-16-4 fertilizer at 0.67 MT/ha. Virgin soil and previously cropped soil artificially infested with pathogens were used in tests determining efficacy of the fungicides.

Field tests were conducted in plots at the Agricultural Research Center, Fort Pierce, FL (ARC-FP), and on commercial farms near the ARC-FP in soils suitable for growing tomatoes. At the ARC-FP, previously cropped, nonfumigated Oldsmar fine sand was disked thoroughly before beds were formed. Preplant fertilizer (4-16-4) was applied to the beds as they were prepared. For tests in which fruit yields were desired, bands of 8-12-20 fertilizer were placed parallel to the bed length under strips of black polyethylene mulch to reduce leaching (4). On commercial farms, virgin soils and soils previously cropped with vegetables were used in the tests. In most cases, the cropped soils had not been used for several years. The uncropped soil was prepared with minimum disturbance of native sod to reduce bed erosion. Beds were generally made and seeds planted by plug mix planting before plant residues were thoroughly decomposed—the common practice with area growers.

Losses to tomato seedling diseases are heaviest during the late summer planting period. The field tests at the ARC-FP were conducted from May through October. Field tests in commercial plantings were conducted from early August through the middle of September, coinciding with periods of high rainfall and high temperatures. At the ARC-FP, the tests followed recent plantings of tomatoes to increase the likelihood of diseases.

**Test plant**—Seeds of tomato (*Lycopersicon esculentum* Walter) were used throughout the tests. The seeds, which were obtained from a commercial supplier, had not been treated with pesticides.

**Planting mix**—Jiffy Mix Plus, a commercial 1:1 mix of horticultural vermiculite and shredded sphagnum peat moss plus nutrients similar to Cornell Peat-Lite Mix A (1), which was used as the planting mix in developing the plug mix technique of planting (3), was used as the carrier for the fungicides in the following tests.

**Preparation of pathogen-infested soil**—Several isolates of *P. aphanidermatum*, *P. myriotylum*, and *R. solani* and one isolate of *P. arrhenomanes* obtained from diseased tomato seedlings were grown for two weeks on twice-autoclaved browntop-millet (*Panicum*

*fasciculatum*) seeds. Thirty infested millet seeds were mixed with 100 g of soil (dry weight). The infested soil was placed in 169.8–339.6-g (6–12-oz) Styrofoam cups for greenhouse tests. Control treatments received the same number of autoclaved but noninfested millet seeds.

**Procedure in greenhouse tests**—Uninfested, limed, fertilized virgin soil was placed in 169.8–339.6-g (6–12-oz) Styrofoam cups for phytotoxicity tests. Infested, limed, fertilized virgin or previously cropped soil was placed in Styrofoam cups for tests to determine efficacy of candidate compounds. The compounds were suspended in distilled water and added to the planting mix. The amount of water used was enough to bring the moisture level of the mix to 60% of its total weight. Depressions were made on the surface of soil in the cups with a No. 7 rubber stopper. One tablespoon of mix containing the candidate compound was placed in the depression and tamped down. Five or ten seeds were placed on top, another tablespoon of mix was added, and it was tamped down again. Planting mix without fungicides served as controls. Standard treatments were 1.25 g ai of captan or 0.67 g ai of ETMT per kilogram of dry weight planting mix. The cups were placed in the greenhouse in a randomized complete block design. One-cup replicates were repeated four to six times. The cups were watered once or twice daily to maintain a high moisture level. In some tests, holes were made in the base of the cups for drainage, but in some trials, holes were not made in order to maintain a high moisture level. No other chemicals were sprayed on or added to plants in the greenhouse.

**Procedure in field tests**—Weighed amounts of seed were mixed with measured amounts of Jiffy Mix Plus in a portable cement mixer to provide an average of about five seeds per hill. This usually resulted in most hills having from two to ten seedlings (3). The planting mix was then divided into smaller batches, and the candidate fungicides were added in the same manner as described for greenhouse tests. Planting mix, 57 cm$^3$, was deposited with a hand-operated planter (3) at 15- or 61-cm (6-in. or 2-ft) intervals, depending on the purpose of the test. The shorter distance was used in tests requiring only disease control data, the 2-ft distance in tests requiring fruit yield. From four to six replications were in a randomized complete block design of 8–20 hills per replicate, depending on the experiment. Foliar sprays consisting of mancozeb, basic copper sulfate, dimethoate, and methomyl were applied at four- to five-day intervals soon after seedling emergence. In addition to seepage irrigation, tests at the ARC-FP were irrigated with about 1 acre-inch of water with a Rainbird sprinkler on days with little or no rainfall. Tests in commercial fields were watered by seepage irrigation only.

**Collecting phytotoxicity data**—All emerged seedlings were counted daily in the greenhouse phytotoxicity tests. The counts were made for five to seven days after the first seedling emerged. Visual ratings of the appearance of the seedlings were made once or twice during each experiment. The rating system used was 1, no difference in appearance from seedlings not exposed to fungicides, to 3, plants severely affected (distorted, dying, or stunted).

**Collecting fungitoxicity data**—Seedlings from infested soil in the greenhouse and field were counted daily for five to seven days after emergence of the first plant. Damped-off seedlings were counted each day for up to 20 days after planting. Each damped-off seedling was replaced with a toothpick to prevent confusion with seedling loss caused by late emergence. By 20 days after planting, seedlings were lifted and rated for root damage. The rating system used in most cases was 1 (healthy root) to 5 (plants damped-off). In some tests, damping-off and root rot were treated separately, and the rating system used for the root rot phase of the disease was 1 (healthy root) to 5 (severe root damage). Sometimes, seedlings were scored merely as diseased or healthy. The number of hills with all seedlings killed was also noted.

If yield data were desired, one or two seedlings per hill were left. Fruits were harvested ripe or mature green one to six times, depending on the experiment. Spot checks for fruit size and quality were made at harvest.

A record of the daily high and low greenhouse temperatures was kept. Daily high and low temperatures and hourly rainfall data were collected for field experiments at the ARC-FP. Rainfall data for tests in commercial fields were obtained when possible.

**Summarization of data**—An emergence index (mean number of days required for seedling emergence) was calculated for seedling emergence in phytotoxicity and fungitoxicity tests. The mean number of days required for emergence (EI) for each replicate was expressed with the following formula:

$$EI = \frac{\sum\limits_{i=1}^{n} N_i D_i}{T}$$

where N is number of seedlings emerging in each 24-hr period; D, days after planting that seedling count was made; and T, total of seedlings emerged in each replicate. The mean emergence ($\overline{EI}$) for a treatment was obtained by determining the mean of the EIs for all replicates of the treatment.

Percentage of emergence per replicate and mean percentage of emergence for treatments were determined in tests in which preemergence damping-off was a problem. When a root rot index was desired, it was calculated for each hill by determining the mean disease rating for all seedlings in the hill. The index for each replicate was the mean of the hill indexes, and for each treatment, the mean of the replicate indexes. When data for damping-off and root rot were separated in the same test, a mean percentage of damping-off was calculated in a similar manner. Likewise, a mean percentage of root rot was calculated for all replicates in some tests. A mean percentage of hills with no plants was calculated for each treatment also.

An appearance index was calculated for each treatment by determining the mean appearance rating for all plants in a hill, then all hills in a replicate, and finally all replicates of a treatment.

All data were subjected to analysis of variance. All percentages were transformed to angles (5) before analysis of variance. Mean comparisons were made by Duncan's multiple range test.

## Results and Discussion

General phytotoxicity symptoms included reduced seed germination, reduced growth of seedlings, and abnormal growth or coloration of cotyledons, leaves, stems, and roots. Different compounds and levels of compounds produced different combinations or intensities of symptoms. Abnormal foliage coloration ranged from dark green to yellow. Stems were twisted in some cases. Roots were smaller than normal; were twisted, swollen, or discolored; or had necrotic tips. Sometimes postemergence seedling death occurred. Usually seedlings in treated planting mix grew more slowly than did seedlings in untreated mix with no other apparent symptoms. In a few cases, the rate of growth was drastically reduced, but plants appeared to be otherwise healthy.

Compounds or rates of compounds that produced marked phytotoxic symptoms on seedlings in greenhouse tests were eliminated from further testing. Rates causing few or no phytotoxic symptoms were evaluated in the greenhouse for control of *Pythium* spp. and *R. solani*.

Under greenhouse conditions, disease incidence was consistent, sufficient for adequate determination of differences between treatments with three or four replicates of one Styrofoam cup each. Disease severity was high in untreated planting mix and in planting mix with a fungicide that was ineffective against the pathogen. Rates of compounds showing moderate to excellent control and little or no phytotoxicity were selected for use in the field.

In tests in the field at the ARC-FP, the most active pathogens were *P. myriotylum* and *P. aphanidermatum*. *R. solani* affected only a few seedlings. The disease severity was such that four to five replications of eight hills each were usually adequate for evaluation of treatments.

Disease incidence in commercial fields was low unless rain occurred at the time of planting or within a few days after planting. Only two of seven test plantings in growers' fields in the last four years had enough disease pressure to determine differences between treatments. The two successful tests were on virgin soil in which the main pathogen was *P. myriotylum*.

Eight- to ten-hill replicates gave consistent results for fruit yield in tests at the ARC-FP, but fruit yields in commercial growers' fields were erratic, even with 14–20 hills per replicate. The lack of uniformity was due to growers' farming practices in the area near the ARC-FP. Commercial fields were prepared with a minimum of land leveling. Compounds and rates of compounds causing decreased yields or delayed fruit production in the field were eliminated. No effect on fruit size or quality has been noticed with any of the materials selected for field testing thus far.

No fungicides are currently registered for use in plug mix planting. Some of the data obtained in the tests above have been submitted to the IR-4 committee to support registration of fungicides for this use.

Phytotoxicity and efficacy data for a fungicide obtained by the procedures outlined above probably will be valid only if the particular combination of planting mix ingredients used in the screening is to be used. The planting mix for most commercial plug mix planting is similar to the mix used in our tests. Due to variance in binding of fungicides by different ingredients in planting mixes, however, changes in the composition of planting mixes require rescreening of fungicides.

## Literature Cited

1. BOODLEY, J. W., and R. SHELDRAKE, JR. 1972. Cornell Peat-lite mixes for commercial plant growing. Information Bulletin, 43. New York College of Agriculture, Cornell University: Ithaca, NY.
2. BRYAN, H. H., N. C. HAYSLIP, P. H. EVERETT, and W. W. DEEN, JR. 1973. Effect of "Plug Mix" seedling and mulch methods on yield and quality of tomatoes grown on calcareous soils. Proc. Am. Soc. Hortic. Sci. 17:333-345.
3. HAYSLIP, N. C. 1973. "Plug Mix" seedling developments in Florida. Proc. Fla. State Hortic. Soc. 86:179-185.
4. HAYSLIP, N. C., and J. R. ILEY. 1966. Use of plastic strips over fertilizer bands to reduce leaching. Proc. Fla. State Hortic. Soc. 79:132-139.
5. SNEDECOR, G. W. 1956. Statistical methods. Ed 5. Iowa State College Press. Ames, IA.
6. SONODA, R. M. 1973. Occurrence of a Pythium disease in virgin sandy soils of south Florida associated with a new method of field seeding tomatoes. Plant Dis. Rep. 57:260-261.
7. SONODA, R. M. 1975. Control of damp-off in tomatoes planted by Plug-Mix method in previously cropped soil. Ft. Pierce ARC Research Report RL-1975-1.
8. SONODA, R. M. 1976. Incorporating fungicides in planting mix to control soilborne seedling diseases of plug-mix seeded tomatoes. Plant Dis. Rep. 60:27-30.

# 35. Control of Seed and Seedling Diseases of Cotton With Seed Fungicides[1]

### Earl B. Minton and C. D. Ranney

Plant pathologist and area director, respectively, USDA, Cotton Research Laboratory, Federal Research, Science and Education Administration, Southern Region, in cooperation with the Texas Agricultural Experiment Station at Lubbock, TX 79401, and Alabama-North Mississippi Area, Starkville, MS 39759

The Cotton Disease Council was established in 1936 in recognition of the economic importance of cotton diseases. Plant pathologists and industry representatives have convened almost every year since 1936 to discuss and plan methods for control of seedling diseases in cotton (*Gossypium hirsutum* L.). The Cotton Disease Council maintains a regional cottonseed treatment program, which coordinates evaluations of new fungicides. Seed treatments reduce seedborne and soilborne diseases, increase plant vigor, eliminate the need to replant, reduce losses from other diseases, and increase the yield and quality of the crop.

Other groups could use this type of program to determine the performance of fungicides and other pesticides. Under the auspices of the Cotton Disease Council, the agricultural chemical industry has cooperated with federal and state agricultural researchers. Through the cottonseed treatment program, they have developed and evaluated new seed-protectant and systemic fungicides. Consequently, several effective seed-treatment fungicides were available to processors when commercial use of all alkyl mercury fungicides was canceled in 1971.

## Techniques

During the early years of the regional beltwide program, each cooperator treated his own seed lots. It soon became apparent, however, that a common seed lot should be used for all evaluations. Each year, one common seed lot with 75–85% germination and infested with seedborne pathogens was used for all evaluations.

Since 1959, the seeds have been grown in Mississippi, and until 1971 were provided without charge by the Stoneville Pedigreed Seed Company, Stoneville, MS.

Each seed lot was machine-delinted; half of each seed lot was then acid-delinted before applying fungicides to both types of delinted seed. Since 1972, half of the seed lot has been machine-delinted, and the other half has been acid-delinted only. During the last three years, a second acid-delinted seed lot, grown in New Mexico or California, has been used for evaluations in Texas, New Mexico, Arizona, and California (the western states program). The seeds used in the regional and western states programs are representative in quality and variety of a high percentage of the seeds planted in their respective regions.

Chemical company representatives nominate fungicide treatments to be included in the regional evaluations during the annual meeting of the Cotton Disease Council. Many candidate treatment nominations, especially combinations of fungicides, involve two or more chemical companies. All treatments nominated must have performed as well as the standards in industry's field and greenhouse programs. Cooperators from Texas eastward, where cotton is grown, select treatments from those nominated, or may request additional treatments for evaluation. The chairman of the Seed Treatment Committee of the Cotton Disease Council evaluates all treatments.

The western states program began about ten years ago. Previously, only selected treatments already included in the regional program were evaluated in the western states program. Today the cooperators of the western states program select for evaluation 16–20 treatments that have increased stands significantly in the regional or other programs.

In both programs, packets of untreated seeds are

included as checks. A standard fungicide, 7.7% *N*-(ethylmercuri)-*p*-toluenesulfonanilide (Ceresan M®), has been used for many years and is still used. Since 1971, 22.8% pentachloronitrobenzene–11.4% 5-ethoxy-3-(trichloromethyl)-1,2-4-thiadiazole (PCNB-ETMT, Terra-Coat L-21®) has been used as a second standard. These entries are used to compare the effectiveness of other fungicides and to determine the progress being made in the development of more effective treatments.

The chairman of the Seed Treatment Committee applies the fungicides to small lots (1.6–6.4 kg, or 3.5–14.1 lb) of seed in a rotating drum treater. This procedure requires smaller quantities of fungicide and less labor than would be needed if each cooperator treated his own seeds. Emulsifiable liquid concentrate formulations are diluted with water to 2% by weight of the seed and then sprayed on the tumbling seed with an atomizer. Oil formulations are applied to the seed with an atomizer without dilution. Wettable powder formulations are mixed with water equivalent to 2% by weight of the seed; the fungicide suspensions are applied to the seed with a small hand sprayer, which disperses them into small droplets. Water is used to reduce the loss of fungicide while the seeds are treated and to insure uniform distribution. The seeds are tumbled in the drum treater for 5 min or more to assure uniform coverage and to aid in water evaporation. Subsamples of each of the treated and untreated seed lots are mailed to each cooperator.

One hundred seeds per plot are planted in a randomized complete block design. Each plot consists of one row 9.14 m (30 ft) long. The seeds are normally planted 10–14 days earlier than the recommended date of planting at each location. Early planting exposes seed and seedlings to conditions more favorable for pathogen than for host development. Maximum disease severity generally occurs under these conditions. Consequently, the most effective candidate fungicides are readily identified. Numbers of surviving and dead seedlings are recorded about 20 and 40 days after planting and are used to calculate seedling emergence and survival. The data are analyzed statistically to determine treatment differences for all test sites. Also in the western states program, a combined analysis of the data across all test sites is calculated to determine the overall average performance of the fungicides for all locations. In the western states program, significant differences can be obtained from both individual and combined analyses; this helps to identify effective treatments. Duncan's multiple range and least significant difference tests are used to determine significant differences at the 0.01 and 0.05 levels of probability among treatments.

The chairman of the Seed Treatment Committee assembles and summarizes the data annually for publication in *The Cotton and Gin Oil Mill Press* and *Proceedings of the Beltwide Cotton Production Research Conferences*. Since the same lots of treated seed are used for several locations, data from both individual and combined test sites are used to select effective treatments for further evaluations, while noneffective treatments are eliminated the first year. Due to extensive evaluations, effective treatments usually can be documented in three years. The policy of the Cotton Disease Council requires that all treatments included in the suggested list for commercial use be evaluated in the regional program for a minimum of three years. Rates of application in experiments are based on formulated products and are therefore easily translated into commercial use.

## Results and Discussion

The impact of these evaluations was apparent in 1971 after alkyl mercury fungicides for seed treatments were canceled. At that time, registered, effective, alternate fungicides were available to replace the alkyl mercuries for cottonseed treatments (1,6). Also, other commodity groups used the data from the cottonseed treatment program to select replacement seed fungicides. Industry uses the results from the regional evaluations to assess the performance of the fungicides across the cotton belt and to register them with the Environmental Protection Agency.

During the early years of this program, another area that received major emphasis was identification of organisms causing seedling diseases. The effects on disease development of seed placement or spacing within the row were also determined.

The level of disease control through seed treatment has been increased by combinations of two or more seed protectants and seed protectants plus systemic fungicides. Also, more effective treatments have been selected by statistical analysis of the data from both individual and combined tests sites (2,3). Treatments that control disease complexes over a wide range of environments, soil types, and organisms are more readily selected from the combined analyses from different test sites than from individual locations. A common seed lot must be used, and the evaluations for each treatment must come from the same batch of treated seed for the combined analyses to be most meaningful.

Conventional planting procedures representative for each area are used when planting seed for evaluation within specific areas. These procedures differ among the test sites. More precise or common planting techniques should allow for more accurate determination of differences among treatments (4). The effects of fungicides on seed germination and seedling vigor also should be determined (5,7). Seed quality can affect plant growth and should not be misinterpreted as a fungicidal response. Some variation in seed quality is desirable, since seeds of wide-ranging quality are used for planting. New planting techniques and other changes certainly will be developed in the future. Three years of evaluations were sufficient in the regional program, however, to test a wide range of climatic conditions, soil types, seed qualities, and microorganisms so that effective treatments could be selected for commercial use.

## Literature Cited

1. MINTON, E. B. 1974. Status of non-mercurial seed treatments, 1967-73. Proceedings of the Western Cotton Production Conference, 1974. pp. 5-9.
2. MINTON, E. B., and G. A. FEST. 1975. Seedling survival from cottonseed treatment experiments at several locations. Crop Sci. 15:509-513.
3. MINTON, E. B., G. A. FEST, and G. L. SCIUMBATO. 1975. Effect of cottonseed treatment on stand. Texas Agric. Exp. Stn. MP-1210c.
4. PINCKARD, J. A., and J. IVEY. 1971. Chemical treatments for cottonseed. La. Agric. Exp. Stn. Bull. 655.
5. PINCKARD, J. A., and D. R. MELVILLE. 1975. Some new

developments in cottonseed dressings. Plant Dis. Rep. 59:262-266.

6. RANNEY, C. D. 1971. Effective substitutes for alkyl mercury seed treatments for cottonseed. Plant Dis. Rep. 55:285-288.

7. RANNEY, C. D. 1972. Multiple cottonseed treatments: Effects on germination, seedling growth and survival. Crop Sci. 12:346-350.

# 36. Field Evaluation of Fungicides as Soybean Seed Treatments

## N. G. Whitney

Research plant pathologist, Texas Agricultural Experiment Station, Texas Agricultural Research and Extension Center, Beaumont, TX 77706

Soybean seed-treatment tests have been conducted at various locations in the United States since 1925 (4). In the southern states, the combination of poor seed quality and cold, wet weather after planting sometimes results in the loss of 20–25% of the potential stand if soybean seeds are not treated with fungicides. Under such conditions, yield increases from seed treatment have been demonstrated (3). This method is intended for field evaluation of the efficacy of fungicides used to control seedling diseases of soybeans. It provides guidelines for establishing test plots, treating and planting the seed, determining the data, and interpreting the results.

## Technique

**Test site**—Soil suitable for growing soybeans and climatic conditions favorable for disease development are necessary. Fungi causing seedling disease should be endemic to the test area. These include *Pythium, Phytophthora, Rhizoctonia,* and *Diaporthe* spp. Naturally occurring inoculum may not be present in sufficient quantities to assure uniform disease development throughout the experimental area. Consequently, culturing selected pathogenic fungi in the laboratory for incorporation into the soil of the research area may be necessary. For example, *Rhizoctonia solani* can be easily increased by transferring a pure fungal culture to an autoclaved moist rice or barley medium. Two to three weeks should be allowed for fungal growth. The inoculum should be incorporated into the test site at the rate of 40 g/m² of soil. Tuite (5) suggests other methods for increasing *Pythium* and *Phytophtora.* Seedborne fungi such as *Diaporthe* are usually found in planting seed (2,6). Cultural practices for the test site should approximate those of commercial producers.

**Seed treatment**—Any locally grown commercial variety is acceptable. Seed germination should range between 60 and 80% to reveal differences in treatments. Running a dosage series for each fungicide is advisable, to ascertain the effective rate for maximum disease control. The dosage series should range between 0.5 and 8.0 g of formulated product per kilogram of seed. Each fungicide is mixed with 10–15 ml of distilled water to form a slurry. The slurry is placed with 1 kg of seed and tumbled in a jar mill or agitated by hand in a plastic bag.

Slurry-treated seed should be allowed to dry for 24 hr, then counted and packaged for planting 40 seed per meter of row for individual plots. In hopper box treatments, the calculated amount of fungicide can be added to individual row packets and shaken before planting. A standard control treatment such as captan or thiram used at the recommended rate and untreated controls must be included to determine disease severity and to provide a basis for treatment comparisons. Precautions should be taken to avoid inhalation of or exposure to fungicide dusts during treatment, packaging, and planting of the seed. Commercial inoculant at 0.25 g per packet of treated seed should be added just before planting, with the following precautions: (i) the fungicide and inoculant should not be combined before being added to the seed, either in a slurry mix or hopper box treatment, and (ii) seed treated with both inoculant and fungicide should not be held for longer than 6 hr. If this occurs, seed should be reinoculated.

**Procedure**—The test design is a randomized complete block with at least four replications per treatment. More replications are needed in areas where disease incidence is light. A precision drop planter should be used to plant the correct number of seeds for the desired length of row. Each plot should be four rows wide and approximately 6 m long. Row width varies, depending on cultural practices of local producers. Under good soil conditions, seedlings should be ready to count in eight to ten days. Plants are then grown to maturity under standard management practices. The two center rows from each plot are harvested and yields calculated for each treatment.

**Data determination**—Emergence counts are determined by counting all healthy seedlings that are free of abnormalities in the two center rows of each plot. Abnormalities include shortened, thickened stems, which may or may not have deep lesions in them; shriveled or weakened stems; damaged or absent primary leaves; missing or moldy cotyledons; or missing terminal buds.

After the stand counts are recorded for each replicated treatment, an analysis of variance of the data is calculated. Duncan's multiple range test (1) is applied for significance at the 5% level. Effective fungicides must have significantly higher stand counts than does the untreated control. Other methods that help to clarify the data include (i) calculations of the percentage of stand for each treatment and /ii) the percentage of change from the untreated control, determined as follows:

$$\text{Percentage of change} = \frac{\left(\begin{array}{c}\text{Number of plants}\\\text{in treatment}\end{array}\right) - \left(\begin{array}{c}\text{Number of plants}\\\text{in control}\end{array}\right)}{\begin{array}{c}\text{Number of plants}\\\text{in control}\end{array}} \times 100$$

Plants should be observed for any type of phytotoxicity and the percentage of severity reported. Common types of phytotoxicity are leaf burning, chlorosis, or stunting. At maturity, yields are statistically analyzed to show significant differences between treatments.

The following information should be included in reporting test results: (i) Product name and EPA registration number, (ii) percentage and chemical identity of active ingredient or ingredients, (iii) treatment rate, (iv) method of application, (v) number of seeds per treatment, (vi) number of replications, (vii) treatment date, (viii) planting date, (ix) date on which emergence count was made, (x) percentage of stand, (xi) percentage of change from untreated control, (xii) emergence significance as determined by Duncan's multiple range test at the 5% level, (xiii) percentage of severity and type of phytotoxicity, and (xiv) yield and significance as determined by Duncan's multiple range test at the 5% level.

## Results and Discussion

In an area where fungal inoculum concentrations are adequate, this test yields a high degree of precision. Tests vary from year to year, however, because of environmental conditions. A dry year produces smaller differences than does a wet year, because some seedling disease fungi require water for movement on or through the soil. Humidity also plays a large role in fungal effects on host plants. Where necessary, overhead irrigation to establish high humidity is desirable. If inoculum levels are not high enough, artificially grown inoculum incorporated into the soil increases the precision of the test.

Data obtained from this type of test are reliable. Treatments showing an increased response over controls can be recommended to producers, provided the chemical has been cleared for use on soybeans with the Environmental Protection Agency.

## Literature Cited

1. DUNCAN, D. B. 1955. Multiple range and multiple F tests. Biometrics 11:1-42.
2. ELLIS, M. A., M. B. ILYAS, and J. B. SINCLAIR. 1975. Effect of three fungicides on internally seed-borne fungi and germination of soybean seeds. Phytopathology 65:553-556.
3. JOHNSON, H. W. 1951. Soybean seed treatment. Soybean Dig. 11(7):17-20.
4. JOHNSON, H. W., and B. KOEHLER. 1943. Soybean diseases and their control. U.S. Dep. Agric. Farmers Bull. 1937.
5. TUITE, J. 1969. Plant pathological methods fungi and bacteria. Burgess Publishing Co.: Minneapolis.
6. WALLEN, V. R., and T. F. CUDDY. 1960. Relation of seed-borne Diaporthe phaseolorum to the germination of soybean. Proc. Assoc. Off. Seed Anal. 50:137-140.

# SECTION IV.

# Nematicide Test Procedures

## 37. Guidelines and Test Procedures for Nematicide Evaluation[1]

Kenneth D. Fisher

Chairman, Joint SON/ASTM E 35.16 Subcommittee on Nematode Control Agents, and director, Life Sciences Research Office, Federation of the American Sciences for Experimental Biology, 9650 Rockville Pike, Bethesda, MD 20014

### Historical Background

In 1972, the Technical Services Division Office of Pesticide Programs, Environmental Protection Agency (EPA), was charged with the responsibility of enforcing rules and regulations related to pesticide use. Because no codified scientific basis existed for enforcing evolving rules and regulations on pesticide use, the Technical Services Division sought assistance in development of methods for assessing infractions of rules for safe, efficient use of pesticides. Since 1973, for various reasons, methods and procedures related to establishing efficacy and safety for registration rather than enforcement alone have become the primary goal.

EPA requested assistance in developing a series of guidelines on test methods from the American Society for Testing and Materials (ASTM). Founded in 1898, ASTM is the world's largest source of voluntary consensus standards. The organization relies on voluntary committees of scientists and engineers to develop standardized procedures of testing that the scientific discipline or industry involved can accept.

EPA approached ASTM instead of the respective scientific discipline societies, because ASTM has more than 75 years of experience as an organization with the sole purpose of developing voluntary consensus standards. With management assistance from ASTM, technical committees composed of producers, users (consumers), and general interest individuals are organized to provide scientific and technical input.

Specialists in specific disciplines prepare ASTM techniques for developing voluntary consensus standards on methods of testing, guidelines on use, or specific test procedures for use in those specific disciplines.

In January 1973, more than 300 specialists met in Philadelphia at the invitation of ASTM and EPA to discuss the formation of an ASTM Committee on Pesticides. The invitees represented most of the disciplines involved with pesticide development and use, from household disinfectants to vertebrate control agents and herbicides. Although ASTM's activities in developing consensus standards for the construction, textile, and various heavy manufacturing industries have been notably successful, these experiences had not prepared ASTM for analogous activities in the biologic sciences. The inherent nature of biologic processes, such as measurement of pesticide efficacy, is not easily delineated by rigid standard methods of testing and evaluation.

While the concept of rigid standard test methods was rejected, the group concluded that development of suggested procedures and recommended guidelines would be a useful activity. On this basis, a majority of those attending voted to establish the E 35 Committee on Pesticides. The organizational structure comprised 18 subcommittees, including 11 technical subcommittees covering each of the plant, animal, or human pest disciplines. The nematologists in attendance formed a Nematode Control Agents Subcommittee (E 35.16), which developed the following statement of its scope and purpose:

Development of standards, definitions, classifications, biological test methods, and

[1]A portion of this paper was delivered at the Symposium on Suggested Guidelines and Test Procedures for Nematicide Evaluation, August 1976, at the 15th Annual Meeting of the Society of Nematologists, Daytona Beach, FL.

recommended procedures for the determination of the effectiveness and usefulness of nematode control agents. These activities will be coordinated with related committees and subcommittees of the ASTM and with appropriate professional organizations and with industry.

After considerable discussion, the Nematode Control Agents Subcommittee concluded that they would ask the Society of Nematologists (SON) to establish an ad hoc committee on nematode control agents and suggested that this SON committee function jointly with the ASTM subcommittee. SON President Bert Endo and the SON Executive Committee responded favorably to this request and indicated that "SON felt an obligation to become involved because the setting of standards of pesticide testing required the experience of those who have used chemicals for the control of pests." Since 1973, the SON and ASTM committees have functioned as a single unit, the Joint SON/ASTM E 35.16 Subcommittee.

This approach of SON was unqiue, and the other disciplinary technical committees of the ASTM disciplinary group did not adopt it. The joint subcommittee concept has functioned well; nematologists who did not join ASTM have worked on task forces and contributed to the genesis of documents that form the basis of the suggested guidelines on evaluation of nematicides.

At its second meeting, the joint subcommittee concluded that a consensus on guidelines should receive the widest possible input. In a questionnaire mailed to all SON members, 158 respondents expressed interest in working on the several task forces of the joint subcommittee. These task forces, varying in size from 6 to more than 35 individuals, were organized to address the following specific areas:

i. Primary screening tests involving laboratory and greenhouse testing techniques

ii. Secondary screening procedures for efficacy trials in the field

iii. Responses of nematode populations

iv. Response of the plant to application of nematicides

v. Site selection factors

vi. Interactions among nematicides and with other pesticides

vii. Environmental considerations, primarily environmental impact and safety to man

Early in 1974, the joint subcommittee made the fundamental decision to confine its activities to development of suggested guidelines and recommended practices. Rigidly prescribed test methods were rejected, because standard methods of nematicide evaluation would not reflect adequately the complexity of the nematicide test situation. Guidelines and recommendations on testing procedures would provide reasonable uniformity in evaluating nematicidal efficacy, yet allow flexibility in evaluation trials as required.

## Current Status and Future Directions

Based on the concept of guidelines in lieu of rigid standards, the joint subcommittee concluded that nematologists should assess periodically the state of the science and prepare guidelines on procedures that reflect the latest available information. The acceptability of such guidelines and procedures for use in developing data on efficacy depends on the extent to which such information reflects a consensus of nematologists in industry, research, extension, and regulatory agencies. This approach should ensure that the best possible information continues to be submitted in support of nematicide registrations.

For these reasons, SON has joined with APS in the preparation of this special volume on laboratory, greenhouse, and field evaluation procedures. This section of the book contains seven task force reports that the joint subcommittee has prepared. In addition, two specific methods still under consideration by the joint subcommittee are included.

The task force reports and guidelines on recommended test methods developed thus far are primarily conceptual. They describe approaches to evaluation of efficacy and suggest useful, efficient alternatives. They also suggest procedures for evaluation that can be replicated and repeated and recommend methods for comparison of data from different tests. In the chapters that follow, emphasis is placed on recommended practices and guidelines, not rigid standard test methods. Publishing these reports in this book provides further opportunity for peer review with regard to their acceptability and utility in efficient evaluation of nematicides. In addition, the several proposed guidelines on test methods have been submitted to the ASTM E 35 Committee on Pesticides for review. Developed by broad consensus review, they provide a useful approach to nematicide evaluation by increasing methodologic precision and allowing valid comparisons among tests whenever possible. They are sufficiently flexible, however, to accommodate variable geographic, nematologic, economic, and other factors that are inherent in any biologic testing situation.

Readers are urged to communicate their comments, suggestions, and criticisms to the individual authors or to the chairman of the Joint SON/ASTM E 35.16 Subcommittee.

## Acknowledgment

The assistance of G. W. Bird, W. W. Osborne, and I. J. Thomason in review of this manuscript is acknowledged.

# 38. Guidelines for Evaluating Nematicides in Greenhouses and Growth Chambers for Control of Root-Knot Nematodes[1]

## C. P. DiSanzo

FMC Corporation, Agricultural Chemical Division, 100 Niagara St., Middleport, NY 14105

## J. Feldmesser

Plant Nematology Laboratory, USDA, ARS, ARC-West, Beltsville, MD 20705

## R. F. Myers

Department of Entomology, Rutgers University, New Brunswick, NJ 08903

## F. C. O'Melia

The Dow Chemical Company, 2800 Mitchell Dr., Walnut Creek, CA 94598

## R. M. Riedel

Department of Plant Pathology, Ohio State University, 1735 Neil Ave., Columbus, OH 43210

## A. E. Steele

USDA, ARS, PO Box 5098, Salinas, CA 93901

In nematology, comparing results obtained by different researchers on the same problem is often impossible because of the susceptibility of plant-parasitic nematodes to the various handling methods. Variation in inoculum preparation, inoculation, and temperature all affect nematode population or plant response or both. The following guidelines should help to minimize these variations in nematicide research.

## Scope

The following guidelines on greenhouse or growth chamber evaluation of nematicides against the root-knot nematode (*Meloidogyne* spp.) are suggested to establish a basis for reproducible methods that will yield data comparable with those that other scientists obtain.

## 1. Guidelines on Preparation Procedures

**1.1 Nematode species**—The root-knot nematode *Meloidogyne incognita* (Kofoid & White, 1919) is the test species of choice. This species is preferred because of the well-pronounced symptoms that develop on the infected root system. Other species may be substituted, but the reasons must be specified.

**1.2 Host plant**—The cultivated tomato, *Lycopersicon esculentum* Mill., is the recommended host plant for this procedure. Suggested cultivars include Rutgers, Pearson A1, Heinz 1350, Marglobe, Bonny Best, or any other recognized susceptible variety. Cucumber (*Cucumis sativus* L.) or other susceptible hosts can be substituted if preferred. Species and variety should be recorded and reported.

**1.3 Conditions**—The optimum average soil temperature during the test is approximately 24–28°C.

**1.4 Nematode culture**—Tomato or other host plant seedlings with two or four large true leaves are transplanted into clay pots (15–20-cm diameter preferred) containing steam-sterilized, sandy soil (30–50% sand). One week after transplanting, galled roots of nematode-infested plants with fully developed egg masses are placed in three holes in the soil around the seedling roots. Holes are then closed with soil. The plants are allowed to grow until fully developed egg masses are formed (six to seven weeks after inoculation).

**1.5 Inoculum preparation**—*1.5.1. Alternative No. 1*—Infected tomato or other host roots containing egg masses are washed under running tap water, cut into short pieces, and comminuted with water in an electric blender for approximately 30 sec. The shredded roots are poured into layers of washed sand in a wooden flat. The flat is covered with plastic sheeting and kept in the shade at greenhouse temperatures (24–28°C) for three to seven days to allow about 50% of the larvae to hatch.

*1.5.2. Alternative No. 2*—Eggs can be secured by the technique that Hussey and Barker (2) developed. *Meloidogyne* eggs are collected by dissolving egg-mass matrixes with 0.53 or 1.05% sodium hypochlorite

[1]Prepared by the Joint SON/ASTM E 35.16 Task Force on Laboratory and Greenhouse Procedures for Evaluation of Nematode Control Agents, C. P. DiSanzo, chairman.

101

(NaOCl) solution. Root-knot nematode-infected tomato or other host-plant roots with egg masses are shaken in the NaOCl solution for 4 min and the free eggs collected by sieving. To obtain uniformity, the sand containing larvae and eggs is mixed in a suitable mixing apparatus (eg, a vee-mixer for 30 sec, a standard soil riffle for five or six passes, or any other such device or technique). If only larvae are required for special use, wet-strength tissue or cloth is put on the layer of sand before the shredded roots are poured onto it.

*1.5.3. Alternative No. 3*—A more rapid but less preferred method for securing root-knot nematode larvae is to cut off the tops of the host plant and keep the soil and roots moist for 10–14 days. This procedure causes breakdown of some roots and about 50% hatch. This method may result in damage to young seedlings due to soil pathogens, but it can be used to inoculate already established plants.

**1.6 Preparation of root-knot nematode-infested soil**—A sample of the infested soil prepared as described is processed for nematodes with the centrifugal flotation extraction technique (3). A 26-$\mu$m (500-mesh) sieve is used to collect the nematodes and eggs, and their number is estimated under a stereomicroscope. Sand containing eggs and larvae is diluted with additional steam-sterilized sandy soil to obtain approximately 2,000 root-knot nematode larvae and eggs per 1,000 cm$^3$ of soil. Depending on the amount of nematode-infested soil needed, mixing should be accomplished in a standardized manner that is identified (eg, a Twin-Shell Mixer [Patterson-Kelley Co., East Stroudsburg, PA] for 60 sec).

**1.7 Soil treatment: Nonvolatile compounds**—Compounds are tested as dry or liquid formulations (eg, dusts, emulsifiable concentrates, solutions, or suspensions). Soil texture should be identified.

*1.7.1. Soil incorporation*—The chemical is mixed with soil at the desired concentration or concentrations. Uniform mixing should be done by the same technique in each test (see 1.5 above). Dry formulations are added directly to the soil, which should be dry to prevent lumping, whereas the liquid formulations are sprayed on the infested soil, which has been previously arranged in a layer not thicker than 2 cm. Fifty milliliters of liquid formulations per square meter are used. If more or less formulation is used, the reason or reasons must be given. The soil is then placed in a mixer. The amount of soil used for testing is usually mixed for at least 30 sec.

*1.7.2. Drench tests*—Nematode-infested soil is prepared as described in 1.6, and four 10-cm pots are filled. Chemical formulation is applied until leaching starts. A tomato seedling with two or four large leaves is transplanted into each pot after 20–30 hr.

*1.7.3. Soil surface treatments*—Pots containing infested soil are prepared as described in 1.7.2. Dry formulations are spread on the entire surface of each pot and the chemical is hand mixed with the top 1 cm of soil, or 5 ml of liquid formulation per pot is sprayed on the soil surface. Tomato seedlings with two or four large leaves are transplanted into each pot soon after treatment.

**1.8 Soil treatment: Volatile compounds**—Infested soil equivalent to four 8-cm pots prepared as described in 1.6 is placed in sealable containers. The soil is tamped firmly to about 2 cm from the top. The test chemical is then drenched into the soil

until 60 bars of water suction (field capacity) is reached. Another method involves injection of 2 ml of chemical formulation with a syringe and cannula into the bottom of the container. The hole is then covered with surrounding soil and the containers sealed. After 96 hr, the soil is transferred to the pots and a seedling tomato or other host plant with two or four large leaves is planted in each pot.

**1.9 Eradication tests**—Tomato seedlings with two or four large leaves are transplanted into 8-cm pots containing nematode-infested soil as described in 1.6 and allowed to grow for 5–15 days. The root systems of these plants are then washed clean under running tap water and transplanted into 15–20-cm pots containing steam-sterilized sandy soil that has been treated as described in 1.7.1, 1.7.2, 1.7.3, and 1.8. After five to six weeks, the root systems are washed, 10–20 swellings are examined for presence of egg masses, and the number is counted under a stereomicroscope. The control is expressed as a percentage of egg masses present on the treatments when compared with the untreated control roots. Thirugnanam (chapter 39) has developed an alternative eradication test procedure.

**1.10 Dip tests**—Host plant seedlings with two to four large leaves, and with their root systems washed clean, are dipped in a dilution of the test chemical for the desired periods of time. The chemical is kept in agitation during treatment. After treatment, the root systems are pressed gently between two layers of paper towel or similar material to remove excess chemical and transplanted in 15-cm pots containing nematode-infested soil. If eradicant activity is to be investigated, plants inoculated as described in 1.9 are used. The activity is determined as described in 1.9.

**1.11 Systemic treatment**—*1.11.1. Systemic control*—The foliage of tomato or other host plants growing in 10-cm pots containing steam-sterilized, sandy soil and having four to six large true leaves is sprayed to runoff by formulating the chemical as wettable powder, water or acetone-water solution (0.5–5% acetone), or other liquid preparation. Each treatment consists of at least four replicates. Just before treatment, the soil is covered to prevent the chemical from reaching it. Covers are removed when the foliage of the treated plants is dry. Half of the plants (two plants if four replicates were used) are inoculated with the nematode one day after treatment, and the remaining plants are inoculated one week later. Sand containing inoculum, which is prepared as in 1.5 but left under greenhouse conditions for 10–15 days to allow most of the larvae to hatch, is deposited at the rate of 2,000 root-knot larvae per 1,000 cm$^3$ of soil around the plant. A thin layer of uninfested soil covers the sand containing the nematodes. The plants are watered afterward.

*1.11.2. Systemic eradicant*—If the purpose of the test is to investigate eradicant activity, plants are grown in 15-cm pots and inoculated with the root-knot nematode ten days before treatment. The plants are allowed to grow for an additional five weeks, and the control is determined as described in 1.9.

**1.12 Control treatments**—Each test should include appropriate controls that are treated alike, including mechanical mixing but excluding the chemicals in test. If a solvent or other material is used for formulating chemicals, controls should be treated with them at the same concentrations. An additional set of controls

## TABLE 38.1. Nomograph of root-knot galling indexes for plant roots infected with Meloidogyne spp.

| Galling index systems[a] | | | | Percentage of total root system galled |
|---|---|---|---|---|
| 0–4 | 0–5 | 1–6 | 0–10 | |
| 0 | 0 | 1 | 0 | 0 |
| | 1 | 2 | 1 | 10 |
| | 2 | 3 | 2 | 20 |
| 1 | | | 3 | 30 |
| | | | 4 | 40 |
| 2 | 3 | 4 | 5 | 50 |
| | | | 6 | 60 |
| | | | 7 | 70 |
| 3 | | | | |
| | 4 | 5 | 8 | 80 |
| | | | 9 | 90 |
| 4 | 5 | 6 | 10 | 100 |

[a]Percentage of swelling present on root system of control plants should be indicated for each test. If percentage of total control plant root system galled is less than 25% (75% of feeder roots uninfected), test should be repeated.

should be included to indicate the effect of the solvent on the performance of the tests.

**1.13 Standards**—Each test should include standard chemicals, the performance of which is known under greenhouse conditions. The standard will be selected according to the characteristics of the chemicals to be tested. If nonvolatile chemicals are to be tested, nonvolatile commercial or experimental materials should be included. If the chemicals are volatile, known fumigants should be tested.

### 2. Determination of Activity

Within a standard number of days after the soil has been inoculated with the root-knot nematode but within two to four weeks after inoculation (except in the case of eradication tests), the soil in which plants are growing is allowed to dry until the plants begin to wilt. The roots are shaken to remove the soil adhering to them, and the extent of swellings developed on the roots of control plants is noted. Roots are scored for the degree of galling using a root-knot nematode root-galling index. The activity of the test chemical is then expressed as a percentage of reduction in the root-galling index (Table 38.1). A portion of the root system should be stained with a biological stain (eg, acid-fuchsin lactophenol) to determine the presence and numbers of nematodes within roots. The sex ratio can also be determined.

**2.1 Galling index**—Roots are scored for the degree of galling using one of the attached root-galling indexes, which are relatively comparable and interchangeable (Table 38.1).

**2.2 Differential root staining**—Portions of root tissue may be stained using any one of several methods that takes advantage of differential stain uptake by nematodes. Because nematodes absorb stains from solutions such as acid-fuchsin-lactophenol or cotton blue-lactophenol more rapidly than does surrounding root tissue, they may be identified and counted in situ under the stereomicroscope. Overstained roots in which nematodes may have been obscured may be destained in lactophenol so as to reveal the presence of nematodes. The staining and counting help to determine the presence of nematodes within roots in the absence of swellings that may be due to conditions of treatment. Differential root staining is advisable to substantiate low root-galling indexes that exhibit chemical efficacy in results that approach 100% control.

**2.3 Statistical treatment of data**—A standard analysis of variance is used to check data for significant differences, plus Duncan's multiple range test to compare treatment means. The data can also be subjected to a probit analysis using Finney's (1) method.

### Literature Cited

1. FINNEY, D. J. 1952. Probit Analysis: A Statistical Treatment of Sigmoid Response Curve. Ed 2. Cambridge University Press: London.
2. HUSSEY, R. S., and K. R. BARKER. 1973. A comparison of methods of collecting inocula of *Meloidogyne* spp., including a new technique. Plant Dis. Rep. 57:1025-1028.
3. JENKINS, W. R. 1964. A rapid centrifugal-flotation technique for separating nematodes from soil. Plant Dis. Rep. 48:692.

# 39. Evaluation of Nematicides for Systemic Eradication of Root-Knot Nematodes[1]

M. Thirugnanam

Nematologist, Rohm and Haas Co. Research Laboratories, Spring House, PA 19477

### Scope and Significance

This procedure is suggested to evaluate a nematicide for systemic eradication properties against root-knot nematodes that exist inside the root tissue prior to

[1]Manuscript prepared in connection with the effort of the Joint SON/ASTM E 35.16 Subcommittee on Nematode Control Agents to develop guidelines on assessing nematicide efficacy.

chemical treatment. This method aids in evaluation of nematicides and their potential for postplanting use.

### 1. Preparatory Procedures

**1.1 Nematode species**—Root-knot nematodes, *Meloidogyne incognita* (Kofoid & White) Chitwood or *M. arenaria* (Neal) Chitwood, are suggested.

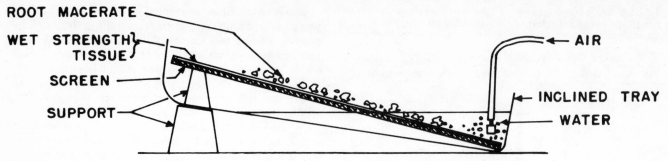

ROOT MACERATE
WET STRENGTH TISSUE
SCREEN
SUPPORT
AIR
INCLINED TRAY
WATER

Fig. 39.1. Modified Baermann apparatus for collection of root-knot inoculum (second stage larvae).

**1.2 Host plant**—The cultivated pepper *Capsicum annuum* L. (chilli, red, or sweet peppers) is the preferred host, because these nematode species produce inconspicuous root galls but conspicuous egg masses lying outside the roots. The variety used should be specified in the report.

**1.3 Conditions**—Keeping greenhouse or growth chamber temperature at 24–28°C is recommended.

**1.4 Nematode culture**—Cultures are easily maintained on susceptible tomato varieties in the greenhouse (see chapter 38).

**1.5 Inoculum procedures**—*1.5.1. Modified Baermann method*—Root-knot–infected tomato roots containing egg masses are washed free of soil in tap water, cut into small pieces, and macerated in a blender for about 30 sec. The macerate is poured evenly onto a screen (15- to 25-mesh) that is already layered with wet-strength tissue and supported by two petri dishes at one end in an inclined shallow tray (Fig. 39.1). Water is added to the tray and maintained at a level sufficient to keep the tissue layer moist during the incubation period. During this period, the larvae hatch on the screen and migrate toward water. Aeration of water helps to keep the larvae alive. After five to seven days of incubation, the aqueous larval suspension is stirred well to get uniform distribution of nematodes for microscopic counting. The aliquot may be further diluted with water to facilitate counting.

*1.5.2. Alternate methods*—DiSanzo et al. (chapter 38) suggest alternative procedures on inoculum preparation. If such procedures are used, reports should describe modification of procedures for inoculum preparation.

## 2. Inoculation Procedures

**2.1**—Three- to four-week-old pepper seedlings grown in steam-sterilized soil and having four to six true leaves are washed in running water to free the roots from soil particles.

**2.2**—The root system of each seedling is placed individually in a petri dish, and 2 ml of aqueous nematode suspension containing about 500 second-stage larvae (see chapter 38) is pipetted directly onto the roots.

**2.3**—Steam-sterilized sand is sprinkled over the entire root system; this soil is maintained moist throughout the inoculation period. After 24 hr, the seedlings are removed and their roots thoroughly washed in running water prior to chemical treatment.

## 3. Chemical Treatment

**3.1 Soil incorporation**—Steam-sterilized soil is treated with test compounds to obtain desired dosages. Chemicals may be applied by thorough incorporation, drench, or surface treatment or by fumigation in the case of volatile compounds (see chapter 38). Seedlings exposed to root-knot infection for 24 hr are planted in treated soils. Each treatment in an experiment should be replicated two to four times. Appropriate checks as well as standard chemical treatments should also be maintained and reported.

**3.2 Foliar application**—Pepper seedlings prepared as in 2.3 above are held horizontally and the foliage sprayed to runoff with test compounds. During spraying, the root system is kept covered to prevent direct contact with the chemical. When the leaves are dry, the plants are potted in steam-sterilized soil. Replications, checks, and standard treatments are required as with soil incorporation.

## 4. Determination of Eradicant Activity

**4.1 Extraction of nematode eggs**—After four to six weeks of treatment, the plants are carefully removed from the soil, the root systems are washed in water, and the number of eggs on each plant is determined by extraction with 0.5% sodium hypochlorite for 10 min and microscopic counting (see chapter 43). The eradicant activity of the test substance is expressed as the percent control of egg production using the following modification of Abbott's formula (1).

$$\text{Percent control of egg production} = \frac{\text{Number of eggs in check plant} - \text{Number of eggs in treated plant}}{\text{Number of eggs in check}} \times 100$$

**4.2 Statistical treatment**—Comparison of test substance eradicant activity with that of known eradicant nematicides and appropriate controls should be analyzed for statistical significance. Methods of analysis should be reported. Analysis of variance and multiple range tests are suggested in addition to comparison of treatment means.

### Literature Cited

1. ABBOTT, W. S. 1925. A method of computing the effectiveness of an insecticide. J. Econ. Entomol. 18:265-267.

# 40. Tests for Nematicidal Efficacy Using Larvae of Heterodera schachtii[1]

**Arnold E. Steele**

Zoologist, Agricultural Research Service, USDA, PO Box 5098, Salinas, CA 93901

## Scope and Significance

These methods are used to hatch larvae of *Heterodera schachtii* for screening chemicals. They should be used to evaluate the effects of chemical treatments on hatching and emergence of larvae, to obtain larvae for in vitro evaluation of chemicals, or for inoculations to evaluate chemical soil treatments.

## 1. Apparatus

The following items are required for the method suggested. In each case, equivalent materials may be substituted.

1.1—Collection cups (small sieves), supplied by Hykro Pet Industries, Esbjerg, Denmark.

1.2—Plastic portion cups (about 12-ml capacity) supplied by Thunderbird Container Corporation, El Paso, TX 79912.

1.3—Sieve made of rigid plastic tubing and nylon mesh (40 per centimeter).

1.4—Long-stemmed glass funnels of convenient size.

1.5—Refrigerator with counter-top cabinet measuring about 0.04 m$^3$ is adequate. Holes large enough to accept the funnel stem are drilled through the top of the refrigerator cabinet.

1.6—Electronic thermoregulator equipped with thermistor probe. Flexible heating tape 11 × 2 mm, 35 W.

1.7—Pinch clamps, pipettes, camel's hair brushes.

## 2. Reagents

Hatching agents, such as aqueous solutions of $4 \times 10^{-6} M$ zinc chloride or sugar beet root diffusate leached from soil of pot-grown sugar beet plants, are required.

## 3. Test Organism

The test organism is *Heterodera schachtii* Schmidt 1871.

## 4. Procedure for Evaluating Chemical Effects on Larval Hatching and Emergence from Cysts

4.1—Newly formed cysts are separated from sugar beet roots and soil by washing, floating, and decanting suspended debris into screens. Newly formed cysts with eggs and larvae are selected, manually separated from washed debris, and stored until needed.

4.2—Two to 5 ml of test solutions are transferred to the portion cups using a precalibrated pipette. Groups of 20 nematode cysts are transferred with a fine-bristle brush to a small wedge-shaped piece of filter paper before going to the collection cups (sieves). This prevents dilution of the test solution during transfer of the cysts. Treatments, including controls with diluent but no added chemicals, are replicated at least four times. The hatching vessels and their contents are incubated at 24°C during the entire test period.

4.3—The cysts are treated with the chemical solutions for one to seven days and then transferred to tap water, which is changed daily, for four days to remove the test materials. Finally, the cysts are placed in a hatching agent for two to four weeks. A host plant such as sugar beet (variety should be reported) is then inoculated with the cysts to evaluate nematode viability.

## 5. Procedure for Hatching Larvae

5.1—Washed and screened root debris containing cysts are added to a large screen (about 10 cm in diameter), which then is placed in a funnel containing a hatching solution. The level of the solution is adjusted so that the debris is wet but not completely covered.

5.2—The stem of the funnel is inserted through the top of the refrigerator cabinet. The cabinet interior is maintained around 8°C. The temperature of the solution bathing the cysts is maintained at 24°C using an electronic thermoregulator equipped with a thermistor and heating tape. If ambient temperatures are above 24°C, supplementary heating of the solution may not be required.

## 6. Evaluating Chemical Effects on Hatched Second-Stage Larvae

6.1—Newly hatched second-stage larvae are treated for 24 hr at 24°C with aqueous solutions of 1, 5, 10, 25, 50, and 100 µg/ml of test chemical.

6.2—Effects of chemical treatments on mobility may be estimated by placing the larvae on tissue paper supported by collection cups (sieves) that are in turn placed in the chemical solution. Larvae remaining on the tissue paper are assumed to be either immobilized or incapable of purposeful movement (disoriented).

6.3—After the initial treatment, the larvae are washed with several liters of tap water. The nematodes are easily concentrated in a small volume of water using a Buchner-type funnel with fritted disk.

6.4—Roots of sugar beet grown in a steam-sterilized sand-soil mixture are inoculated with treated and nontreated larvae.

6.5—At 18 and 30–35 days after inoculation, the plants are harvested and the roots and soil examined for adult sugar beet nematodes.

[1]Manuscript prepared in connection with the effort of the Joint SON/ASTM E 35.16 Subcommittee on Nematode Control Agents to develop guidelines on assessing nematicide efficacy.

## 7. Procedure for Bioassay of Nematicidal Efficacy

**7.1**—The numbers of larvae added per plant or unit weight of soil should be specified; however, not less than 2,000 larvae or eggs and larvae from 20 selected viable cysts should be added per plant. If cysts are used, the numbers of eggs and larvae may be estimated by counting them or by hatching the eggs in a solution containing a hatching agent. At least four, and preferably six to ten, plant replicates should be used for each treatment or untreated check.

**7.2 Mechanics of inoculation**—*7.2.1*—If plants are inoculated at the time of transplanting, the inoculum is placed in a hole just large enough to accept the transplant roots.

*7.2.2*—If established plants are inoculated, the inoculum should be evenly distributed in three holes large enough to accept the pipette tip (1.5–3.0 cm deep) and located 1.5–3.0 cm from the plant stem.

## 8. Assessing Nematode Populations

**8.1**—Nematicide efficacy may be evaluated one or more times during the plant's growth period. At least one count should be obtained before production of the second nematode generation.

**8.2**—Counts may be made on nematodes extracted from soil or plant tissues and may include any or all stages. Counts of adult males may be obtained by harvesting plants 18 days after inoculation and placing roots on funnels in a moist chamber for five to ten days. For counts of adult females, plants are harvested 30–35 days after inoculation and the roots and potting soil washed and examined for nematodes.

**8.3**—Barker et al. (chapter 43) have described preferred methods of population assessment.

**8.4**—Counts of accumulated larva hatches should be analyzed for statistical significance by analysis of variance; the statistical procedures used should be reported.

# 41. Test Materials and Environmental Conditions for Field Evaluation of Nematode Control Agents[1,2]

## A. W. Johnson[3]

Research nematologist, ARS-USDA, Coastal Plain Experiment Station, Tifton, GA 31794

### Scope

Test materials and environmental and cultural conditions are important factors that influence the results of secondary and field evaluation of nematicides. The scope of this chapter is to describe the information on these factors that should be recorded during field tests of experimental nematicides to achieve meaningful evaluation.

### 1. Test Materials

All test materials should be compared with an untreated control and with a known standard, usually one of the materials currently recommended. For evaluation of field trials of experimental nematicides, one should know as much as possible about the biologic activity and chemical and physical properties of the test material. Pertinent literature or technical reports should be reviewed before designing field trials.

**1.1 Formulation**—The records for data should

include formulation type—emulsifiable concentrate, wettable powder, flowable, water soluble, and granular (mesh size). The names and percentage of every ingredient in the formulation, lot number on the package label, and dates sent and received should be recorded. If a nematicide is diluted before application, the amount of diluent used should be specified by common and chemical names.

**1.2 Rate or rates of application**—Rates should be clearly, precisely stated as formulation and active ingredient in one or more of the following terms: the quantity per unit of area if treated overall (broadcast), the quantity per linear distance and row spacing if row treated, and width of band and row spacing if band treated.

**1.3 Number and timing of applications**—*1.3.1*—Dates (month, day, year) of preplant or postplant applications or both should be recorded.

*1.3.2. Timing of applications*—Proper timing of application is critical. Information should specify the time of application in terms of crop planting date, emergence date, growth stage, preharvest interval, and intervals between applications. Target-pest population levels and incidence should be specified.

1.3.2.1—Preplant—before seeds or seedlings are planted.

[1]Prepared by the Joint SON/ASTM E 35.16 Task Force on Test Material and Environmental Conditions in Field Evaluation of Nematode Control Agents, A. W. Johnson, chairman.

[3]See addendum for names and addresses of cooperating authors.

1.3.2.2—At planting—at the time seeds or seedlings are planted.

1.3.2.3—Postplant—after seeds or seedlings are planted and on established plantings. Plant size, stage of growth, or number of days since emergence for postplant applications or a combination of these should be recorded.

**1.4 Method of application**—Method of application, including specialized equipment, should be specified. Such descriptive terms as spraying, injecting, spacing, soaking, rinsing, and flooding should be used when appropriate. Soil applications should include such information as band width, row spacing, chisel spacing, depth of application, and time interval between application and incorporation. If applied on the surface, the method and depth of incorporation, if any, should be stated. Description of row applications should include whether they were in furrow, band over row, or side dressed (preplant, at planting, postplant, postemergence). For side-dressed applications, placement in relation to seed or plant should be given. Treatment may be broadcast, strip, row, site, root dip, or foliar spray, and may be injected into the soil or applied to the soil surface as a drench, spray, granule, or solution in irrigation water. Granules may be incorporated into the top few centimeters of soil, or the active ingredient may be washed from the granules by irrigation or rainfall. Care should always be taken to prevent recontamination of treated areas by cultivation or other means whereby soil from nontreated areas is blended with soil from treated areas. Methods of application in irrigation water include overhead sprinkler, flood, trickle, row or furrow, and basin.

## 2. Environmental and Cultural Conditions

Information about pretreatment, at-treatment, and posttreatment environmental and cultural factors that could affect nematicidal efficacy should be recorded. Erratic results of incomplete experimentation may be due, at least in part, to effects of such factors as relative humidity, wind, rainfall, and air temperature during the test period. The relationship of all environmental and cultural factors to crop, pathogen, and nematicide should be considered and explained in any evaluation of results.

Soil factors include the identity of target and nontarget nematodes and their relative density before, during, and after testing; temperature; soil types, including textural variations with depth; pH; field capacity; nutrient levels; percentage of organic matter; presence or absence of crop refuse or trash; percentage of soil moisture and, if possible, an estimate of drainage; amounts and frequency of rainfall or irrigation or both and type of irrigation (eg, flood, sprinkle, row, basin); and other data that may affect the application and performance of the nematicide being tested. Occurrence and quantity of other organisms that affect crop growth and nematode populations should be included. Applications of fertilizers, lime, or other soil amendments such as herbicides, fungicides, and insecticides should be recorded. Previous cropping history and pesticide usage are also important.

# ADDENDUM

## Test Materials and Environmental Conditions for Field Evaluation of Nematode Control Agents

**Task Force Chairman**
Dr. A. W. Johnson
Research Nematologist
USDA
Tifton, Georgia 31794

Dr. Jack Altman
Botany and Plant Pathology Department
Colorado State University
Fort Collins, CO 80521

Fred W. Bistline
The Coca-Cola Company
PO Box 3189
Orlando, FL 32802

Dr. Bill B. Brodie
USDA, ARS
Department of Plant Pathology
Cornell University
Ithaca, NY 14850

Dr. R. A. Chapman
Department of Plant Pathology
University of Kentucky
Lexington, KY 40506

Dr. D. W. Dickson
Entomology Research Laboratory
Archer Rd.
Gainesville, FL 32601

James M. Epps
USDA, ARS
605 Airways Blvd.
Jackson, TN 38301

Dr. G. D. Griffin
Crops Research Laboratory
Utah State University
Logan, UT 84321

Dr. C. M. Heald
USDA, ARS
PO Box 267
Weslaco, TX 78596

Dr. L. R. Hodges
Union Carbide Corporation
Box 1906
Salinas, CA 93901

Dr. L. R. Krusberg
Botany Department
University of Maryland
College Park, MD 20740

Dr. S. A. Lewis
Department of Plant Pathology and Physiology
Clemson University
Clemson, SC 29631

Dr. D. H. MacDonald
Department of Plant Pathology
University of Minnesota
St. Paul, MN 55108

Dr. Michael McKenry
Department of Nematology
University of California
Riverside, CA 92502

Dr. N. A. Minton
USDA, ARS
Coastal Plain Experiment Station
Tifton, GA 31794

Dr. Calvin Orr
Southern Plains Research and Entomology Center
RFD 3
Lubbock, TX 79401

Prof. A. J. Overman
AFED, IFAS
5007 60th St. E.
Bradenton, FL 33505

Dr. G. C. Smart
Department of Entomology and Nematology
University of Florida
Gainesville, FL 32601

Dr. E. J. Spyhalski
6713 Pontiac Dr.
North Little Rock, AR 72116

A. E. Steele
PO Box 5098
Alisal Branch
Salinas, CA 93901

Dr. I. J. Thomason
Department of Nematology
University of California
Riverside, CA 92502

Dr. E. J. Wehunt
PO Box 87
Byron, GA 31008

Dr. L. V. White
Great Lakes Chemical Company
PO Box 2200
West Lafayette, IN 47906

# 42. Site Selection Procedures for Field Evaluation of Nematode Control Agents[1]

**G. W. Bird**

Department of Entomology, Michigan State University, East Lansing, MI 48824

**R. A. Davis**

Environmental Protection Agency, Beltsville, MD 20705

**P. W. Johnson**

Agriculture Canada, Vineland, Ont. L0R 2E0

**D. MacDonald**

Department of Plant Pathology, University of Minnesota, St. Paul, MN 55108

**M. McKenry**

San Joaquin Valley Research and Extension Center, University of California, 9240 Riverbend Ave., Parlier, CA 93648

**R. B. Malek**

Department of Plant Pathology, University of Illinois, Urbana, IL 61801

**C. Orr**

USDA, Lubbock, TX 79401

**W. M. Powell**

Department of Plant Pathology, University of Georgia, Athens, GA 30602

**R. M. Riedel**

Department of Plant Pathology, Ohio State University, Columbus, OH 43210

**J. L. Ruehle**

USDA Forest Service, Forestry Sciences Laboratory, Athens, GA 30601

**J. L. Townshend**

Agriculture Canada, Vineland, Ont. L0R 2E0

The process of site selection is an important subsystem within any well-planned experiment for field evaluation of nematode control agents. Inadequate consideration of this subsystem is frequently responsible for the failure of field investigations. The objective of this chapter is to describe and discuss the five major components of site selection: (i) preliminary experimental procedure, (ii) selection of provisional test site, (iii) evaluation of the nematode parameters of the provisional test site, (iv) evaluation of the biological, physical, and chemical parameters of the provisional test site, and (v) final selection of the test site and experimental design (Fig. 42.1). The third and fourth components should be monitored throughout evaluation of the nematode control agent to determine its long-term

---

[1]Prepared by the Joint SON/ASTM E 35.16 Task Force on Site Selection for Field Evaluation of Nematode Control Agents.

influence on the ecology of the test site. These procedures for site selection are recommended for use in conjunction with those that the SON/ASTM Task Forces have developed for assessment of nematode population responses, test materials, and environmental considerations; evaluation of plant responses and interactions of nematodes with other pathogens and pesticides; and assessment of safety to humans (chapters 41, 43, 44).

## Preliminary Experimental Procedure

The first two steps in the preliminary experimental procedure are to define the problem clearly and concisely and state the experimental objectives (Fig. 42.2). Without fulfilling these prerequisites, a successful investigation is almost impossible. In most cases, the preliminary experimental procedure includes selection of test species, test host, and experimental treatments. Next, the definition of the problem, including a review of the biological, chemical, and physical properties of the control agent, and the experimental objectives are critically analyzed (4,8). Adjustments in the definition of the problem and experimental objectives, as well as desired test species, test host, or experimental treatments, must be made before the experiment begins.

## Provisional Site Selection

Suitable test sites for field evaluation of nematode control agents may be naturally or artificially infested with the test species. The investigator may make provisional selection of a test site after considering the experimental objectives, experimental design, replication, experimental units for observation (plot size, shape, general characteristics), and suitability of a site for experimental purposes (Fig. 42.3A). Success of

TABLE 42.1. Site selection checklist for field evaluation of nematode control agents[a]

1. Definition of problem

2. Experimental objectives

3. Proposed test species

4. Proposed test host (cultivar)

5. Proposed treatments

6. Proposed experimental unit

7. Proposed experimental design and replication

[a]Information to be recorded and evaluated before analysis of provisional sites.

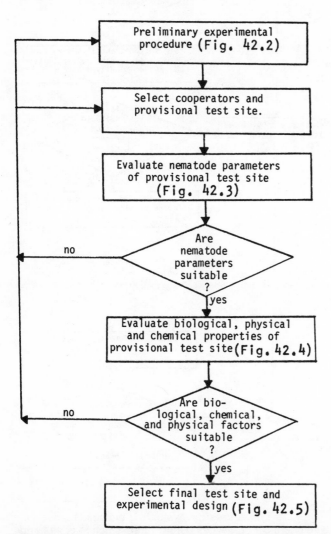

Fig. 42.1. Generalized procedure for selection of experimental sites and designs for evaluation of nematode control agents.

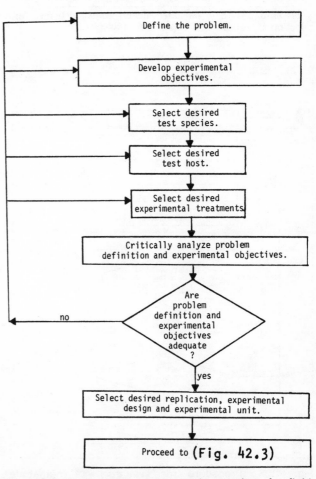

Fig. 42.2. Preliminary experimental procedure for field evaluation of nematode control agents. Information required in Table 42.1 should be recorded.

provisional site selection depends largely on the investigator's experience and desire to locate and acquire the most suitable test site. Initial population density of the test species in the site must be stable and equal to or above the economic threshold level for the specific host-parasite relationship. Having relatively uniform horizontal and vertical distributions of nematode populations is desirable. In most locations,

however, soil texture is not uniform, and nematode species can be expected to favor zones of habitation within the soil profile, depending on the biology and physics of the soil environment. Since uniform testing conditions may not be attainable under field conditions, researchers must take steps to identify, measure, and record variables. It is helpful if the site has a history of uniform cropping sequence and productivity. Any chemicals that have recently been applied to the site should not be of such a nature that they will interfere with the experimental objectives.

The control that an investigator has over a test site is important. Investigator-controlled sites, such as those at experiment stations, and grower-owned sites both have advantages and disadvantages. Good cooperators are indispensable assets. While the use of small plots and experimental equipment is often adequate for fulfillment of experimental objectives, experimental

**TABLE 42.2. Site selection checklist for field evaluation of nematode control agents**[a]

1. Location or proposed site

2. Sampling date and procedure

3. Size of unit sampled

4. Number of units sampled

5. Number of cores per sample

6. Sample processing procedure and date

7. Population density of proposed test species

8. Horizontal distribution of proposed test species

9. Vertical distribution of proposed test species

10. Concomitant nematode species

11. Population densities of concomitant nematode species

[a]Nematode data to be collected and analyzed before final selection of test site and experimental design.

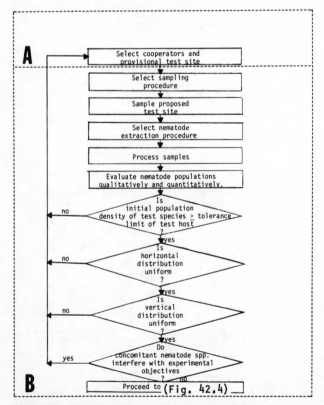

Fig. 42.3. Procedure for evaluation of nematode parameters in site selection for evaluation of nematode control agents. Information required in Table 42.2 should be recorded.

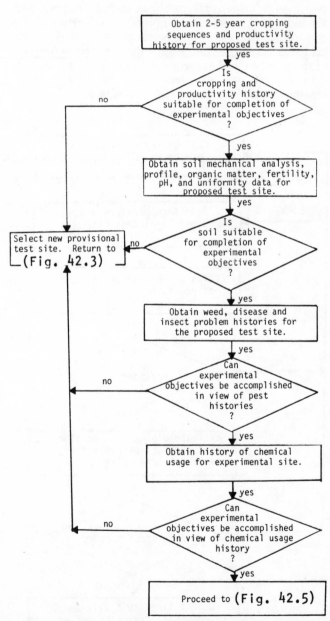

Fig. 42.4. Procedures for evaluating biological, physical, and chemical properties before final selection of site and experimental design for evaluation of nematode control agents. Information required in Table 42.3 should be recorded.

sites (not demonstration plots) maintained under commercial agricultural conditions often can minimize the number of years required for complete evaluation of a nematode control agent.

## Nematode Evaluation

The nematode parameters of a proposed test site must be evaluated in detail before final site selection (Fig. 42.3B). The first step in this component is to select a sampling procedure suitable for the proposed test site, test host, and test species. Many experiments fail because of inadequate sampling procedures. Samples must evaluate both the horizontal and vertical distribution of the nematode population (chapter 43). Samples should be processed as soon as possible after collection or stored in a suitable environment (10–15°C) to prevent nematode mortality and population alteration. The assay technique varies for different test species, sites, and hosts (chapter 43) (7).

Sample assay is followed by identification of the nematodes and analysis of the population density. The initial population of the test species should be greater than the tolerance limit of the specific cultivar of the proposed host. A site with relatively uniform horizontal and vertical distribution of the test species is most desirable, and the population should be in a relatively stable condition. Concomitant plant-parasitic nematode species also should be distributed uniformly, their populations being of a type and density that will not interfere with experimental objectives. If any of the nematode factors are not suitable for successful evaluation, a new provisional test site should be selected and evaluated in respect to its nematode parameters (Fig. 42.3B).

## Biological, Physical, and Chemical Factors

The biological, physical, and chemical properties of the provisional site must be evaluated in relation to the experimental objectives (Fig. 42.4). The site should be capable of supporting normal growth and development of the desired cultivar of the proposed test host. Its suitability can be determined only through investigation of the cropping sequence and production records of the site for the previous two or more growing seasons. The potential for disease, insect, and weed problems should be evaluated and measures considered for their control. Information about soil texture, profile, organic matter, fertility, and pH uniformity should be analyzed in relation to the experimental objectives. A record of all edaphic chemical applications during the previous two years should be obtained. Although soil fertility and pH can be adjusted, the site should not be used unless all other edaphic chemical applications during the previous year were applied uniformly to the

---

**TABLE 42.3. Biological, physical, and chemical factor checklist for field evaluation of nematode control agents**[a]

1. Cropping sequence and productivity history (two or more years)

2. Mechanical analysis of soil, including organic matter

3. Soil profile and uniformity

4. Soil fertility

5. Soil pH

6. History of weed problems

7. History of disease problems

8. History of insect problems

9. History of chemical use

[a]Data to be collected and analyzed before selection of test site and experimental design.

---

**TABLE 42.4. Site selection checklist for evaluation of nematode control agents**[a]

1. Description of site location (include coordinates and local landmarks

2. Size and shape of test area

3. Test cultivars

4. Experimental treatments

5. Experimental design

6. Replication

7. Experimental units (size and shape)

8. Initial population density of the nematode test species for each experimental unit

9. Proposed statistical analysis of data

[a]Information to be recorded during the final selection of test site and experimental design.

---

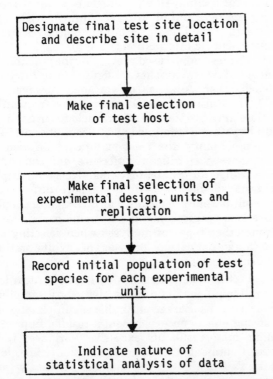

Fig. 42.5. Final site and experimental design selection for evaluation of nematode control agents. Information required in Table 42.4 should be recorded.

entire proposed experimental area. If any of these biological, physical, or chemical factors are unsuitable for completion of the experimental objectives, a new provisional site should be selected and evaluated for its nematological, chemical, physical, and biological properties.

## Final Site and Experimental Design Selection

If a provisional test site meets the requirements for evaluation of a nematode control agent, it should be selected as test site and described in detail (Fig. 42.5). The description should contain its size, shape, and location, including coordinates and local landmarks. Nematologists should become familiar with use of coordinates for geographic descriptions, since this technique most likely will be used in future computerized management of ecosystems.

The investigator should evaluate some of his initial decisions and make the final selection of the test host species and cultivar, experimental units, number of replications, and experimental design (1–4,6,8). The initial population density of the nematode test species should be recorded for each experimental unit and the nature of statistical analyses of data described.

**Experimental unit and replication**—Final selection of plot size, plot shape, and the number of replications must be related to the precision desired for the experiment. Although precision in an experiment can usually be increased through additional replication, the extent of improvement decreases as the number of replications is increased beyond a certain level. Generally, five to eight replications provide reasonable precision for evaluation of nematode control agents in a properly selected field site.

Size and shape of field plots also influence experimental precision. Increasing plot size usually minimizes variability and increases precision. Once a certain size has been reached, however, the increase in precision decreases rapidly if the plot size is enlarged further. The nature of the nematode control agents being evaluated also must be considered during selection of plot size and shape. In addition to available references on plot size and shape for specific crops, experience from similar tests is important in selecting adequate experimental units and the appropriate number of replications.

**Experimental design**—The most commonly used experimental designs and statistical analyses involve random placement of treatments within the experimental units. The completely randomized, randomized complete block, Latin square, and several types of factorial designs are usually used in evaluation of nematode control agents. The suitability of a particular experimental design depends in part on the test site and the desired statistical analysis of the data to be collected. Generally, the best design is the simplest one possible for the test site, experimental objectives, and desired analyses of data (1–4).

The completely randomized design is the simplest of the designs most commonly used in evaluation of nematode control agents. It is organized by assigning treatments at random to a previously determined number of experimental units. Any number of treatments and experimental units can be used with this simple, flexible design. Its main disadvantage is a low level of precision.

The randomized complete block is most commonly used in field trials. The test site is divided into blocks of experimental units, and the treatments are assigned at random to the units within each block. Each treatment is usually assigned once to each block. This arrangement minimizes variability among the experimental units within each block. The number of treatments should be as few as possible to meet the objectives of the experiment.

The Latin square is useful when two major sources of variation are present (eg, soil variability in two directions), because the randomization of treatments is grouped by both columns and rows. The design makes it possible to remove variability of experimental error associated with columns and rows. If variation is associated with the columns, this design allows more precision than does the randomized block design. Each treatment is assigned to one experimental unit in each block, and to one in each column. A Latin square design requires at least as many replications as there are treatments. Locating a suitable site is difficult if many treatments are required.

In field tests of nematode control agents, evaluating two or more factors is sometimes necessary. This is usually accomplished through a factorial experiment. Split-plot designs and the split-plot variation are frequently used for factorial experiments. A completely randomized, randomized complete block, or Latin square design is used, with treatments of the second factor assigned to subplots within each initial plot. These designs have numerous variations. The split-plot principle also can be used in experiments in which successive observations are made over a period of time, or to compare data statistically from multiple-year tests.

**Statistical data analysis**—Although statistical analysis of data is not a component of the site selection subsystem, it must be considered in relation to experimental design. All data should be analyzed and the results evaluated in accordance with the experimental conditions, original objectives, and previous scientific information (1,2,4–6).

Student's t test can be used to evaluate the probability that means of two treatments are significantly different from each other. Analysis of variance, however, is probably the statistical procedure used most frequently when more than two treatments are included in a test. If the null hypothesis is rejected at ≤0.05 or other desired level of probability, the least significant difference, Tukey's honestly significant difference, and Duncan's multiple range or Student-Newman-Kuels' multiple range tests should be used to compare differences among multiple pairs of means. The investigator should not overlook regression, correlation, probit, and numerous other types of analyses when selecting an appropriate statistical tool for assistance in interpreting experimental results.

Several other factors must be considered in regard to the overall approach of analysis. Methods of sampling for nematodes as Barker et al. discuss in chapter 43 have strong effects on experimental results. Improper sampling may result in excessive variance, thus masking treatment effects. In fumigation trials, low population densities of nematodes frequently are associated with certain treatments. One basic assumption in analysis of variance is that variability is much the same over the whole experiment. When population densities vary markedly among treatments, however, this assumption is probably invalid. If the populations follow a Poisson or negative binomial distribution, the variance increases with the mean. If this occurs, analyzing a simple function of the data and not the original population densities is desirable. Variance can be stabilized by transforming the data.

Another problem in data analysis occurs if nematode control agents have differential effects on several target nematode species. A given control agent may reduce populations of one pathogen successfully, but fail to control a second one that continues to parasitize and damage the host plants. An analysis of growth or yield data may fail to indicate control of the test species because of the damage caused by the second. Conversely, a control agent that reduces populations in general may result in growth and yield responses that cannot be attributed solely to control of the test species.

These comments tend to enforce the axiom in statistics that the manner in which the data are obtained dictates the analysis of any set of data. The analyses are exact only if the underlying assumptions are satisfied. Since this is often difficult in field trials involving nematodes, much depends on the skill of the researcher in selecting the method of analysis that best fits the circumstances of the experimental situation.

## Literature Cited

1. COCHRAN, W. G., and G. M. COX. 1964. Experimental Designs. Ed 2. John Wiley & Sons, Inc.: New York.
2. FEDERER, W. T. 1955. Experimental Design: Theory and Application. The MacMillan Co.: New York.
3. Le CLERG, E. L., W. H. LEONARD, and A. G. CLARK. 1972. Field Plot Technique. Ed 2. Burgess Publishing Co.: Minneapolis.
4. LITTLE, T. M., and F. J. HILLS. 1972. Statistical Methods in Agricultural Research. Agricultural Extension Service, University of California: Davis.
5. PROCTOR, J., and C. F. MARKS. 1974. The determination of normalizing transformations for nematode count data from soil samples and of efficient samples schemes. Nematologica 20:395-406.
6. SNEDECOR, G. W., and W. G. COCHRAN. 1967. Statistical Methods. Ed 6. Iowa State University Press: Ames.
7. SOUTHEY, J. F. 1970. Laboratory Methods for Work With Plant and Soil Nematodes. Tech. Bull. No. 2. Ministry of Agriculture, Fisheries and Food. Her Majesty's Stationery Office: London.
8. UNTERSTENHOFER, G. 1963. The basic principles of crop protection field trials. Pflanzenschutz-Nachr. Bayer (Am. Ed.) 16(3):81-117.

# 43. Determining Nematode Population Responses to Control Agents[1]

### K. R. Barker
Department of Plant Pathology, North Carolina State University, Raleigh, NC 27607

### J. L. Townshend
Canada Department of Agriculture, Box 185, Vineland Station, Ont. L0R 2E0

### R. E. Michell
Environmental Protection Agency, Room 1143, S. Agriculture Bldg., Washington, DC 20250

### D. C. Norton
Department of Botany and Plant Pathology, Iowa State University, Ames, IA 50010

### J. L. Ruehle
Southeastern Forest Experiment Station,Forestry Sciences Laboratory, Carlton Street, Athens, GA 30602

### G. W. Bird
Department of Entomology, Michigan State University, East Lansing, MI 48823

### R. A. Chapman
Department of Plant Pathology, University of Kentucky, Lexington, KY 40506

### R. A. Dunn
Entomology and Nematology Department, University of Florida, Gainesville, FL 32611

### M. Oostenbrink
Department of Nematology, Agricultural University, Duivendaal 1, Wageningen, The Netherlands

### W. M. Powell
Department of Genetics and Plant Pathology, University of Georgia, Athens, GA 30601

### I. J. Thomason
Department of Plant Nematology, University of California, Riverside, CA 92502

### K. D. Fisher
American Society of Experimental Biologists, 9650 Rockville Pike, Bethesda, MD 20014

### C. J. Southards
Agricultural Biology Department, PO Box 1071, University of Tennessee, Knoxville, TN 37901

### W. F. Mai
Department of Plant Pathology, New York State College of Agriculture, Cornell University, Ithaca, NY 14850

### G. C. Smart, Jr.
Department of Entomology and Nematology, Archer Road Laboratory, University of Florida, Gainesville, FL 32611

### R. A. Rohde
Department of Plant Pathology, University of Massachusetts, Amherst, MA 01002

### D. W. Dickson
Department of Entomology and Nematology, Archer Road Laboratory, University of Florida, Gainesville, FL 32611

### R. B. Malek
Department of Plant Pathology, 106 Horticulture Field Laboratory, University of Illinois, Urbana, IL 61801

### G. D. Griffin
Crops Research Laboratory, UMC 63, Utah State University, Logan, UT 84321

### P. L. Taylor
SIATSA, La Lima, Honduras

### W. H. Thames, Jr.
Plant Science Department, Texas A&M University, College Station, TX 77840

### J. L. Murad
College of Life Sciences, Louisiana Tech University, Ruston, LA 71270

Chemical control agents are necessary for efficient production of many crops. If nematicides are to be tested and used properly, there must be an informed consensus on reliable, standardized sampling procedures and extraction methods. In selecting extraction procedures, special consideration must be given to target species and soil type. Regardless of methods used in evaluating nematode responses to nematicides, results reflect any weaknesses in the experiments. For nematode population data to be of value, soil samples must be collected at the proper time in an appropriate probing pattern. Samples also must be handled properly after collection and the appropriate extraction procedure

[1]Prepared by the Joint SON/ASTM E 35.16 Task Force on Nematode Population Response, K. R. Barker, chairman.

must be used.

No one method of collecting nematode samples for assay or extracting nematodes to evaluate nematicides can be used in all situations. Therefore, the procedures described in this chapter are presented only as guidelines. Specific procedures for sampling and extracting nematodes, however, are recommended for various combinations of crops, soil textures, sampling periods, and nematodes. If any of these procedures is to be used reliably to evaluate responses to nematode

**TABLE 43.1. Extraction procedures for combinations of nematode and soil types[a]**

| Type of soil | Types of Nematodes | | | | |
| | Sedentary forms | | | Migratory forms | |
| | Meloidogyne[b] | Heterodera spp.[c] | Criconemoides spp. | Endoparasites[d] | Ectoparasites[e] |
|---|---|---|---|---|---|
| Sandy | CF, BT, SAE-CF, SBT, SAE-BT, SFS, EE | CF, BT. SAE-CF, SBT, FC, SAE-BT, CFHS | CF, SAE-CF | CF, BT, SBT SAE-BT, SAE-CF, SFS, MC, SK | CF, BT, SBT, SFS, SAE-BT, SAE-CF |
| Clay | CF, BT, SBT, SAE-CF, EE, SAE-BT | CF, SAE-CF, BT, SAE-BT, FC, CFHS | CF, SAE-CF | CF, BT, SAE-CF, SBT, SAE-BT, MC, SK, BBT | CF, SAE-CF, BT, SBT, SAE-BT |
| Organic | CF, BT, EE | CF, BT, CFHS | CF, SAE-CF | CF, BT, MC, SK, BBT | CF, BT |

[a]See Table 43.2 for selection of specific techniques for given sampling time and Appendix 43.A for abbreviations.
[b]Also for other nematodes—*Nacobbus, Rotylenchulus,* and *Tylenchulus* spp.
[c]If hatching factor is required, use FC only.
[d]*Pratylenchus, Radopholus, Hoplolaimus;* for some plants, *Helicotylenchus, Tylenchorhynchus, Ditylenchus,* and *Aphelenchoides.*
[e]*Belonolaimus, Dolichodorus, Helicotylenchus, Hemicycliophora, Longidorus, Paratylenchus, Rotylenchus, Scutellonema, Trichodorus, Tylenchorhynchus,* and *Xiphinema.* (SFS is most efficient for large forms such as *Belonolaimus* and *Xiphinema.*)

**TABLE 43.2. Extraction procedures for use at various sampling times for nematode parasites of annual and perennial crops[a]**

| Sampling time | Types of nematodes | | | | |
| | Sedentary forms | | | Migratory forms | |
| | Meloidogyne[b] | Heterodera spp.[c] | Criconemoides[d] | Endoparasites[e] | Ectoparasites[f] |
|---|---|---|---|---|---|
| **Annuals** | | | | | |
| P_i | CF, BT, SAE-CF, SBT, SFS, SAE-BT | BT, CFHS, SBT, FC, SAE-CF *or* BT | CF, SAE-CF | CF, BT, SBT, SFS, SAE-BT | CF, BT, SAE-CF, SBT, SFS, SAE-BT |
| P_ie | BT, SBT, SAE-BT | BT, SBT, SAE-BT | CF (+VS)[d] | BT, SBT, SAE-BT | BT, SBT, SAE-BT |
| P_m | CF(?), BT, SAE-CF, SBT + EE, SAE-BT | CF(?), BT, SBT, SAE-BT | CF, SAE-CF (+VS)[d] | CF(?), BT, SBT, SAE-BT, MC, SK | CF(?), BT, SBT, SAE-BT |
| P_f | CF, SBT + EE + GI | CF, SAE-CF, SBT, FC, SAE-BT, CFHS | SAE-CF,CF | CF, SBT, SAE-CF, SFS, SAE-BT, MC, SK | CF, SAE-CF, BT, SBT, SFS, SAE-BT |
| **Perennials** | | | | | |
| P_i | CF, SAE-CF, SBT + EE + GI, SAE-BT | ... | CF, SAE-CF | CF, SAE-CF, BT, SAE-BT, SBT, SFS, MC, SK | CF, SAE-CF, BT, SBT, SFS, SAE-BT |
| P_ie | BT, SBT, SAE-BT | ... | SAE-CF,CF (+VS)[d] | BT, SBT, SAE-BT, MC, SK | BT, SBT, SAE-BT |
| P_m and P_f | Same as with annuals | | | | |

[a]See Table 43.1 for consideration of soil type, Table 43.3 for data required for sampling times; and Appendix 43.A for abbreviations.
[b]Also for other nematodes—*Nacobbus, Rotylenchulus,* and *Tylenchulus* spp.
[c]If hatching factor is required, use FC only.
[d]Need vital stain to identify living specimens for P_ie and sometimes P_m sampling.
[e]*Pratylenchus, Radopholus, Hoplolaimus;* for some plants, *Helicotylenchus, Tylenchorhynchus, Ditylenchus,* and *Aphelenchoides.*
[f]*Belonolaimus, Dolichodorus, Helicotylenchus, Hemicycliophora, Longidorus, Paratylenchus, Rotylenchus, Scutellonema, Trichodorus, Tylenchorhynchus,* and *Xiphinema.* (SFS is most efficient for large forms such as *Belonolaimus* and *Xiphinema.*)

control agents, investigators must understand the population dynamics of the target nematode on the particular crop and the specific geographic region involved. Procedures for extracting nematodes are given in the section on procedures and in Tables 43.1 and 43.2.

In evaluating the efficacy of nematode control agents, characterization of nematode population responses is as important as determining plant responses. Most nematicides exert their maximum effect shortly after application, but long-term changes in soil biology often occur. For example, the proportions of species in nematode communities often shift after chemical treatments of soil. In addition, some nematicides stimulate plant growth in the absence of nematodes (1); others affect insect, fungal, or bacterial populations or a combination of these (21, chapters 44, 45).

The most important considerations in determining nematode population responses to control agents are collection of representative soil or root samples from each plot or both, adequate mixing of these samples, and extraction of nematodes by the most appropriate procedure. Such procedures depend on the kinds and numbers of nematodes present, their characteristics, and the nature and condition of the samples, including soil texture and the time of collection. If nematode numbers are below a detectable level, an appropriate bioassay may also be used.

Appendix 43A lists abbreviations and symbols and Appendix 43B is a glossary of descriptive terms used in this chapter.

## Sampling

Sampling procedures vary, depending on the crop, nematode species, host-parasite relationships, and soil type. Ideal times for sampling (when differences in population densities between treated and nontreated plots are greatest) also vary with crop, nematode, and geographic location. *The primary objective in sampling is to collect a sample that truly represents the population in a given plot or field at a given time.* Nematode populations vary greatly both horizontally and vertically. Many species occur as aggregates (egg masses or cysts) or are skewed in distribution because of plant or soil influences. Therefore, sampling is frequently the weakest link in field evaluation of nematicides. Important considerations in soil sampling are (i) the number, diameter, and depth of cores needed to provide an adequate sample, (ii) a representative pattern of sampling to obtain reliable data on population densities, (iii) sampling at a time that reflects population density at critical stages during the growing season, (iv) condition of the soil, and (v) proper handling and storage of samples.

For most sampling, soil should be moist but not wet, preferably below 60-cm tension (percentage) of water suction (field capacity). The percentage of moisture should be determined from three to five representative samples per sampling time. After collection, certain clay or silt soils (which set hard when dry) should be dried slowly and crumbled daily during the drying process (16).

To evaluate fully the effects of chemical soil treatments on nematode populations and plant growth, four samplings should be made: a pretreatment ($P_i$) (within one week prior to treatment), an intermediate posttreatment ($P_{ie}$), a midseason ($P_m$), and a final sampling ($P_f$). To avoid excessive spring planting delays in northern areas, nematicide treatments are often applied in the fall. In such situations, sampling schedules would need to be modified accordingly ($P_i$ before treatment, $P_{ie}$ in spring). To determine nematicide effects on perennials, taking posttreatment samples in late fall and early spring is also necessary.

To standardize sampling procedures, one should use a 2.0-cm ID sampling tube or a cone-shaped tube (11) and sample to a depth of 20 cm unless otherwise indicated. In areas with hot, dry summers, cores should be taken to a depth of 30–45 cm for $P_i$. Each core (2.0 cm in diameter by 20 cm deep) provides about 62 cm³ of soil. Minimum numbers of soil cores should be collected in a systematic procedure (Fig. 43.1A) to cover each plot according to its size as follows: (i) small plots (1–5 m²), 10 cores, (ii) medium size plots (5–100 m²), 20 or more cores, (iii) large plots (greater than 100 m²), 30 or more cores.

Careful collection of soil samples by well-trained, responsible individuals is of utmost importance for reliable nematode assays. The sampling procedure, however, must be modified for different kinds of crops. With established annual row crops, sampling should be done in the row, with cores coming from the root zone, 5–12 cm from the stems (Fig. 43.1B). Special care must

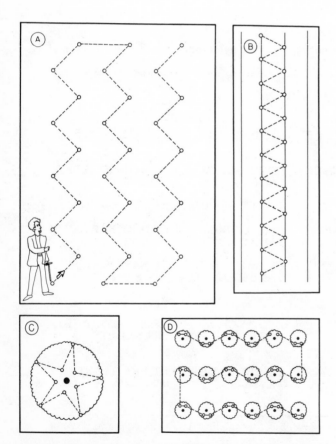

Fig. 43.1. Diagrams for collecting soil samples. A) Recommended pattern for collecting minimum of 20-30 cores for pretreatment nematode assay of large test field or plot. B) Sampling pattern for collecting soil in two center rows in four-row plot (performance is determined only from these two rows). C) Procedures for sampling single-plant plot. D) Pattern for sampling in feeder-root zone of established perennials (for small plots, additional cores should be collected from each plant).

be given to single-plant plots (Fig. 43.1C). For deep-rooted perennials such as grape and citrus, samples should be collected at two or three depths (15, 30, and 60–100 cm). For shallow-rooted perennials, eg, turfgrasses, sampling to a depth of 8–12 cm is sufficient. For ornamentals and other perennials such as fruit trees, borings should be collected in the drip line to a depth of 20 cm (Fig. 43.1D).

With certain endoparasites such as *Pratylenchus* spp. (Tables 43.2, 43.3), extracting nematodes from roots and soil at $P_m$ and $P_f$ samplings is often necessary. In these instances, a larger sampling tube (5.0-cm ID) should be used to obtain roots with the soil sample. Alternatively, one may dig and collect root samples from a minimum of ten plants per plot. A brief description of each type of sampling period follows:

$P_i$—Each plot should be sampled in a systematic manner within one week before treatment. This sampling is particularly important to establish a base population density for each individual plot, because nematode populations may vary greatly from plot to plot. By using these base population densities, the effects of various nematicides in each plot can be measured more precisely.

$P_{ie}$—This sampling can be used for relating numbers of nematodes to crop performance for fast-acting nematicides.

**CAUTION**: Do not collect soil treated with highly toxic, slow-acting materials such as organic phosphates or carbamates within 14 days after application.

*Fumigant nematicides*—These should be sampled two to three weeks after application in the same pattern as initial sampling if the treatment is broadcast. For row treatment, the sampling should be done in the row.

*Nonfumigant nematicides*—Sampling should be done four to six weeks after application, because most of these compounds are slow acting. One may use methods of extraction that depend on nematode motility or use a vital stain (12, 30) to recognize living nematodes at this sampling time.

$P_m$—The optimum time for this *critical sampling* for *determining* the *efficacy of nematicides, varies with the life cycle of the target nematode species, host-parasite relationships, crop, climate,* and *type of nematicide.* Normally it is best done when the population density in nontreated controls is high (6–12 weeks after treatment). If too much time elapses after treatment, nematode numbers in treated plots often exceed those in controls because of the larger root system and lack of competitors or predators or both in the treated plots. For many perennials, eg, turf grasses, or when little is known about the population dynamics of the target species, sampling monthly for five or six months is desirable. Final population densities ($P_f$) are often more useful than $P_m$ densities in northern geographic areas that have short growing seasons.

$P_f$—The final population is usually determined at the time of harvest for annual crops. This sampling is most useful for determining residual effects of nematicides. Nematode numbers in treated plots often exceed those in nontreated plots at this sampling. For *Meloidogyne* spp., root-knot indexes should be obtained at this time. (See standard root-knot indexes in Table 43.4.)

### Care and Conditioning of Soil Samples

Soil samples for nematode assays should be regarded as perishable and handled accordingly. Exposure to temperatures of 40°C *or above,* even for a short time, kills some species. All soil and root samples should be placed in plastic bags to prevent drying, kept out of the sun, and transported in an insulated chest.

Ideally, nematodes should be extracted no later than two days after collecting soil samples. Since storage is often necessary, however, samples should be stored at 10–15°C to keep the nematodes physiologically young

**TABLE 43.3. Nematode population data required from standardized-volume soil samples from annual and perennial crops[a]**

| Sampling time | Types of nematodes | | | | |
| | Sedentary forms | | | Migratory forms | |
| | Meloidogyne spp. | Heterodera spp. | Criconemoides spp. | Endoparasites | Ectoparasites |
|---|---|---|---|---|---|
| **Annuals** | | | | | |
| $P_i$ | Larvae in soil | Cysts and larvae in soil | Nemas in soil | Nemas in soil | Nemas in soil |
| $P_{ie}$ | Living larvae in soil | Living larvae | Living nemas in soil | Living nemas in soil | Living nemas in soil |
| $P_m$ | Larvae and eggs (soil and roots) | Larvae and cysts | Living nemas in soil | Living nemas in soil and roots | Nemas living in soil |
| $P_f$ | Above + GI | As above | Nemas in soil | Nemas in soil and roots | Nemas in soil |
| **Perennials** | | | | | |
| $P_i$ | Larvae and eggs (soil and roots) | ... | Nemas in soil | Nemas in soil and roots | Nemas in soil |
| $P_{ie}$ | Living larvae | ... | Living nemas in soil | Living nemas in soil and roots | Living nemas in soil |
| $P_m$ and $P_f$ Same as for annuals | | | | | |

[a]See Tables 43.1 and 43.2 for specific extraction procedures for specific combinations of types of nematodes and soil.

and active (38). Lower storage temperatures may be desirable in northern regions. High temperatures allow hatching of eggs and rapid aging of motile forms; low temperatures may cause chilling injury (27).

Adequate mixing of composite soil samples before removing an aliquant (usually 100 cm³) for extracting nematodes is also important when the entire sample (500 cm³ or greater) cannot be processed. Such samples should be screened first and then mixed thoroughly by a suitable procedure, eg, running them through a soil sample splitter two to four times. Also, soil samples may be mixed effectively by coning and quartering. Mixing the composite sample avoids undue variation caused by nematode aggregation. Root or other plant tissues should be cut and mixed when selecting an aliquant for extraction. Chopping tissues in blenders may be satisfactory in some instances, but this is not recommended for general use (due to release of toxic substances in some plants such as peach).

**CAUTION:** Although mixing soil is necessary, special care (gentle mixing or use of flotation procedures) must be given to samples containing *Xiphinema, Longidorus,* and *Trichodorus* spp., which are often injured and lose their motility by mechanical mixing of soil.

## Calibration and Standardization

Concentrations of all chemicals used for extracting nematodes should be expressed on a molar basis—micrograms per milliliter or gram (see specific procedures). In procedures using the centrifuge, the relative force times gravity should be determined rather than simply determining the number of revolutions per minute (rpm). When stirrers are used, rpms are sufficient. Subsamples should be based on volume of soil rather than weight, because weight varies greatly with soil moisture. The water content should be determined and reported as a percentage of oven-dry weight of soil.

TABLE 43.4. Nomograph of root-knot galling indexes for Meloidogyne spp.

| Galling index systems[a] | | | | Percentage of total root system galled |
|---|---|---|---|---|
| 0-4 | 0-5 | 1-6[b] | 0-10 | |
| 0 | 0 | 1 | 0 | 0 |
| | 1 | 2 | 1 | 10 |
| | 2 | 3 | 2 | 20 |
| 1 | | | 3 | 30 |
| | | | 4 | 40 |
| 2 | 3 | 4 | 5 | 50 |
| | | | 6 | 60 |
| 3 | | | 7 | 70 |
| | 4 | 5 | 8 | 80 |
| | | | 9 | 90 |
| 4 | 5 | 6 | 10 | 100 |

[a]Roots are scored for degree of galling using one of several root-galling indexes, all of which are comparable and relatively interchangeable. Minimum of ten root systems should be evaluated per treatment.

[b]We recommend this scheme for use in evaluating nematicides (see Fig. 43.2 for examples similar to those of Zeck [40]).

## Safety and Precautions With Specific Procedures

Certain hazards prevail for some of the methods described. When using the NaOCl extraction procedure for eggs produced by *Meloidogyne* spp. (4) and other nematodes that produce external egg masses, work should be done in a fume hood to avoid inhaling the vapors. Soil samples should not be collected within 14 days after application of highly toxic nematicides (organic phosphates or carbamates). When soil assays are made within two to four weeks after applying chemicals at any residual concentration, appropriate precautions (rubber or plastic gloves) should be taken in handling and mixing soil.

When choosing methods for extracting various nematode species, investigators must be familiar with their biology and population dynamics to select the most appropriate method for each sampling time (Tables 43.1, 43.2). Root samples are best used for endoparasites (*Meloidogyne, Pratylenchus*) and semiendoparasites (*Hoplolaimus, Helicotylenchus*) in evaluating chemical soil treatments, because a significant portion of these nematode populations may exist in the roots. Root sampling may not be necessary shortly after a chemical treatment, because the fractions of the populations in the soil reflect the relative efficacy of the test material. Many sedentary (immobile) ectoparasites can be extracted only by flotation procedures. *For soil samples collected within one to three weeks after chemical soil treatments, methods that yield only motile nematodes are best.* Vital stains (Phloxine B and new Blue R) may be used with procedures such as the centrifugal-flotation method, which yields dead as well as live specimens. Many nonfumigant nematicides act over a period of six weeks or more, causing nematode starvation and slow disappearance from the soil (18, 25). Time of early posttreatment sampling requires adjustment for such materials.

Several factors may affect the efficiency of specific extraction procedures. Certain problems, such as losing nematodes through sieve openings, occur with numerous techniques. Procedures and potential major difficulties are:

i. *Baermann methods (BF), sieving-Baermann trays (SBT)*—Excessive soil or debris, inappropriate temperature used for incubation, nematode motility, excessive microbial activity, and excessive water over soil.

ii. *Centrifugal flotation (CF)*—Inappropriate type or use of centrifuge or sieves, high amount of organic matter, nematodes primarily in roots, and incorrect sugar concentration.

iii. *Sugar-flotation sieving (SFS)*—High moisture, especially with clay soils, high clay or organic content of soil, incorrect sugar or Separan® concentration, incorrect mesh of sieves, and dark color of molasses, which if used in lieu of sugar, interferes with decanting.

iv. *Elutriators*—Incorrect rate of water flow, wrong sieve size, and soil texture (high organic matter or clay).

v. *Mist techniques*—Evaporation in dry climates, which may reduce water temperature below optimum (22–25°C at root level) for some nematodes, and toxic substances in roots (eg, peach if combined with blender).

vi. *Fenwick can (FC)*—Excessive moisture in cysts and high organic matter.

vii. *Shaker incubation*—Excessive microbial activity

and inappropriate incubation temperature (22–25°C optimum for nematode species).

viii. *Semiautomatic elutriator (SAE)*—Incorrect water, air flow, or sieves or all three and high organic matter.

ix. *Elutriator-NaOCl egg extraction (EE)*—Erratic root distribution and egg hatching before assay.

### Procedures

Selection of assay procedures depends on the kinds and numbers of nematodes present, host, the nature and condition of the samples (including soil texture), time of collection, and chemical soil treatments used. Nematode populations may consist of relatively motile to quiescent forms—cysts, individual eggs, eggs in masses, or various combinations of these. Ectoparasitic forms occur primarily in the soil, whereas high percentages of endoparasites are in roots, root fragments, or other plant parts during certain periods of the year. Actively moving larvae or adults (in roots or soil) can be extracted by (i) Baermann trays or modifications thereof, (ii) a combination of flotation or sieving or both and Baermann methods, or (iii) flotation (or elutriation) and cotton wool methods. Methods adapted to both motile and nonmotile forms include (i) sieving, (ii) centrifugal flotation, (iii) sugar-flotation sieving, and (iv) Seinhorst's two-Erlenmeyer flask sedimentation apparatus. The flotation methods are of limited use for endoparasites when most are inside the roots or other plant parts, but they are superior for ectoparasites. For the extraction of nematodes from roots, the Seinhorst-mist apparatus is ideal, but incubation on a gyratory shaker (2) is also satisfactory. Methods for extracting free eggs and egg masses are now being developed. Extraction of cysts of *Heterodera* spp. also requires special techniques, such as the Fenwick can or "heavy sugar" centrifugation (9). If population densities are below a detectable level, an appropriate bioassay may be used (eg, Rutgers tomato for most *Meloidogyne* spp.; Lee soybean for *Heterodera glycines*).

Specific outlines for each recommended method of extracting nematodes follows. Many methods that require special equipment or excessive labor inputs or both are not described. As indicated earlier, many nematodes are lost during sieving in any procedure. *Such losses often can be minimized by allowing the nematodes to settle out of suspension and decanting the excess water instead of sieving.* A second passage of the suspension may be necessary when population densities are low. Required equipment and chemicals are listed for each of the following methods.

### A. Extraction of nematodes, motile or nonmotile in soil only

1. *Centrifugal flotation (CF) (19, modified)*—This is an excellent method for routine assays, and is the best procedure when *Criconemoides* is the target genus.

a. Equipment—Centrifuge with horizontal (swinging bucket) head with 50-ml or larger tubes; operation to 420 × *g*; mechanical stirrers; 35-, 325-, and 400-mesh sieves; 100-, 150-, and 1,000-ml beakers; and soil-sample splitters (W. S. Tyler Co., Mentor, OH 44060) or coarse sieves (for mixing).

b. Chemicals—Sucrose solution, 454 g, in sufficient water to make 1 L of solution (sp gr, 1.18 or 1.33 M).

NOTE: Commercial detergents may enhance recovery by this method (39).

c. Procedure

i. Mix the soil.

ii. Place a 100-cm³ aliquant in a 1,000-ml beaker and add sufficient water to bring the total volume to 600 ml.

iii. Stir for 20 sec and allow the soil to settle for 60 sec (maximum time of 20–30 sec is best for *Criconemoides* spp.).

iv. Decant onto a 40-mesh sieve over a 325-mesh sieve.

v. Using a wash bottle, rinse the 40-mesh sieve while still over the 325-mesh sieve (excessive rinsing washes small nematodes through both sieves). For *Heterodera* spp., add a 60-mesh sieve between the 40- and 325-mesh sieves to collect cysts.

vi. Wash the debris and nematodes from the 325-mesh sieve into a 150-ml beaker.

vii. Pour the washings (step vi) into 50-ml centrifuge tubes.

viii. Place the tubes in the centrifuge (*be sure to balance the tubes*).

ix. Centrifuge at 420 × *g* for 5 min.

x. Decant water from the tubes (nematodes are in the soil pellet in the bottom of the tubes).

xi. Refill centrifuge tubes with sucrose solution and mix with stirring rod or vibrator mixer.

xii. Centrifuge for 30 sec at 420 × *g* (nematodes remain suspended in sugar solution). *Do not* use the brake on certain centrifuges, as it may cause sufficient vibration to dislodge the pellet.

xiii. Decant the sugar solution-nematode suspension onto a 400-mesh sieve (pour slowly with small nematodes).

xiv. Rinse the residue and nematodes from the 400-mesh sieve into a 150-ml beaker (about 20 ml of water, suitable for making counts).

2. *Flotation-sieving method (SFS) (5)*—This method is well suited for extracting large nematodes from sandy soils. It generally is not as efficient as CF, but is acceptable if centrifuges are not available.

a. Equipment—Mechanical stirrers; 35-, 325-, and 400-mesh sieves; 100-, 150-, and 1,000-ml beakers; wash bottles or mist hoses; and soil sample splitter or sieves (for mixing).

b. Chemicals—Sucrose solution (0.7*M*); Separan NP10® (The Dow Chemical Company, Midland, MI 48640), final concentration of 12.5 µg/ml. NOTE: Molasses (26) can be used in lieu of sucrose (sp gr, 1.1).

c. Procedure

i. Mix soil.

ii. Place a 100-cm³ subsample in a 1,000-ml beaker and add sufficient 0.7*M* sucrose and Separan (12.5 µg/ml) solution to bring total volume to 500 ml.

iii. Stir with a motorized stirrer for 20 sec.

iv. Allow the soil to settle for about 2 min.

v. Decant the liquid onto a 40-mesh sieve over a 325-mesh sieve.

vi. Rinse the 40-mesh sieve while it is still over the 325-mesh sieve.

vii. Using a wash bottle, rinse the nematodes and debris from the 325-mesh sieve into a 150-ml beaker (about 50 ml of water).

viii. Swirl the 150-ml beaker and allow the contents to settle for 5–10 sec.

ix. Decant the nematode suspension onto a 400-mesh sieve and rinse the residue into a 150-ml beaker (about 20 ml of water).

3. *Semiautomatic elutriator (SAE) (3)*—This new approach to nematode extraction includes an elutriator similar to Oostenbrink's (22), plus a sample splitter and sieve shaker. It may be used in combination with Baermann trays (BT) or centrifuge (CF).

a. Equipment—Simple elutriator, Oostenbrink type; aqueous sample splitter; water and air supplies; motorized sieve shaker; 10-, 24-, 40-, 60-, 400-, and 500-mesh sieves. (May be semiautomated with time clocks, etc. [3]).

b. Procedure

i. Add 500 cm$^3$ of nonmixed soil to the elutriator (with air and water flowing at desired rates).

ii. Run the elutriator for 3 min, catching roots on the 40-mesh sieve over a sample splitter and "free" nematodes on the 400-mesh sieve on the motorized shaker.

iii. Rinse the sieves.

—For *Meloidogyne* spp., process the roots from the 40-mesh sieve by NaOCl method (4).

—For *Pratylenchus* and other migratory endoparasites, roots are trapped on the 40-mesh sieve and incubated in the mist chamber.

—Cysts of *Heterodera* may be collected on 60-mesh sieve under 10- and 24-mesh sieves.

—Any fraction (1/15, 1/5, or all) of nematodes in the soil are collected on sieves on the shaker. *Criconemoides* and related genera may be cleaned by centrifugation (first method described), and other species by BT, SFS, or CF with 500-mesh sieves.

4. *Decanting and sieving (CDS)*—The simple, modified version of Cobb's (7) sifting and gravity method, which is described below, is useful in extracting nematodes for inoculation purposes and routine assays when combined with BT. Addition of Separan eliminates the need for numerous sievings. Thorne (35) has described the original method, and Townshend (36) has illustrated it.

a. Equipment—10-L pails; 10-, 25-, 50-, 100-, 200-, 325-, and 400-mesh sieves; 50-, 150-, 250-, and 500-ml beakers.

b. Chemicals—Separan with sucrose can be used to reduce the steps involved. A stirrer and 600-ml suspension with 100 cm$^3$ of soil should be used as under SFS.

c. Procedure (short, modified method)

i. Place 500 cm$^3$ of soil in a large pan or pail and cover well with water containing Separan (12.5 $\mu$g/ml). Thoroughly break all lumps and allow them to settle about 2 min.

ii. Decant onto a 40-mesh sieve over a 325-mesh sieve.

iii. Resuspend the original soil in water and repeat step ii if maximum recovery is desired.

iv. Combine the washings from steps ii and iii, stir, and allow them to settle for 10 sec.

v. Decant through a 400-mesh sieve and rinse the residue (nematodes) into a clean beaker for counting (may place on BT for cleaner samples).

**B. Extraction of motile species of nematodes**

1. *Baermann trays (BT) (37)*—This procedure is useful for extracting nematodes from small soil samples, root fragments, larvae of *Heterodera* (in the spring), and debris coming from elutriators.

a. Equipment—17.5-cm diameter (20-mesh) plastic screen (sieve-type) support with 4-mm high support legs or similar metal unit, wet-strength tissues, epoxy resin-coated 20-cm diameter aluminum pie pans.

b. Procedure

i. Mix the soil.

ii. Place a 100-cm$^3$ aliquant of soil uniformly over the tissue, which is superimposed on a plastic or stainless steel screen (do not use copper screens).

iii. Place the screen with soil in a 20-cm pie pan and add water just to cover the soil.

iv. Incubate the samples at 21–24°C (cover or stack to reduce evaporation). Add water as needed.

v. Collect the nematodes from the pans after three days. If dirty, clean by pouring through a 500-mesh sieve. For maximum recoveries, nematodes should be collected over a 1–14-day period.

2. *Combination of BT with elutriation (BTE) or sieving (BTS)*—The combination of Baermann trays with elutriation (or sieving), a modification of the Christie-Perry (6) procedure, is still invaluable for extracting nematodes from soil and root fragments.

a. Equipment—10-, 24-, and 400-mesh sieves; elutriator; screen supports; 15–20-cm diameter pie pans (stainless steel or resin treated); wet-strength facial tissue, milk filter, or muslin filter.

b. Procedure

i. Add a nonmixed 500-cm$^3$ soil sample to the elutriator.

ii. Collect the root and soil fractions from 10- and 400-mesh sieves, respectively (see semiautomatic elutriator), or use the modified Cobb's decanting-sieving (CDS) method.

iii. Place these fractions on facial tissues supported on screen.

iv. Carefully add sufficient water just to cover the residue. Add more water as needed with time.

v. Collect the nematodes after incubation at 21–24°C for three days (for maximum recoveries, collect them over a 10–14-day period).

3. *Semiautomatic elutriator (SAE)*—See procedure A3.

4. *Seinhorst elutriator (29)*

5. *Oostenbrink elutriator (22)*

**C. Extraction of cysts**—In assays for cysts of *Heterodera,* a homogenizer (Ten-Broeck) (32) should be used to crush extracted cysts to determine numbers of eggs.

1. *Centrifugation with heavy sugar (CFHS) (9, modified)*—This procedure is useful for isolating cysts and larvae of *Heterodera* spp., but problems may be encountered with some fine clay soils.

a. Equipment—Same as for CF for routine use, except 25- and 100-mesh sieves are needed.

b. Chemicals—1.8$M$ sucrose solution (615 g of sucrose) dissolved in sufficient warm water to make 1 L of solution (1.23 sp gr).

c. Procedure

i. Wash 100 cm$^3$ of soil through a 25-mesh sieve and collect it in a beaker (use about 1.0 L of water).

ii. Mix the solution thoroughly and allow it to settle for 10–15 sec.

iii. Pour the supernatant through a 100-mesh screen (add a 325-mesh sieve for larvae).

iv. Wash any residue from the screen into a centrifuge tube or tubes with 1.8$M$ sucrose solution.

v. Centrifuge at 420 $\times$ $g$ for 2.5 min.

vi. Collect the supernatant on a 100-mesh screen (add a 325-mesh sieve for larvae).

vii. Rinse thoroughly.

viii. Wash the sample into a beaker, using about 20 ml of water.

ix. Crush the cysts with a Ten-Broeck homogenizer and count the eggs and larvae.

2. *Modified Fenwick can (FC) (13,14,31,32)*—This procedure is useful for extracting cysts from dry soil, but centrifugation (CFHS) and other methods are becoming more widely used.

a. Equipment—Modified Fenwick can, 60–80-mesh sieve with bowl, camel's hair brush (No. 00, 0, or 1), homogenizer (Ten-Broeck) plain glass slide, 100-ml bottle, pipette, aquarium pump, 400-ml Erlenmeyer flask, 250-ml volumetric flask, funnel, petri dish, 18.5-cm diameter filter papers, acetone or acetone-carbon tetrachloride 3:1.

b. Procedure

i. Mix the soil thoroughly.

ii. Fill the modified Fenwick can with water. Place the sample of 100 $cm^3$ of well-mixed soil in the top sieve (20.5-cm diameter 18- or 24-mesh sieve).

iii. Wash the sample into the apparatus via the funnel. The coarse material is retained on the top sieve, heavy soil particles such as sand sink to the bottom of the apparatus, and the floating cysts are carried off over the overflow collar.

iv. Cysts, root debris, and other particles are collected on a 20-cm diameter sieve (60–80 mesh). Particles that are 175 $\mu$m and smaller pass with water through the sieve.

v. After washing, dry the debris at room temperature. Transfer the somewhat dried debris retained on the sieve to a 250-ml capacity flask.

vi. Pour technical acetone, or a mixture of three parts acetone and one part carbon tetrachloride, into a volumetric flask up to level 1 (neck of flask). Shake the flask and fill it completely. **CAUTION**: Use the exhaust hood.

vii. After 1 min, decant the floating cysts and debris through a filter paper (18.5-cm diameter) in a glass funnel into a second, or Erlenmeyer flask while rotating the original flask. The acetone passes through the filter.

viii. Place the filter in a petri dish and view it through a dissecting microscope (magnification, ×50) with overhead light. Pick up the cysts with a camel's hair brush (No. 00, 0, or 1) and transfer them to a small watch glass with moist filter paper. Identify the cysts under the dissection microscope using an overhead light. Use a camel's hair brush to transfer the cysts of the desired species into a small drop of water in the glass tube of the homogenizer. Place the piston in the tube and *carefully* rotate it by hand. Pour the eggs and larvae that were released from the cysts into a bottle. Fill the bottle with water, up to 100 ml. Mix the suspension carefully using compressed air. Pipette out two 10-ml aliquants and place them in 10-ml Perspex dishes for counting.

3. *Semiautomatic elutriator (SAE)*—This new method has promise for recovering cysts of *Heterodera* (see description above).

4. *Seinhorst's extraction procedure for Heterodera cysts from moist soil (29)*

### D. Separation of nematodes from plant tissues

1. *Modified Seinhorst mist apparatus (mist chamber [MC]) (28)*—This is the most widely used method for obtaining nematodes from plant tissues.

a. Equipment—Time clock, water mixer-warmer,

water regulator and filter, solenoid switch, cover (fiberglass, Plexiglas, or other suitable material) with doors to funnel racks or supports, superfine nozzles, plastic petri dishes, 10-cm diameter glass funnels with rubber tubing, 1-L plastic cups with holes in the bottom, clamps, 500-mesh sieve (specifications available from K. R. Barker).

b. Procedure

i. Place a representative sample of roots or other plant tissues in plastic cups superimposed over an open Baermann funnel. This funnel is supported over a second Baermann funnel by a plastic petri dish with a 2.5-cm hole in the center. Use wet-strength facial tissues in the cups to reduce debris.

ii. Set the time clock to regulate desired mist (on 1 min, off 2 min). Adjust the water mixer to the flow rate that gives a temperature of 24°C.

iii. Collect the nematodes from the funnels every three to five days through a 14-day period (nematodes such as *Pratylenchus* spp. will continue to emerge for weeks).

iv. Concentrate and clean the nematode suspensions if needed with a 500-mesh sieve.

2. *Shaker (SK) (2,33)*—Several researchers have found this shaker procedure to yield numbers of *Pratylenchus* and other species from roots similar to those obtained from mist chambers.

a. Equipment—Gyratory shaker, 125-ml flasks.

b. Chemicals—Ethoxyethyl mercuric chloride (Aretan®) and streptomycin sulfate or other suitable antibiotics.

c. Procedure

i. Wash roots and cut into 1–2-cm segments.

ii. Place a representative sample of root tissue (0.5–5.0 gm) in a 125-ml flask.

iii. Cover the tissue with a mixture of 10 $\mu$g/ml of ethoxyethyl mercuric chloride and 50 $\mu$g/ml of streptomycin sulfate.

iv. Incubate the mixture at 100 rpm for 48 hr.

v. Collect nematodes on a 325- or 400-mesh sieve; rinse them into a 150-ml beaker and count them.

3. *Blender-Baermann tray (BBT) (34)*—Although many problems are encountered with this procedure, it is useful for limited situations.

a. Equipment—Blender, beakers, 325-mesh sieve, and Baermann funnels or pans.

b. Chemicals—Ethoxyethyl mercuric chloride (Aretan) in a 10 $\mu$g/ml solution (other materials such as antibiotics or certain fungicides [captan] may be used instead of Aretan).

c. Procedure

i. Rinse the plant tissues until they are free of soil.

ii. Weigh the tissues to be processed and place them in the blender. Use not more than 50 g of tissue (fresh weight basis) with 200 ml of water in 1.9-L (2-qt) blender.

iii. Homogenize the mixture for 15 sec.

iv. Decant the suspension from the blender, including the rinse water, over a 325-mesh sieve.

v. Using a wash bottle containing antibiotic solution (see SK method), wash the debris from the sieve into a beaker.

vi. Pour the liquid that passed through the sieve over the sieve again.

vii. Combine the material collected on the sieve in the second passage with that from the first passage.

viii. Gently decant the suspension of material collected on the sieve over the filter of the Baermann apparatus.

ix. Fill the Baermann apparatus with sufficient antibiotic solution to barely submerge the debris on the filter.

x. Replenish the incubation solution with water as needed.

xi. Collect the nematodes after incubation at 21–24°C for two to three days.

In addition to the combined procedure described above, blending followed by wet sieving is useful for certain nematodes such as *Radopholus similis* on banana.

### E. Extraction of nematode eggs

1. *Elutriation-dissolution of gelatinous matrices of egg masses and staining (EE) (4)*—Variation of egg numbers in the field sometimes causes problems with this assay method. It is useful, however, for midseason to late-season assays of *Meloidogyne* spp. and other nematodes that form egg masses.

a. Equipment—Sample splitter or semiautomatic elutriator, 150- and 600-ml beakers, stirrers, 15-cm household sieve, 40- and 500-mesh sieves, exhaust hood, 5-ml dipper, compressed air (for cleaning sample splitter and use with elutriator).

b. Chemicals—Sodium hypochlorite (NaOCl), antifoam spray, acid fuchsin, and lactic acid.

c. Procedure

i. Mix the soil.

ii. With water flow adjusted to 350 ml/sec, or 60–80 ml/sec if air-water mixture is used, place a 500-cm³ aliquant of soil in the elutriator. Turn the water on for 2–3 min, trapping root fragments on a 40-mesh sieve.

iii. With a spray nozzle, wash the residue off the sieve into a 600-ml beaker and add water to 200 ml.

iv. Add 20 ml of 5.25% NaOCl and spray with an antifoam agent.

v. Stir the mixture under an exhaust hood for 10 min.

vi. Using a household sieve to retain debris, take a 5-ml sample with a dipper and rinse into a 150-ml beaker.

vii. Pour this subsample onto a 500-mesh sieve and wash the eggs from the sieve into a clean 150-ml beaker (should have 20–25-ml suspension).

viii. Add two drops of 0.35% acid fuchsin in 25% lactic acid and boil for 1 min under the exhaust hood (a microwave oven is useful).

ix. Allow the mixture to cool before counting.

2. *Extraction of eggs by centrifugation (8,10,15,17)*

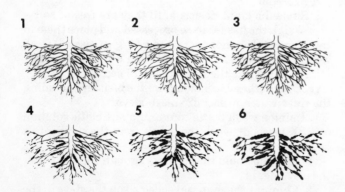

Fig. 43.2. Scheme for rating field and greenhouse infestation and bioassay evaluation of *Meloidogyne* spp. See Table 43.4 (GI 1-6) for descriptions of each root-knot class.

### F. Bioassays of nematode populations
(24)—With low natural populations of nematodes such as *Meloidogyne, Heterodera,* and *Ditylenchus* spp., bioassays are the most reliable procedures. The following is a typical bioassay for *Meloidogyne*.

1. *Equipment*—Fumigated sandy loam soil, 10-cm clay pots, 15-cm plastic pots, pot labels, three-week-old Rutgers or other susceptible seedlings, nutrient solution, and greenhouse space. Other recommended tomato cultivars are Person A1, Heinz 1350, Marglobe, Bonny Best, and Manapal.

2. *Procedure*

a. Fill the bottom 2 cm of the clay pots with sterile soil.

b. Add 250 cm³ of test soil to a given clay pot.

c. Transplant the tomato seedling to a 10-cm pot, and fill the remainder of the pot with sterile soil (place the 10-cm pot in the 15-cm plastic pot, which minimizes contamination).

d. Grow plants for five to six weeks at 24–28°C, providing nutrients and water as needed. **Do not overwater.**

e. Harvest the plants by washing the roots out carefully. Rate the nematode development by use of the gall index (Table 43.4). One may also determine the numbers of eggs (see NaOCl method).

The third root-knot index (1-6) in Table 43.4 (nomograph) is recommended for evaluating effects of nematicides on *Meloidogyne* spp. treated in the field or greenhouse. Figure 43.2 illustrates examples of each class. The other schemes in the nomograph (Table 43.4), which DiSanzo et al. (chapter 38) prepared, are acceptable.

Some confusion may be encountered in rating galling caused by different *Meloidogyne* spp., since the number and size of galls incited by a given number of larvae vary with species and host plant. Concentrating primarily on the proportion of roots galled rather than the size of galls, one can minimize this problem. When the cause of galls is uncertain, roots should be stained with 0.05% acid fuchsin in lactophenol and cleared in lactophenol (20,35). The contents of galls (numbers of larvae, eggs, and adults) can then be determined.

Gall indexes for *Meloidogyne* spp. should be limited to galling. If the rating of necrosis is desired, a separate rating system such as Powell et al. (23) developed is suitable for rating necrosis associated with root-knot. Their classification system is 0, no necrosis; 1, less than 10% of the root system necrotic; 2, 11–25% necrotic; 3, 25–50% necrotic; 4, 51–75% necrotic; and 5, 76–100% necrotic. Based on all root systems per treatment, the disease or necrosis index is as follows:

$$\frac{\begin{bmatrix} \text{Number of} \\ \text{plants in} \\ \text{class } 1 \times 1 \end{bmatrix} + \begin{bmatrix} \text{Number of} \\ \text{plants in} \\ \text{class } 2 \times 2 \end{bmatrix} + \cdots \begin{bmatrix} \text{Number of} \\ \text{plants in} \\ \text{class } 5 \times 5 \end{bmatrix} \times 100}{\text{Number of plants in treatment} \times 5}$$

A similar lesion or necrosis index is also useful for rating banana roots for infection by *Radopholus similis*.

### Reporting Results

Replicate and treatment mean results for the following should be determined:

i. Mean numbers of nematodes per 100 cm³ or greater volume of soil (use of a mechanical stage on the microscope to control counting dish increases efficiency in making nematode counts).

ii. Numbers of nematodes per gram of root or other tissue (fresh or dry weight) when endoparasites are extracted from same plots. *Total nematodes per plant for greenhouse experiments should be included.*

iii. For *Heterodera* spp., numbers of larvae and cysts as well as numbers of eggs per cyst.

iv. For *Meloidogyne* spp., numbers of eggs per unit volume of soil or weight of roots (fresh or dry weight).

v. For *Meloidogyne* spp., root-knot indexes based on the index given under bioassay procedure. (Table 43.3 summarizes the data required for various types of nematodes and host plants.)

vi. Plant response (chapter 44).

Precision in determining nematode population responses to nematicides depends on how the samples are collected, handled, stored, extracted, and counted. With proper execution in each of these areas, variation usually ranges from 10–30%. Oostenbrink (personal communication) found that the usual coefficient of variation for this error is about 25%. This error varies, however, with nematode species and population densities. In estimating egg numbers for *Meloidogyne* spp., this coefficient of variation is considerably greater. *Thus, proper statistical analyses of all data must be used to estimate the variance for a given experiment, including coefficient of variation and standard error.* Since some zero readings and considerable variation are normally encountered, the $\log_{10}$ (P + 1) transformation should be used for nematode population data.

## Literature Cited

1. ALTMAN, J., and K. M. TSUE. 1965. Changes in plant growth with chemicals used as soil fumigant. Plant Dis. Rep. 49:600-602.
2. BIRD, G. W. 1971. Influence of incubation solution on the rate of recovery of *Pratylenchus brachyurus* from cotton roots. J. Nematol. 3:378-385.
3. BYRD, D. W., Jr., K. R. BARKER, H. FERRIS, C. J. NUSBAUM, W. E. GRIFFIN, R. H. SMALL, and C. A. STONE. 1976. Two semi-automatic elutriators for extracting nematodes and certain fungi from soil. J. Nematol. 8:206-212.
4. BYRD, D. W., Jr., H. FERRIS, and C. J. NUSBAUM. 1972. A method for estimating numbers of eggs of *Meloidogyne* spp. in soil. J. Nematol. 4:266-269.
5. BYRD, D. W., Jr., C. J. NUSBAUM, and K. R. BARKER. 1966. A rapid flotation-sieving technique for extracting nematodes from soil. Plant Dis. Rep. 50:954-957.
6. CHRISTIE, J. R., and V. G. PERRY. 1951. Removing nematodes from soil. Proc. Helminthol. Soc. Wash. 18:106-108.
7. COBB, N. A. 1918. Estimating the nema population of soil. U.S. Dep. Agric. Tech. Circ. 1:1-48.
8. COOLEN, W. A., and C. J. D'HERDE. 1972. A method for the quantitative extraction of nematodes from plant tissue. Min. Agric. Ghent. Belgium.
9. DUNN, R. A. 1969. Extraction of cysts of *Heterodera* species from soils by centrifugation in high density solutions. J. Nematol. 1:7 (Abstr.).
10. DUNN, R. A. 1973. Extraction of eggs of *Pratylenchus penetrans* from alfalfa callus and relationship between age of culture and yield of eggs. J. Nematol. 5:73-74.
11. ESSER, R. P., J. B. MacGOWAN, and H. M. VAN PELT. 1965. Two new nematode subsampling tools. Plant Dis. Rep. 49:265-267.
12. FENNER, L. M. 1962. Determination of nematode mortality. Plant Dis. Rep. 46:383.
13. FENWICK, D. W. 1952. Sampling techniques and the limits of their applicability. Proc. Int. Nematol. Symp. FAO Rome. pp. 8-12.
14. FENWICK, D. W. 1952. The estimation of the cyst contents of the potato-root eelworm *H. rostochiensis.* J. Helminthol. 26:55-68.
15. FLEGG, J. J. M., and D. G. McNAMARA. 1968. A direct sugar-centrifugation method for the recovery of eggs of *Xiphinema, Longidorus* and *Trichodorus* from soil. Nematologica 14:156-157.
16. GOODEY, J. B. 1963. Laboratory methods for work with plant and soil nematodes. (Ed. 4) Min. Agric. Fish. Food Tech. Bull. 2.
17. GOORIS, J., and C. J. D'HERDE. 1972. A method for the quantitative extraction of eggs and second stage juveniles of *Meloidogyne* spp. from soil. Min. Agric. Ghent. Belgium.
18. HOUGH, A., and I. J. THOMASON. 1975. Effects of aldicarb on the behavior of *Heterodera schachtii* and *Meloidogyne javanica.* J. Nematol. 7:221-229.
19. JENKINS, W. R. 1964. A rapid centrifugal-flotation technique for separating nematodes from soil. Plant Dis. Rep. 48:692.
20. McBETH, C. W., A. L. TAYLOR, and A. L. SMITH. 1941. Note on staining nematodes in root tissue. Proc. Helminthol. Soc. Wash. 8:26.
21. NUSBAUM, C. J., and F. A. TODD. 1970. The role of chemical soil treatments in the control of nematode-disease complexes of tobacco. Phytopathology 60:7-12.
22. OOSTENBRINK, M. 1960. Estimating nematode populations by some selected methods. In SASSER, J. N., and W. R. JENKINS (eds.). Nematology—Fundamentals and Recent Advances With Emphasis on Plant Parasitic and Soil Forms. University of North Carolina Press: Chapel Hill. pp. 85-102.
23. POWELL, N. T., P. L. MELENDEZ, and C. K. BATTEN. 1971. Disease complexes in tobacco involving *Meloidogyne incognita* and certain soil-borne fungi. Phytopathology 61:1332-1337.
24. POWELL, W. M., and C. J. NUSBAUM. 1963. Investigations on the estimation of plant parasitic nematode populations for advisory purposes. N.C. Agric. Exp. Stn. Tech. Bull. 156.
25. RICH, J. R., and G. W. BIRD. 1973. Inhibition of *Rotylenchulus reniformis* penetration of tomato and cotton roots with foliar applications of oxamyl. J. Nematol. 5:221-224.
26. RODRIGUEZ-KABANA, R., and P. S. KING. 1972. The use of sugarcane molasses as an economical substitute for sugar in the extraction of nematodes from soil by the flotation-sieving technique. Plant Dis. Rep. 56:1093-1096.
27. SAYRE, R. M. 1964. Cold-hardiness of nematodes. I. Effects of rapid freezing on the eggs and larvae of *Meloidogyne incognita* and *M. hapla.* Nematologica 10:168-179.
28. SEINHORST, J. W. 1950. De betekenis van de toestand van de grond voor het optreden van aantasting door het stengelaaltje (*Ditylenchus dipsaci* [Kuhn] Filipjev). Tijdschr. Plantenziekten 56:289-348.
29. SEINHORST, J. W. 1964. Methods for the extraction of *Heterodera* cysts from not previously dried soil samples. Nematologica 10:87-94.
30. SHEPHERD, A. M. 1962. New Blue R, a stain that differentiates between living and dead nematodes. Nematologica 8:201-208.
31. s'JACOB, J. J., and J. V. BEZOOIJEN. 1975. A Manual for Practical Work in Nematology. (Ed. 2) Wageningen.
32. SOUTHEY, J. F. (ed.). 1970. Laboratory methods for work with plant and soil nematodes. Tech. Bull. Min. Agric. Fish. Food 2.
33. SZCZYGIEL, A. 1962. Methods of extracting nematodes of the genus *Pratylenchus* Filipjev, 1936, from the roots of plants. Ekologia Polska, Ser. B. 8:153-159.
34. TAYLOR, A. L., and W. Q. LOEGERING. 1953. Nematodes associated with root lesions in abaca. Turrialba 3:8-13.
35. THORNE, G. 1961. Principles of Nematology. McGraw-Hill Inc.: New York.
36. TOWNSHEND, J. L. 1962. An examination of the efficiency of the Cobb decanting and sieving method. Nematologica 8:293-300.
37. TOWNSHEND, J. L. 1963. A modification and evaluation of the apparatus for the Oostenbrink direct cottonwool filter extraction method. Nematologica 9:106-110.
38. VAN GUNDY, S. D., A. F. BIRD, and H. R. WALLACE. 1967. Aging and starvation in larvae of *Meloidogyne*

*javanica* and *Tylenchulus semipenetrans.* Phytopathology 57:559-571.

39. WEHUNT, E. J. 1973. Sodium-containing detergents enhance the extraction of nematodes. J. Nematol. 5:79-80.

40. ZECK, W. M. 1971. A rating scheme for field evaluation of root-knot nematode infestations. Pflanzenschutz-Nachr. (Am. Ed.) 24:141-144.

# Bibliography

ANDERSSON, S. 1970. A method for the separation of *Heterodera* cysts from organic debris. Nematologica 16:222-226.

ANSCOMBE, F. J. 1950. Soil sampling for potato root eelworm cysts. A report presented to the Conference of Advisory Entomologists. Ann. Appl. Biol. 37:286-295.

AYALA, A., J. ROMAN, and A. C. TARJAN. 1963. Comparison of four methods for isolating nematodes from soil samples. J. Agric. Univ. P. R. 47:219-225.

BARKER, K. R., G. V. GOODING, Jr., A. S. ELDER, and R. E. EPLEE. 1972. Killing and preserving nematodes in soil samples with chemicals and microwave energy. J. Nematol. 4:75-79.

BARKER, K. R., and C. J. NUSBAUM. 1971. Diagnostic and advisory programs. In ZUCKERMAN, B. M., W. F. MAI, and R. A. ROHDE (eds.). Plant parasitic nematodes. Vol. 1. Academic Press: New York. pp. 281-301.

BARKER, K. R., C. J. NUSBAUM, and L. A. NELSON. 1969. Effects of storage temperature and extraction procedure on recovery of plant-parasitic nematodes from field soils. J. Nematol. 1:240-247.

BARKER, K. R., C. J. NUSBAUM, and L. A. NELSON. 1969. Seasonal population dynamics of selected plant-parasitic nematodes as measured by three extraction procedures. J. Nematol. 1:232-239.

CAIRNS, E. J. 1960. Methods in nematology: A review. In SASSER, J. N., and W. R. JENKINS (eds.). Nematology—Fundamentals and Recent Advances With Emphasis on Plant Parasitic and Soil Forms. University of North Carolina Press: Chapel Hill. pp. 33-84.

CAVENESS, F. E., and H. J. JENSEN. 1955. Modification of the centrifugal-flotation technique for the isolation and concentration of nematodes and their eggs from soil and plant tissue. Proc. Helminthol. Soc. Wash. 22:87-89.

CHAPMAN, R. A. 1957. The effects of aeration and temperature on the emergence of species of *Pratylenchus* from roots. Plant Dis. Rep. 41:836-841.

CHITWOOD, B. G., and J. FELDMESSER. 1948. Golden nematode population studies. Proc. Helminthol. Soc. Wash. 15:43-55.

DAS, P. K., and Y. S. RAO. 1971. On the optimal sampling time for assessment of nematode populations in rice soils. Curr. Sci. 40:17-18.

GOOD, J. M., and J. FELDMESSER. 1967. Plant nematicides and soil fumigants. Dev. Ind. Microbiol. 8:117-123.

GOOD, J. M., J. N. SASSER, and L. I. MILLER. 1963. A suggested guide for reporting experiments on nematicidal chemicals. Plant Dis. Rep. 47:159-163.

GOODEY, T. 1963. Soil and freshwater nematodes. (Rewritten by J. B. Goodey). John Wiley and Sons: New York.

GOWEN, S. R., and J. E. EDMUNDS. 1973. An evaluation of some simple extraction techniques and the use of hydrogen peroxide for estimating nematode populations in banana roots. Plant Dis. Rep. 57:678-681.

HARRISON, M. B. 1967. Influence of nematicidal treatments on nematode populations. Phytopathology 57:650-652.

HUSSEY, R. S., and K. R. BARKER. 1973. A comparison of methods of collecting inocula of *Meloidogyne* spp., including a new technique. Plant Dis. Rep. 57:1025-1028.

JONES, F. G. W. 1955. Quantitative methods in nematology. Ann. Appl. Biol. 42:372-381.

JONES, F. G. W. 1956. Soil populations of beet eelworm (*Heterodera schachtii* Schm.) in relation to cropping. II. Microplot and field plot results. Ann. Appl. Biol. 44:25-56.

KABLE, P. F., and W. F. MAI. 1968. Overwintering of *Pratylenchus penetrans* in a sandy loam and a clay loam at Ithaca, NY. Nematologica 14:150-152.

KIMPINSKI, J., and H. E. WELCH. 1971. Comparison of Baermann funnel and sugar flotation extraction from compacted and non-compacted soils. Nematologica 17:319-320.

KLEYBURG, P., and M. OOSTENBRINK. 1959. Nematodes in relation to plant growth. I. The nematode distribution pattern of typical farms and nurseries. Neth. J. Agric. Sci. 7:327-343.

LEWIS, T., and L. R. TAYLOR. 1967. Introduction to experimental ecology. Academic Press: New York.

METLITSKII, O. Z. 1971. A rapid method for isolating nematodes from soil. Zashch. Rast. 9:41-42.

MEYER, K. B., and B. V. NELSON. 1971. An improved method for concentrating nematodes from the Baermann apparatus. Lab. Exp. Biol. Univ. Mass. pp. 899-900.

MORIARTY, F. 1960. Laboratory errors associated with the estimation of the population density of *Heterodera* species in soil. Ann. Appl. Biol. 48:665-680.

MURPHY, P. W. 1962. Extraction methods for soil animals. II. Mechanical methods. In MURPHY, P. W. (ed.). Progress in Soil Zoology. Butterworths: London. pp. 115-155.

NUSBAUM, C. J., and K. R. BARKER. 1971. Population dynamics. In ZUCKERMAN, B. M., W. F. MAI, and R. A. ROHDE (eds.). Plant Parasitic Nematodes. Vol. 1. Academic Press: New York. pp. 303-323.

OOSTENBRINK, M. 1966. Major characteristics of the relations between nematodes and plants. Meded. Landbouwhogesch. Wageningen 66:1-46.

OOSTENBRINK, M. 1970. Comparison of techniques for population estimation of soil and plant nematodes. In PHILLIPSON, J. (ed.). Methods of Study in Soil Ecology. UNESCO: Paris. pp. 249-255.

OOSTENBRINK, M. 1972. Evaluation and integration of nematode control methods. In WEBSTER, J. M. (ed.). Economic Nematology. Academic Press: New York. pp. 497-514.

PATEL, G. J., M. V. DESAI, and H. M. SHAH. 1969. A sampling method for estimation of nema population in case of bajri (*Pennisetum typhoideum* S & H) grown in sandy loam soils of Gujarat. All India Nematol. Symp. New Delhi, Aug. 21-22, p. 6 (Abstr.).

PITCHER, R. S., and J. J. M. FLEGG. 1968. An improved final separation sieve for the extraction of plant-parasitic nematodes from soil debris. Nematologica 14:123-127.

REED, J. F. 1953. Sampling soils for chemical tests. Better Crops with Plant Food Magazine. Reprint DD-10-53.

SASSER, J. N., K. R. BARKER, and L. A. NELSON. 1975. Correlations of field populations of nematodes with crop growth responses for determining relative involvement of species. J. Nematol. 7:193-198.

SEINHORST, J. W. 1956. The quantitative extraction of nematodes from soil. Nematologica 1:249-267.

SEINHORST, J. W. 1962. Extraction methods for nematodes inhabiting soil. In MURPHY, P. W. (ed.). Progress in Soil Zoology. Butterworths: London. pp. 243-256.

SEINHORST, J. W. 1962. Modifications of the elutriation method for extracting nematodes from soil. Nematologica 8:117-128.

SEINHORST, J. W. 1966. Killing nematodes for taxonomic study with hot f.a. 4:1. Nematologica 12:178.

SEINHORST, J. W. 1970. Dynamics of populations of plant parasitic nematodes. Annu. Rev. Phytopathol. 8:131-156.

SEINHORST, J. W. 1970. Separation of *Heterodera* cysts from organic debris in ethanol 96%. Nematologica 16:330.

SEINHORST, J. W. 1973. Principles and possibilities of determining degrees of nematode control leading to maximum returns. I. Protection of one crop sown or planted soon after treatment. Nematologia Med. 1:93-105.

SEINHORST, J. W., and H. DEN OUDEN. 1966. An improvement of Bijloo's method for determining the egg content of *Heterodera* cysts. Nematologica 12:170-171.

SKELLAM, J. G. 1962. Estimation of animal populations by extraction processes considered from the mathematical standpoint. In MURPHY, P. W. (ed.). Progress in Soil Zoology. Butterworths: London. pp. 26-36.

SPEARS, J. F. 1968. The Golden Nematode Handbook. (Survey, Laboratory, Control, and Quarantine Procedures).

Agric. Handbook 353. U.S. Dep. Agric.

TARJAN, A. C. 1972. Observations on extracting citrus nematodes, *Tylenchulus semipenetrans,* from citrus roots. Plant Dis. Rep. 56:186-188.

TARJAN, A. C., W. A. SIMANTON, and E. E. RUSSELL. 1956. A labor-saving device for the collection of nematodes. Phytopathology 46:641-644.

THISTLETHWAYTE, B., and R. M. RIEDEL. 1969. Expressing sucrose concentration in solutions used for extracting nematodes. J. Nematol. 1:387-388.

TOWNSHEND, J. L. 1967. Plant-parasitic nematodes in grape and raspberry soils of Ontario and a comparison of extraction techniques. Can. Plant Dis. Surv. 47:83-86.

TSENG, S. T., K. R. ALLRED, and G. D. GRIFFIN. 1968. A soil population study of *Ditylenchus dipsaci* (Kühn) Filipjev in an alfalfa field. Proc. Helminthol. Soc. Wash. 35:57-62.

VAN DER PLANK, J. E. 1963. Plant diseases: Epidemics and Control. Academic Press: New York.

WALLACE, H. R. 1963. The biology of plant parasitic nematodes. Edward Arnold: London.

WILLARD, J. R., and M. S. PETROVICH. 1972. A direct centrifugal-flotation method for extraction of nematodes from clay soils. Plant Dis. Rep. 56:808-810.

YOUNG, T. W. 1954. An incubation method for collecting migratory endoparasitic nematodes. Plant Dis. Rep. 38:794-795.

## APPENDIX 43.A

### Abbreviations and symbols

| | | | |
|---|---|---|---|
| BBT | Blender-Baermann tray combination | $P_i$ | Initial nematode density |
| BF | Baermann funnels | $P_{ie}$ | Posttreatment nematode density |
| BT | Baermann trays | $P_m$ | Midseason nematode density |
| CDS | Cobb's decanting-sieving method | SAE | Semiautomatic elutriator |
| CF | Centrifugal flotation | SAE-BT | Combination of semiautomatic elutriator with |
| CFHS | Centrifugal flotation with heavy sugar | | Baermann trays |
| EE | Elutriation-dissolution of gelatinous matrices of egg masses and staining | SAE-CF | Combination of semiautomatic elutriator with centrifugal flotation |
| FC | Fenwick can | SBT | Sieving-Baermann trays |
| GI | Gall index | SFS | Sugar-flotation sieving |
| MC | Mist chamber | SK | Shaker extraction |
| $P_f$ | Final nematode density | VS | Vital stain |

## APPENDIX 43.B

### Glossary of descriptive terms

**Bioassay**—use of appropriate hosts to determine relative population levels (has been used primarily for endoparasites such as root-knot and cyst nematodes, but can be used for any nematode under suitable propagation conditions, usually done in greenhouse)

**Detection level**—minimum population density of a nematode species that can be identified routinely from a soil or plant tissue sample (varies with each sampling and extraction technique)

**Economic threshold density**—minimum number of nematodes per unit volume of soil (or weight of plant tissue) required to cause an economically significant loss in crop production

**Fumigant nematicide**—gas, volatile liquid, or solid that diffuses through the soil pore spaces as a vapor and is particularly water soluble

**Grid or systematic sampling** (stratified)—collection of specified number of borings over entire plot area (one for each square meter or other area measure) for $P_i$. If crops are established, collection of borings from root zone (in the rows for row crops)

**Horizontal head**—for centrifuge, head is so structured that individual tubes (swinging bucket) are in horizontal position when centrifuge is operating

**Molar solution**—1 g mol wt in sufficient water to make 1 L of solution

**Multipurpose chemicals**—of three basic types:
*Fumigant*—volatile chemical that disperses throughout the soil and controls nematodes, fungi, bacteria, insects, or weeds or a combination of these
*Nonfumigant nematicide* (contact nematicides)—nonvolatile chemical that does not dispense readily except through soil water but controls nematodes (and in some cases, soil insects) that come in contact with it
*Systemic nematicide*—chemical that is absorbed by a plant and translocated throughout the plant and controls nematodes

**Nonfumigant nematicide**—one that has little or no volatility and that must be mixed thoroughly in soil as a granular formulation or applied in water (water-soluble formulation)

$P_f$—final population density (per 100–500 cm$^3$ of soil near or at harvest or the end of the growing season)

$P_i$—pretreatment initial population density (per 100–500 cm$^3$ of soil as determined within one week before treatment)

$P_{ie}$—intermediate (two to four weeks), posttreatment population density (per 100–500 cm$^3$ of soil)

$P_m$—midseason population density (per 100–500 cm$^3$ of soil)

**Population density**—number of nematodes per unit volume of soil or weight of plant tissue

**ppm**—parts per million (1 mg/L [1 lb/120,000 gal] of H$_2$O equals 1 ppm). With water, is equivalent to $\mu$g/ml (w/v) or $\mu$g/g (w/w). Recommend that $\mu$g/ml or $\mu$g/g be used in lieu of ppm. The solvent should be specified when nonwater solutions are used

**Random sampling**—points for collecting borings determined by random chance (not usually suitable for nematodes such as *Meloidogyne* spp., which occur in aggregates)

**Relative g force** (relative centrifugal force)—for centrifuge, force (minimum, average, maximum) times that of gravity: Rel. $g$ force $= 0.00001118 \times r \times RPM^2$ (r, radius in centimeters)

**Sampling tube**—metal (all or part) tool for collecting soil cores of constant diameter and volume. Major types are:
*Oakfield soil sampling tube*—2.0 cm ID, open side, 38 cm long (good for routine sampling to depth of 20 cm)
*Viemeyer soil sampling tube*—2.5 cm ID, 120 cm long (with reinforced head, tube can be driven into soil to a depth of 90 cm)
*Can or bucket auger*—steel auger with 5–7.5-cm diameter × 15-cm bucket fitted with cutting edges at the base (useful for obtaining large soil samples and roots to depths of 90–100 cm)
*Cone-shaped sampling tube*—useful for routine sampling of loose sandy soils (11)

**Sieve-mesh number**—number of wires (filaments running in each direction) per inch. Recommend that size of openings be given in micrometers and that mesh number be given in parentheses

**Soil sterilant**—a chemical (eg, methyl bromide) that kills all soil organisms

**Tolerance limit**—minimum number of nematodes that inhibit plant growth (number per unit volume of soil or weight of plant tissue). May not be of economic significance

# 44. Plant Responses in Evaluation of Nematode Control Agents[1]

## C. C. Orr

Zoologist, Agricultural Research Service, USDA, Texas Agricultural Experiment Station, R 3, Lubbock, TX 79401

## C. M. Heald

ARS-USDA, Weslaco, TX 78596

## E. Kerr

University of Nebraska, Mitchell, NE 69357

## R. Kinloch

University of Florida, Gainesville, FL 32601

## R. E. Michell

EPA, Washington, DC 20460

## M. Norris

Dow Chemical Co., Midland, MI 48640

## W. W. Osborne

Virginia Polytechnic Institute and State University, Blacksburg, VA 24060

## R. Riggs

University of Arkansas, Fayetteville, AR 72701

## Background and Scope

Nematicide registration must be supported by efficacy data that demonstrate control of the target nematode or nematodes and no adverse effect or effects on the plant being treated. In situations in which symptoms of nematode feeding are definitive and diagnostic, reduction of such symptoms in treated plants can demonstrate control of the target nematode.

The major reason for using a nematode control agent is to prevent, arrest, or mitigate adverse plant reactions attributable to plant-parasitic nematodes. Use of a nematicide may result in a beneficial, adverse, or undetectable plant response. Such responses are generally due to such factors as control or lack of control of the target nematode or nematodes, one or more nontarget pests on the experimental plot, or a stimulative or phytotoxic effect on the plant.

Chemicals and pests are each capable of inducing beneficial and adverse plant responses. The type and degree of plant response observed may vary under different conditions. Variables include host plant, climate, soil, soil organisms, method of application, other chemicals, and plant nutrition.

The scope of this chapter is to describe the major types of plant responses found in efficacy and phytotoxicity tests associated with use and development of nematode control agents, and some of the factors affecting these responses.

## Gross Reactions

**Plant symptoms**—Responses of plants to nematode control agents are determined from leaves, stems, and roots. Symptoms on above-ground parts may be stunting, chlorosis, or malformations from nematode injury; chlorosis, scorch, stunting, abnormal growth, or death from control agent injury; accelerated vigor and growth; or no symptoms due to treatment.

Estimates of plant response include stand count, yield, weight, height, and visual ratings of growth and development (2,4–10). Root reactions are measured by infection sites, lesions, galls, nematodes per gram of root, root pruning, root proliferation, or malformation (1,3,11). The survival of sedentary nematodes depends on the long life of surrounding host cells. In spite of this accommodation of the host at the infection site, the overall growth of the plant may be severely reduced, and crop failure is a common result.

[1]Prepared by the Joint SON/ASTM E 35.16 Task Force on Plant Responses in Evaluation of Nematode Control Agents, C. C. Orr, chairman.

**Disease complexes and synergism**—Nematode injury in plants is often accompanied by fungal invasion of plant tissue, with substantially increased damage. Use of other pesticides also may result in synergistic reactions in plants when used with nematode control agents. Biologic or chemical combinations often cause unexpected reactions, and may even by phytotoxic.

**Soil properties**—Inherent properties of the soil influence plant responses to nematode control agents. Insufficient fertilizer and trace elements can result in stunting or chlorosis similar to symptoms of nematode injury. The type, texture, moisture content, temperature, and percentage of organic matter may affect plant response by altering the fate, availability, and movement of chemicals; vigor of host plants; or degree of nematode activity. Edaphic factors influence activity of mycorrhiza, microflora, microfauna, and insects, which in turn influence plant response.

## Growth and Development

**Emergence**—Seedling response to nematode control agents may be measured by the rate of emergence and final stand counts. Increased stand counts or rate of emergence may indicate nematode control. Decreased growth may indicate inactivity of the control agent or phytotoxicity. Good stands resulting from nematode control frequently have less seedling disease.

**Early growth**—As plants mature, nematode control may be expressed as larger plants, expanded leaves, increased internode length, or other evidence of increased growth and vigor. Conversely, injury symptoms—stunting, root pruning, discoloration, or morphologic abnormalities—can be observed and rated.

**Senescence and maturity**—Yield is perhaps the most common measure for evaluating benefits derived from nematode control agents. Yield can be measured by weight, size, quality, or aesthetic value of plants or plant parts. Yield of field crops is usually expressed in terms of kilograms per hectare, while other crop responses may be expressed in terms of grade of fruit or root commodity, quality, smoothness, color, or eye appeal.

## Yield and Economic Response

**Chemical**—The amount and type of plant response attributable to nematode control depends on the genera and population density of nematodes and the effectiveness or phytotoxicity of the control agent. Nematodes vary in susceptibility to chemicals. The differences may be due to inherent resistance of the nematode to the chemical, the stage in its life cycle, or whether it was in the soil or roots when the control agent was applied. At certain stages in the life cycle, the nematode may have protective coverings, as in the egg stage, or the larvae may have entered plant tissue and thus be protected. In addition, nematode susceptibility to control agents may be influenced by the inherent action and concentration of the agent, duration of exposure to the agent, temperature, and moisture content of the soil. The efficacy of materials is also influenced by method and timing of application, placement, and degree of incorporation into the soil mass.

**Plant**—The return for dollars invested in nematode control agents is the most important factor governing their use. Cost of nematode control is influenced by the

material chosen, number and timing of applications, and method of application. Increased yields make the cost of control agents a minor production input for many intensively cultivated high-profit crops.

## Nutritional and Flavor Effects

Responses of plants to nematode control agents can be measured readily by quality, size, and yield of produce. Quality of nematode-infected underground plant parts is often severely reduced, even though a corresponding reduction in yield may not be seen. If chemical control agents leave harmful residues or metabolites in the harvested crop or in the soil, they may require specialized equipment or techniques to be detected. Changes in the flavor or nutritional content of the harvested commodity may be attributed to residues or breakdown products of nematode control agents, or alteration of the plant's metabolism. Occasionally, related problems occur with treated commodities intended for processing rather than for fresh market use. For example, altered baking and milling characteristics, inhibition of fermentation, and liberation of excessive gas in canned goods have resulted from use of chemical control agents.

## Varietal Differences

Plant varieties differ in response to nematode control agents. The inherent characteristics of a crop variety dictate its level of resistance to a pathogen or phytotoxicity. Therefore, varieties differ in growth, development, and productivity when nematode control agents are used. Similarly, physiologic and genetic mechanisms may control the levels of residue and metabolites of the control agent found in plants.

## Literature Cited

1. BEINGEFORS, S. 1971. Resistance to nematodes and the possible value of induced mutations. Reprint from "Mutation breeding for disease resistance." IAEA-PL-412/21. Vienna. pp. 209-235.
2. BERGESON, G. B. 1968. Evaluation of factors contributing to the pathogenicity of *Meloidogyne incognita*. Phytopathology 58:49-53.
3. ENDO, B. Y. 1975. Pathogenesis of nematode-infected plants. Annu. Rev. Phytopathol. 13:213-238.
4. HEALD, C. M. 1967. Pathogenicity of five root-knot nematode species on *Ilex crenata* 'Helles.' Plant Dis. Rep. 51:581-585.
5. HOESTRA, H., and M. OOSTENBRINK. 1962. Nematodes in relation to plant growth. IV. *Pratylenchus penetrans* (Cobb) on orchard trees. Neth. J. Agric. Sci. 10:4, 286-296.
6. JONES, J. E., L. D. NEWSOM, and E. L. FINLEY. 1959. Effect of reniform nematode on yield, plant characters, and fiber properties of upland cotton. Agron. J. 51:353-346.
7. OOSTENBRINK, M. 1966. Major characteristics of the relation between nematodes and plants. Meded. Landbouwhogesch. Wageningen 66:1-46.
8. SEINHORST, J. W. 1973. Principles and possibilities of determining degrees of nematode control leading to maximum returns. I. Protection of one crop sown or planted soon after treatment. Nematologia Med. 1:93-105.
9. TAYLOR, A. L., and A. M. GOLDEN. 1954. Preliminary trials of D-D Hi-Sil as a soil fumigant. Plant Dis. Rep. 38:63-64.
10. THORNE, G. 1961. Principles of Nematology. New York: McGraw-Hill Book Company, Inc., p. 553.
11. TOWNSHEND, J. L. 1958. The effect of *Pratylenchus penetrans* on a clone of *Fragaria vesca*. Can. J. Bot. 36:683-685.

# 45. Nematicide Evaluation as Affected by Nematode and Nematicide Interactions With Other Organisms[1]

## G. B. Bergeson

Department of Botany and Plant Pathology, Purdue University, West Lafayette, IN 47907

## G. D. Griffin

Crops Research Laboratory, UMC, Utah State University, Logan, UT 84321

## H. W. Lembright

Dow Chemical Company, USA, Drawer H, Walnut Creek, CA 94596

## B. F. Lownsbery

Department of Nematology, University of California, Davis, CA 95616

## D. C. Norton

Department of Botany and Plant Pathology, Iowa State University, Ames, IA 50010

The application of a nematicide often has direct or indirect effects on organisms other than nematodes that influence plant growth. A complete, accurate evaluation of a nematicide must cover these factors. Before evaluating a nematicide in a field, the investigator should know the history of past infestations by soilborne fungal and bacterial pathogens, nematode-borne viruses, and soil insects. If available, a selective pesticide for these pests should be included to measure more precisely the nematicidal effectiveness of the test compound.

## Scope

This chapter lists examples of microbial interaction that may complicate evaluation of nematicide effects on nematodes. The purpose is to remind the investigator that many diverse interactions occur. Methods of investigating associated organisms and the diseases they cause are not described in this chapter. Short statements for illustrative purposes summarize documented associations that have been published.

## 1. Indirect Effects of Nematicides

**1.1 Fungi**—Nematodes increase the severity of fungal diseases by acting as vectors and wounding agents, by modifying host tissue and rhizosphere, and by overcoming varietal resistance. Some of the more important, well-known associations are:

*Belonolaimus spp.*
● Greatly increases the development of Fusarium wilt in cotton and overcomes resistance to *Fusarium* in Fusarium-resistant varieties (19,36,63)

●Increases the severity of *Pythium aphanidermatum* on chrysanthemums (41,48).

*Helicotylenchus spp.*
●Penetrates the mycorrhizal mantle of short-leaf pine seedlings and creates infection courts for *Phytophthora cinnamoni* (5)
●Causes superficial lesions that are important as infection courts for *Rhizoctonia* sp. in banana roots (86)

*Ditylenchus dipsaci*
●Potato stems injured by this nematode are more susceptible to infections with *Phoma solanicola* (34)

*Heterodera glycines*
●Greatly increases the severity of Fusarium wilt in soybeans (80)

*Heterodera schachtii*
●Facilitates the entrance of *Rhizoctonia solani* in sugar beets
●Increases damping-off of sugar beets by *Pythium ultimum* (93)

*Heterodera rostochiensis*
●Increases severity of *Verticillium dahliae* infection of potato (31) and *Rhizoctonia solani* infection of tomato (20)

*Heterodera tabacum*
● Increases the severity of *Verticillium albo-atrum* infection in tomato (60)

*Hoplolaimus spp.*
●Prolongs the susceptibility of cotton seedlings to postemergence damping-off by *Rhizoctonia solani* (12)
●Penetrates the mycorrhizal mantle of pine seedlings and increases their exposure to fungal pathogens (81)
●Associated with *Fusarium oxysporum* in reduced growth of peach seedlings (92)
●Facilitates entry of *Fusarium oxysporum* f. sp. *pisi*

[1]Prepared by the Joint SON/ASTM E 35.16 Task Force on Side Effects of Nematicide Application, G. B. Bergeson, chairman.

into the vascular tissue of peas (45)

*Meloidogyne*

●Species of this genus are important in predisposing host plants to fungal diseases. Host tissues are modified to create a better substrate for the fungus

*M. arenaria*

●Interacts synergistically with *Pythium myriotylum* to increase peanut root rot (26)

*M. hapla*

●Causes significantly greater incidence and density of *Aspergillus flavus* in kernels of peanut (62)

*M. incognita*

●Greatly increases severity of Fusarium wilt in tomato, muskmelon, tobacco, cotton, cabbage, and mimosa (3,9,19,23,39,42,54,94)

●Interacts synergistically in cotton seedling diseases with *Alternaria tenuis, Glomerella gossypii, Fusarium oxysporum* f. sp. *vasinfectum,* and *Rhizoctonia solani* (15)

●Predisposes tobacco to infection by *Pythium ultimum, Trichoderma harzianum, Curvularia trifolii, Botrytis cinerea, Aspergillus ochraceus, Penicillium martensii* (76)

●Predisposes cotton seedlings to damping-off caused by *Pythium debaryanum* (12)

●Increases severity of *Rhizoctonia solani* and *Phytophthora* of tobacco (7)

●Overcomes resistance of tobacco varieties to *Phytophthora parasitica* (77) and of tomato varieties to *Fusarium oxysporum* f. sp. *lycopersici* (11)

*M. javanica*

●Increases the severity of Fusarium wilt in tomato, mimosa, tobacco, cowpea, sweet potato, and chrysanthemum (27,40,57,89)

●Interacts synergistically with *Macrophomina phaseoli* in kenaf foot rot

*Paratylenchus hamatus*

●Associated with *Rhizoctonia* in root rot of celery (51)

*Pratylenchus sp.*

● Nematodes of this genus usually predispose hosts to fungal infection by creating wounds that facilitate fungal penetration

*P. brachyurus*

●Interacts synergistically with *Rhizoctonia solani* to increase damping-off of soybean seedlings (46)

*P. minyus*

●Reduces incubation time and increases the incidence and severity of *Verticillium dahliae* in peppermint (24)

●Is consistently associated with *Rhizoctonia solani* in root rot of wheat (8)

*P. penetrans*

●Overcomes resistance in Wisconsin Perfection pea to *Fusarium oxysporum* f. sp. *pisi,* and enhances Fusarium wilt in Rondo pea (71,84)

●Increases severity of Verticillium wilt in tomato, cotton, eggplant, and potato (18,58,65)

●Increases severity of *Aphanomyces euteiches* infection in peas when the inoculum level of the fungus is low (72)

*P. pratensis*

●Interacts synergistically with *Cylindrocarpon radicicola* in reducing the growth of potato, carrot, red clover, tomato, spinach, and violet seedlings (32)

*P. scribneri*

●Interacts synergistically with *Fusarium moniliforme* to decrease corn growth (73)

*P. vulnus*

● Reduced growth of peach trees when associated with *Fusarium oxysporum* (92)

*P. zeae*

●Interacts synergistically with *Pythium graminicola* to reduce growth of sugarcane (37,82)

*Radopholus similis*

●Increases by 100% the incidence of *Fusarium oxysporum* f. sp. *cubense* infection in banana and increases the incidence and severity of *Fusarium* sp. in citrus trees (25,69)

*Rotylenchulus reniformis*

●Interacts synergistically to increase severity of Fusarium wilt in cotton. Also, was able to overcome resistance to wilt in some varieties (61,68)

●Increases the susceptibility of cotton to *Rhizoctonia solani* infection when the nematode population is high (12)

*Trichodorus christiei*

●Associated with *Fusarium moniliforme* in reducing root growth of sugarcane (49)

●Increases the infection incidence of *Verticillium albo-atrum* on tomato (17)

*Tylenchorhynchus sp.*

*T. claytoni*

●Increases the incidence of Fusarium wilt in tobacco (35)

*T. martini*

●Increases the incidence of root rot of peas caused by *Aphanomyces euteiches* (30)

*T. dubius*

●Increases the severity of Fusarium blight of turf (91)

*Tylenchulus semipenetrans*

●Increases the incidence of *Fusarium oxysporum* in rough lemon by 33% and that of *F. solani* by 350% (90)

*Tylenchus agricola*

●Increases penetration of the vascular stele of corn by *Fusarium roseum* (44)

**1.2 Bacteria**—Nematicides may affect some bacterial diseases of plants because nematodes are vectors for bacterial pathogens of apical meristems, and are vectors or predisposing agents for endophytic soft rot bacteria, vascular wilt pathogens, or bacteria that cause cankers (75). Nematodes may also be antagonists of beneficial bacteria such as *Rhizobium* spp. (6).

*Anguina tritici*

●Is a vector of *Corynebacterium tritici,* which causes tundu or yellow slime disease of wheat (16,87)

*Aphelenchoides ritzema-bosi*

●Is a vector of *Corynebacterium facians,* which together with the nematode causes typical cauliflower disease of strawberry (75)

*Criconemoides xenoplax*

●Increases susceptibility of peach and plum trees to bacterial canker caused by *Pseudomonas syringae,* probably by reducing the vigor of the host (50,64)

*Ditylenchus dipsaci*

●Wounds rhubarb plants and is a vector of *Erwinia rhaponticum,* which causes crown rot of rhubarb (59)

●Is a vector of *Corynebacterium insidiosum,* the cause of bacterial wilt of alfalfa (33)

*Heterodera glycines*

●Certain races inhibit the formation of nitrogen nodules by *Rhizobium japonicum* in soybean (6)

*Meloidogyne sp.*

●Assists by wounding to increase the severity of carnation wilt caused by *Pseudomonas caryophylli* (85)

*Meloidogyne incognita*

●Increases the severity of bacterial canker caused by *Corynebacterium michiganense* in tomato (66)

●Assists by wounding to increase markedly the incidence and severity of Granville wilt caused by *Pseudomonas solanacearum* in tobacco

*M. hapla*

●Greatly increases the incidence of bacterial wilt caused by *Corynebacterium insidiosum* in alfalfa (38,70)

*Helicotylenchus dihystera*

●Assists by wounding to cause an increase in severity of carnation wilt caused by *Pseudomonas caryophylli* (85)

*Xiphinema americanum* and *Criconemoides xenoplax*

●Parasitism is associated with Cytospora canker of spruce (22)

**1.3 Viruses**—The following species are vectors of the viruses indicated. Control of these nematodes could result in crop improvement by lowering the incidence of virus infection.

*Longidorus sp.*

*L. attenuatus*

● Transmits lettuce ring spot virus (88)

*L. elongatus*

●Transmits raspberry leafcurl, beet ring spot, and tomato black ring viruses (88)

*L. macrosoma*

●Transmits raspberry ring spot virus (88)

*Paratrichodorus sp.*

*P. allius, P. christiei, P. nanus, P. pachydermus, P. porosus,* and *P. teres*

●Transmits tobacco rattle virus (88)

*P. anemones, P. pachydermus,* and *P. teres*

●Transmits pea early browning virus (88)

*Trichodorus sp.*

*T. cylindricus, T. primitivus,* and *T. similis*

●Transmits tobacco rattle virus (88)

*T. viruliferus*

●Transmits pea early browning virus (88)

*Xiphinema sp.*

*X. americanum*

●Transmits tobacco ring spot and tomato ring spot viruses (88)

*X. coxi*

●Transmits cherry leaf roll and arabis mosaic viruses (88)

*X. diversicaudatum*

●Transmits strawberry latent ring spot, arabis mosaic, carnation ring spot, and cherry leaf roll viruses (88)

*X. index*

●Transmits grapevine fanleaf virus (88)

*X. italiae*

●Transmits grapevine fanleaf virus (88)

## 2. Beneficial Effects

**2.1 Fungi**—Several fumigant nematicides have fungicidal properties, especially at high rates of application. Therefore, a nematicide may affect a nematode-fungus interaction directly, indirectly, or both.

*Telone and D-D*

●*Pythium arrhenomanes* infection was reduced 60% in pineapple fields treated with 351 L/ha, and was reduced below detection level by 702 L/ha (1)

●Kills *Phytophthora cinnamoni, Rhizoctonia solani,* and *Thielaviopsis basicola* at 2.1 ml/ft$^3$ in a sealed container held at 24°C (95)

●Reduces *Phytophthora citrophthora* and *P. parasitica* in the 30–60-cm soil horizon when applied at 455 kg/ha (4).

*DBCP*

●Reduced the inoculum potential of *Rhizoctonia solani* for 106 days in a peanut field treated at 14 L/ha applied in the row (2)

●Reduced *Rhizoctonia solani* infection on Kentucky bluegrass with few plant parasitic nematodes when applied at the rate of 174 ml/1,000 ft$^2$ (52)

●Prevented damping-off of cotton seedlings due to *Pythium ultimum* in greenhouse experiments when applied at 14 L/ha (12)

●Reduced *Pythium* sp. propagules in soil by 66% when applied at 22.5 L/ha (14)

**2.2 Insects**—Insects may disseminate (10,29,53) and prey (13,21,47,67,79) on nematodes. Many insects and nematodes have common hosts, but the comparative damage that insects or nematodes do after direct or indirect application of nematicides has not been determined. This is surprising, because many nematicides also have insecticidal properties. Some granular nematicides were tested, registered, and marketed first as insecticides; their use in nematode control came later. The knowledge that certain nematicides are also insecticides may help to interpret experimental results. Other pesticides in commercial use, however, have not been tested thoroughly as nematicides or insecticides, and the possibility remains that they have greater biocidal activity than the manufacturer specifies.

A partial list of insecticides being used or tested as nematicides appears in Table 45.1. Registration of chemicals for insect control implies that they also may have biocidal or biostatic nematicidal properties. Regardless of registration status, the manufacturer should be consulted for biotoxic properties of compounds. Most manufacturers issue data sheets concerning biotoxic properties, suggested uses, and preliminary results of products being developed.

## 3. Harmful Effects

**3.1 Bacteria**—Fumigation at high dosages or in cold wet soils or both can suppress nitrifying bacteria so that plant growth is retarded by ammonia toxicity or nitrate starvation. Symptoms might be confused with direct phytotoxicity of the fumigant, but sometimes the suppression of nitrification by a nematicide may have these beneficial effects: (i) reduced nitrate loss by leaching, (ii) reduced buildup of nitrates in ground water, and (iii) prolonged nitrogen availability during crop growth.

*1,3-dichloropropene (D-D and Telone)*

●The rate of 204 kg/ha of technical material suppressed nitrifying bacteria for six to ten weeks (28)

**TABLE 45.1. List of pesticides active against nematodes, insects, and related pests**

| Common name | Trade name | Insects and related pests controlled |
|---|---|---|
| *Fumigants* | | |
| Chloropicrin | Picfume®, Larvacide® | Grubs, wireworms |
| Ethylene dibromide | Dowfume W-85®, Soilbrom-85® | Wireworms, garden centipedes |
| 1,3-Dichloropropene-chloropicrin mixes | Telone C-17® | Wireworms, garden centipedes |
| Ethylene dibromide-chloropicrin mixes | Terrocide® series | Wireworms, garden centipedes, white grubs |
| Methyl bromide | Dowfume MC₂®, Brom-o-gel®, Brozone® | Many insects and mites |
| Chloropicrin-methyl bromide mixes | Terr-o-gas®, Dowfume MC-33® | Soilborne insects, wireworms |
| 1,3-Dichloropropenes and related chlorinated hydrocarbons | D-D®, Telone®-II | Garden centipedes, wireworms |
| Sodium methyl dithiocarbamate | Vapan® | Garden centipedes |
| *Nonfumigants* | | |
| Aldicarb | Temik® | Aphids, boll weevil, Colorado potato beetle, cotton leaf perforator, flea beetles, leafhoppers, leafminers, lygus bugs, mealybugs, mites, sugar beet root maggot, thrips, white flies |
| Carbofuran | Furadan® | Alfalfa weevil, armyworms, corn borers, corn rootworms, fleabeetles, lygus bugs, pea aphid, snout beetles, sugarcane borer, thrips, wireworms |
| Diazinon | Diazinon®, Sarolex® | Ants, armyworms, clover mites, cutworms, diggerwasps, earwigs, springtails, sod webworms, sowbugs |
| Fensulfothion | Dasanit®, Terracur P® | Cabbage maggot, corn rootworms, European corn borer, lesser cornstalk borer, onion maggot, seed corn beetle, seed corn maggot, sugar beet root maggot, thrips, tuber flea beetle larva, wireworms |
| Ethoprop | Mocap® | Corn rootworms, flea beetles, wireworms |
| Oxamyl | Vydate® | Flea beetles |
| *Miscellaneous* | | |
| Parathion | Parathion® | Many insects |
| Phorate | Thimet® | Soil and systemic insecticide |
| Phenamiphos | Nemacur® | Aphids, mites, thrips, fleahoppers, mealybugs, adult cucumber beetles |
| *Experimental* (not labeled as nematicides) | | |
| CGA-12223 | | Chinch bugs, corn rootworms, cutworms, European chafer, Japanese beetle, onion maggot, sod webworm, southern masked chafer, wireworms, seed corn maggot |
| Disulfoton | Di-Syston® | Aphids, flea beetles, leafhoppers, mites, rootworms, southern potato wireworm, thrips, whiteflies |
| Fonofos | Dyfonate® | Corn rootworms, cutworms, garden centipedes |

*1,2-dibromo-3-chloropropane and dibromoethane*

● At the rate of 29 and 74 kg/ha, respectively, nitrifying bacteria were suppressed for a period of two to four weeks (28)

## 3.2 Fungi
*D-D*

● Row treatment at 93 L/ha nearly doubled the number of tomato plants killed by *Sclerotina sclerotiorum*. D-D may have inhibited some natural antagonist of the fungus (78)

● Dosage levels ranging from 505–1,010 L/ha significantly increased the stipe formation of *Sclerotinia sclerotiorum* and consequently increased the severity of the disease on lettuce (74)

**3.3 Mycorrhizae**—Mycorrhizal fungi are highly specialized, symbiotic, nonpathogenic parasites of most plant species (55,56). They function as "fungus roots" for plants, and plant species may vary from highly mycorrhiza dependent (absence may result in plant mortality) to apparently independent. Nematicides with fungicidal properties can adversely affect the normal development and infection process of mycorrhizae by suppressing the initial population levels in the soil (43,83). Soil fumigants are generally fungicidal at some concentration, and therefore may affect mycorrhizae.

Control of mycorrhizae, especially those in the soil surface, is generally necessary to observe plant growth suppression. This is most commonly observed in fumigation with methyl bromide under a tarpaulin. Most other tarpaulined fumigants leave sufficient mycorrhizae in the surface layers to inoculate the soil adequately during normal cultural practices. Tarpaulined fumigation, however, with 1,3-D or chloropicrin and certain other special techniques of applying these two chemicals also can suppress drastically mycorrhizae populations.

This effect is most likely to be obvious with endomycorrhizae, which, like nematodes, are obligate parasites and do not produce surface fruiting bodies and airborne spores. Suppression of ectomycorrhizae may result in initial suppression of plant growth, but these fungi produce surface fruiting bodies and airborne spores, and fumigated soil is rapidly reinfested with them.

Little is known about the ectendomycorrhizae.

*3.3.1. Suggested guidelines for evaluating mycorrhizal effects*—Because certain fumigant nematicides are transient, they can be used effectively to allow reinoculation with species of mycorrhizae that are preferred over the native species. Furthermore, field experience with fumigant nematicides suggests that stimulation follows initial suppression. Methods for testing the effects of a candidate nematicide against mycorrhizae are similar to those used for nematicides, except for the following modifications:

i. Use closed containers with infested soil

ii. Include mycorrhizae in the soil instead of nematodes. Rates of application should be ten times that required for nematodes to predict whether an effect on mycorrhizae is likely to occur

iii. Add nematicides and roll on roller mill for 10–30 min

iv. Incubate as with nematodes

v. Following incubation, seed with the mycorrhizae host and grow for 12 weeks, or 15 weeks with pine species. Remove roots gently, rinse with water, and examine with at least ×10 magnification. Compare with control for suppression or stimulation

*3.3.2. Species of mycorrhizal fungi to consider for testing purposes*

i. Endomycorrhizae—*Glomus fasciculatus,* because it is ubiquitous, has a high level of symbiosis and wide host range

ii. Ectomycorrhizae—*Pisolithus tinctorius* or *Thelephora terrestris,* for similar reasons

## Literature Cited

1. ANDERSON, E. J. 1966. 1,3-Dichloropropene 1,2-dichloropropane mixture found active against Pythium arrhenomanes in field soil. Down Earth 22(3):23.
2. ASHWORTH, L. S., Jr., B. C. LANGLEY, and W. H. THAMES, Jr. 1964. Long-term inhibition of Rhizoctonia solani by a nematocide, 1,2-dibromo-3-chloropropane. Phytopathology 54:187-191.
3. ATKINSON, G. F. 1892. Some diseases of cotton. Alabama Agric. Exp. Stn. Bull. 41:61-65.
4. BAINES, R. C., L. J. KLOTZ, T. A. DEWOLFE, R. H. SMALL, and G. O. TURNER. 1966. Nematocidal and fungicidal properties of some soil fumigants. Phytopathology 56:691-698.
5. BARHAM, R. O. 1972. Effect of nematodes and Phytophthora cinnamoni on mycorrhiza of shortleaf pine seedlings. Phytopathology 62:801 (Abstr.).
6. BARKER, K. R., D. HUISINGH, and S. A. JOHNSTON. 1972. Antagonistic interaction between Heterodera glycines and Rhizobium japonicum on soybeans. Phytopathology 62:1201-1205.
7. BATTEN, C. K., and N. T. POWELL. 1971. The Rhizoctonia-Meloidogyne disease complex in flue-cured tobacco. J. Nematol. 3:164-169.
8. BENEDICT, W. G., and W. B. MOUNTAIN. 1954. Studies on the association of Rhizoctonia solani and nematodes in a rootrot disease complex of winter wheat in southwestern Ontario. Proc. Can. Phytopathol. Soc. 21:12.
9. BERGESON, G. B. 1975. The effect of Meloidogyne incognita on the resistance of four muskmelon varieties to Fusarium wilt. Plant Dis. Rep. 59:410-413.
10. BLAIR, G. P., and H. M. DARLING. 1968. Red ring disease of the coconut palm, inoculation studies and histopathology. Nematologica 14:395-403.
11. BOWMAN, P., and J. R. BLOOM. 1966. Breaking the resistance of tomato varieties to Fusarium wilt by Meloidogyne incognita. Phytopathology 56:871 (Abstr.).
12. BRODIE, B. B., and W. E. COOPER. 1964. Relation of parasitic nematodes to postemergence damping-off of cotton. Phytopathology 54:1023-1027.
13. BROWN, W. L. 1954. Collembola feeding upon nematodes. Ecology 35:421.
14. BUMBIERIS, M. 1970. Effect of DBCP on pythiaceous fungi. Plant Dis. Rep. 54:622-624.
15. CAUQUIL, J., and R. L. SHEPHERD. 1970. Effect of root knot nematode-fungi combinations on cotton seedling disease. Phytopathology 60:448-451.
16. CHEO, C. C. 1946. A note on the relation of nematode (Tylenchus tritici) to the development of the bacterial disease of wheat caused by Bacterium tritici. Ann. Appl. Biol. 33:446-449.
17. CONROY, J. J., and R. J. GREEN, Jr. 1974. Interactions of the rootknot nematode Meloidogyne incognita and the stubby root nematode Trichodorus christiei with Verticillium albo-atrum on tomato at controlled inoculum densities. Phytopathology 64:1118-1121.
18. CONROY, J. J., R. J. GREEN, Jr., and J. M. FERRIS. 1972. Interaction of Verticillium albo-atrum and the root lesion nematode Pratylenchus penetrans in tomato root at controlled inoculum densities. Phytopathology 62:362-366.
19. COOPER, W. E., and B. B. BRODIE. 1962. Correlation between Fusarium wilt indices of cotton varieties with root-knot and with sting nematodes as predisposing agents. Phytopathology 52:6 (Abstr.).
20. DUNN, E. 1970. Interactions of Heterodera rostochiensis

Woll., and Rhizoctonia solani Kuhn on the tomato plant. 10th International Nematology Symposium of the European Society of Nematologists, Pescara, Italy, September 1970. p. 68 (Abstr.).

21. ESSER, R. P. 1963. Nematode interactions in plates of nonsterile water agar. Proc. Soil Crop Sci. Soc. Fla. 23:121-138.

22. EPSTEIN, A. H., and G. D. GRIFFIN. 1962. The occurrence of Cystospora canker on spruce in the presence of Xiphinema americanum and Criconemoides xenoplax. Plant Dis. Rep. 46:17.

23. FASSULIOTIS, G., and G. J. RAU. 1969. The relationship of Meloidogyne incognita acrita to the incidence of cabbage yellows. J. Nematol. 1:219-222.

24. FAULKNER, L. R., W. J. BOLANDER, and C. B. SKOTLAND. 1970. Interaction of Verticillium dahliae and Pratylenchus minyus in Verticillium wilt of peppermint: Influence of the nematode as determined by a double root technique. Phytopathology 60:100-103.

25. FEDER, W. A., and J. FELDMESSER. 1961. The spreading decline complex: The separate and combined effects of Fusarium spp. and Radopholus similis on the growth of Duncan grapefruit seedlings in the greenhouse. Phytopathology 51:724-726.

26. GARCIA, R., and D. J. MITCHELL. 1975. Synergistic interactions of Pythium myriotylum with Fusarium solani and Meloidogyne arenaria in pod rot of peanut. Phytopathology 65:832-833.

27. GILL, D. L. 1958. Effect of root-knot nematodes on Fusarium wilt of Mimosa. Plant Dis. Rep. 42:587-590.

28. GOOD, J. M., and R. L. CARTER. 1965. Nitrification lag following soil fumigation. Phytopathology 55:1147-1150.

29. GRIFFITH, R. 1968. The relationship between the red ring nematode and the palm weevil. J. Agric. Soc. Trin. 68:342-356.

30. HAGLUND, W. A., and T. H. KING. 1961. Effect of parasitic nematodes on the severity of common root rot of canning peas. Nematologica 6:311-314.

31. HARRISON, J. A. C. 1971. Association between the potato cyst-nematode, Heterodera rostochiensis Woll., and Verticillium dahliae Kleb. in the early-dying disease of potatoes. Ann. Appl. Biol. 67:185-193.

32. HASTINGS, R. J., and J. E. BOSHER. 1938. A study of the pathogenicity of the meadow nematode and associated fungus Cylindrocarpon radicicola Wr. Can. J. Res. 16:225-229.

33. HAWN, E. J. 1963. Transmission of bacterial wilt of alfalfa by Ditylenchus dipsaci (Kuhn). Nematologica 9:65-68.

34. HIJINK, M. J. 1963. A relation between stem infection by Phoma solanicola and Ditylenchus dipsaci on potato. Neth. J. Plant Pathol. 69:318-321.

35. HOLDEMAN, Q. L. 1956. The effect of the tobacco stunt nematode on the incidence of Fusarium wilt in flue-cured tobacco. Phytopathology 46:129.

36. HOLDEMAN, Q. L., and T. W. GRAHAM. 1953. The sting nematode breaks resistance to cotton wilt. Phytopathology 43:475 (Abstr.).

37. HOLTZMANN, V., and G. S. SANTO. 1971. Effect of temperature on the interrelationship of Pratylenchus zeae and Pythium graminicola on sugarcane. Phytopathology 61:1321 (Abstr.).

38. HUNT, O. J., G. D. GRIFFIN, J. J. MURRAY, M. W. PEDERSEN, and R. N. PEADEN. 1971. The effects of rootknot nematodes on bacterial wilt in alfalfa. Phytopathology 61:256-259.

39. JENKINS, W. R., and B. W. COURSEN. 1957. The effect of rootknot nematodes, Meloidogyne incognita acrita and M. hapla, on Fusarium wilt of tomato. Plant Dis. Rep. 41:182-186.

40. JOHNSON, A. W., and R. H. LITTRELL. 1969. Effect of Meloidogyne incognita, M. hapla, and M. javanica on the severity of Fusarium wilt of chrysanthemum. J. Nematol. 1:122-125.

41. JOHNSON, A. W., and R. H. LITTRELL. 1970. Pathogenicity of Pythium aphanidermatum to chrysanthemum in combined inoculations with Belonolaimus longicaudatus or Meloidogyne incognita. J. Nematol. 2:255-259.

42. JONES, J. P., and A. J. OVERMAN. 1971. Interaction of five soil-borne pathogens of tomato. Phytopathology 61:897 (Abstr.).

43. KLEINSCHMIDT, G. D., and J. W. GERDEMANN. 1972. Stunting of citrus seedlings in fumigated nursery soils related to the absence of endomycorrhizae. Phytopathology 62:1447-1453.

44. KISIEL, M., K. DEUBERT, and B. M. ZUCKERMAN. 1969. The effect of Tylenchus agricola and Tylenchorhynchus claytoni on root rot of corn caused by Fusarium roseum and Pythium ultimum. Phytopathology 59:1387-1390.

45. LABRUYERE, R. E., H. DEN OUDEN, and J. W. SEINHORST. 1959. Experiments on the interaction of Hoploaimus uniformis and Fusarium oxysporum f. pisi race 3 and its importance in "early yellowing" of peas. Nematologica 4:336-343.

46. LINDSEY, D. W., and E. J. CAIRNS. 1971. Pathogenicity of the lesion nematode, Pratylenchus brachyurus, on six soybean cultivars. J. Nematol. 3:220-226.

47. LINFORD, M. B., and J. M. OLIVEIRA. 1938. Potential agents of biological control of plant-parasitic nematodes. Phytopathology 28:14 (Abstr.).

48. LITTRELL, R. H., and A. W. JOHNSON. 1969. Pathogenicity of Pythium aphanidermatum to chrysanthemum in combined inoculations with Belonolaimus longicaudatus or Meloidogyne incognita. Phytopathology 59:115-116 (Abstr.).

49. LIU, L. J., and A. AYALA. 1970. Pathogenicity of Fusarium moniliforme and F. roseum and their interaction with Trichodorus christiei on sugar cane in Puerto Rico. Phytopathology 60:1540 (Abstr.).

50. LOWNSBERY, B. F., H. ENGLISH, E. H. MOODY, and F. SCHICK. 1973. Criconemoides xenoplax experimentally associated with a disease of peach. Phytopathology 63:994-997.

51. LOWNSBERY, B. F., and J. W. LOWNSBERY. 1952. Paratylenchus hamatus Thorne & Allen associated with celery disease in Connecticut. Phytopathology 42:13 (Abstr.).

52. MADISON, J. H. 1961. The effect of pesticides on turf grass disease incidence. Plant Dis. Rep. 45:892-893.

53. MAMIYA, Y., and N. ENDA. 1972. Transmission of Bursaphelenchus lignicolus (Nematoda: Aphelenchoididae) by Monochamus alternatus (Coleoptera:Cerambycidae). Nematologica 18:159-162.

54. MARTIN, W. J., L. D. NEWSOM, and J. E. JONES. 1955. Relationship of nematodes of the genera Meloidogyne, Tylenchorhynchus, Helicotylenchus and Trichodorus to the development of Fusarium wilt in cotton. Phytopathology 45:349 (Abstr.).

55. MARX, D. H. 1975. Mycorrhizae of forest nursery seedlings. USDA Agric. Handbook No. 470, Forest Nursery Diseases in the United States. pp. 35-40.

56. MARX, D. H., W. C. BRYAN, and W. A. CAMP. 1970. Effect of endomycorrhizal formed by Endogyne mossea on growth of citrus. Mycologia 63:1222-1226.

57. McCLURE, T. T. 1949. Mode of infection of the sweet potato wilt Fusarium. Phytopathology 39:876-886.

58. McKEEN, C. D., and W. B. MOUNTAIN. 1960. Synergism between Pratylenchus penetrans (Cobb) and Verticillium albo-atrum R. & B. in eggplant wilt. Can. J. Bot. 38:789-794.

59. METCALFE, G. 1940. Bacterium rhaponticum (Millard) Dowson, a cause of crown-rot disease of rhubarb. Ann. Appl. Biol. 27:502-508.

60. MILLER, P. M. 1975. Effect of the tobacco cyst nematode, Heterodera tabacum, on severity of Verticillium and Fusarium wilts of tomato. Phytopathology 65:81-82.

61. MINTON, E. B., A. L. SMITH, and E. J. CAIRNS. 1964. Effects of 7 nematode species on 10 cotton selections. Phytopathology 54:625 (Abstr.).

62. MINTON, N. A., D. K. BELL, and B. DOUPNIK, Jr. 1969. Peanut pod invasion by Aspergillus flavus in the presence of Meloidogyne hapla. J. Nematol. 1:318-320.

63. MINTON, N. A., and E. B. MINTON. 1966. Effect of root knot and sting nematodes on expression of Fusarium wilt of cotton in three soils. Phytopathology 56:319-322.

64. MOJTAHEDI, H., B. F. LOWNSBERY, and E. H. MOODY. 1975. Ring nematodes increase development of bacterial cankers in plums. Phytopathology 65:556-559.

65. MORSINK, F., and A. E. RICH. 1968. Interactions between Verticillium albo-atrum and Pratylenchus penetrans in the Verticillium wilt of potatoes. Phytopathology 58:401 (Abstr.).

66. MOURA, R. M. de, E. ECHANDI, and N. T. POWELL. 1975. Interaction of Corynebacterium michiganense and Meloidogyne incognita on tomato. Phytopathology 65:1332-1335.

67. MURPHY, P. W., and C. C. DONCASTER. 1957. A culture method for soil meiofauna and its application to the study of nematode predators. Nematologica 2:202-214.

68. NEAL, D. C. 1954. The reniform nematode and its relationship to the incidence of Fusarium wilt of cotton at Baton Rouge, Louisiana. Phytopathology 44:447-450.

69. NEWHALL, A. G. 1958. The incidence of Panama disease of banana in the presence of the root knot and the burrowing nematodes (Meloidogyne and Radopholus). Plant Dis. Rep. 42:853-856.

70. NORTON, D. C. 1969. Meloidogyne hapla as a factor in alfalfa decline in Iowa. Phytopathology 59:1824-1828.

71. OYEKAN, P. O., and J. E. MITCHELL. 1971. Effect of Pratylenchus penetrans on the resistance of a pea variety to Fusarium wilt. Plant Dis. Rep. 55:1032-1035.

72. OYEKAN, P. O., and J. E. MITCHELL. 1972. The role of Pratylenchus penetrans in the root rot complex of canning pea. Phytopathology 62:369-373.

73. PALMER, L. T., D. MacDONALD, and T. KOMMEDAHL. 1967. The ecological relationship of Fusarium moniliforme to Pratylenchus scribneri in seedling blight of corn. Phytopathology 57:825 (Abstr.).

74. PARTYKA, R. E., and W. F. MAI. 1958. Nematocides in relation to sclerotial germination in Sclerotinia sclerotiorum. Phytopathology 48:519-520.

75. PITCHER, R. S., and J. E. CROSSE. 1958. Studies in the relationship of eelworms and bacteria to certain plant diseases. II. Further analysis of the strawberry cauliflower disease complex. Nematologica 3:244-256.

76. POWELL, N. T., P. L. MELENDEZ, and C. K. BATTEN. 1971. Disease complexes in tobacco involving Meloidogyne incognita and certain soil-borne fungi. Phytopathology 61:1332-1337.

77. POWELL, N. T., and C. J. NUSBAUM. 1958. The effect of root-knot nematode resistance on the incidence of black shank in tobacco. Phytopathology 48:344 (Abstr.).

78. RANKIN, H. W., and J. M. GOOD. 1959. Effect of soil fumigation on the prevalence of southern blight of tomatoes. Plant Dis. Rep. 43:444-445.

79. RODRIGUEZ, J. G., C. F. WADE, and C. N. WELLS. 1962. Nematodes as a natural food for Macrocheles muscaedomesticae (Acarina: Macrochelidae), a predator of the house fly egg. Ann. Entomol. Soc. Am. 55:507-511.

80. ROSS, J. P. 1965. Predisposition of soybeans to Fusarium wilt by Heterodera glycines and Meloidogyne incognita. Phytopathology 55:361-364.

81. RUEHLE, J. L., and D. H. MARX. 1969. Parasitism of pine mycorrhizae by lance nematodes. J. Nematol. 1:303 (Abstr.).

82. SANTO, G. S., and O. V. HOLTZMAN. 1970. Interrelationships of Pratylenchus zeae and Pythium graminicola on sugarcane. Phytopathology 60:1537 (Abstr.).

83. SCHENK, N. C., and D. P. H. TUCKER. 1974. Endomycorrhizal fungi and the development of citrus seedlings in Florida fumigated soils. J. Am. Soc. Hortic. Sci. 99:284-287.

84. SEINHORST, J. W., and K. KUNIYASU. 1971. Interaction of Pratylenchus penetrans and Fusarium oxysporum forma pisi race 2 and of Rotylenchus uniformis and F. oxysporum f. pisi race 1 on peas. Nematologica 17:444-452.

85. STEWART, R. N., and A. F. SCHINDLER. 1956. The effect of some ectoparasitic and endoparasitic nematodes on the expession of bacterial wilt in carnations. Phytopathology 46:219-222.

86. STOVER, R. H. 1966. Fungi associated with nematode and nonnematode lesions on banana roots. Can. J. Bot. 44(12):1703-1710.

87. SWARUP, G., and N. J. SINGH. 1962. A note on the nematode-bacterium complex in tundu disease of wheat. Indian Phytopathol. 15:294-295.

88. TAYLOR, C. E. 1972. Transmission of viruses by nematodes. In KADO, C. I., and J. O. AGRAWAL (eds.) Principles and techniques in plant virology. New York: Van Nostrand Reinhold Co. pp. 226-247.

89. THOMASON, I. J., D. C. ERWIN, and M. J. GARBER. 1959. The relationship of the root-knot nematode, Meloidogyne javanica, to Fusarium wilt of cowpea. Phytopathology 49:602-606.

90. VAN GUNDY, S. D., and P. H. TSAO. 1963. Growth reduction of citrus seedlings by Fusarium solani as influenced by the citrus nematode and other soil factors. Phytopathology 53:488-489.

91. VARGAS, J. M., Jr., and C. W. LAUGHLIN. 1972. The role of Tylenchorhynchus dubius in the development of Fusarium blight of Merion Kentucky bluegrass. Phytopathology 62:1311-1314.

92. WEHUNT, E. J., and D. J. WEAVER. 1972. Effect of nematodes and Fusarium oxysporum on the growth of peach seedlings in the greenhouse. J. Nematol. 4:236 (Abstr.).

93. WHITNEY, E. D. 1971. Synergistic effect between Heterodera schachtii and Pythium ultimum on damping-off of sugarbeet vs. additive effect of H. schachtii and P. aphanidermatum. Phytopathology 61:917 (Abstr.).

94. YOUNG, P. A. 1939. Tomato wilt resistance and its decrease by Heterodera marioni. Phytopathology 29:871-879.

95. ZENTMYER, G. A., and J. B. KENDRICK. 1949. Fungicidal action of volatile soil fumigants. Phytopathology 39:864 (Abstr.).

## Bibliography

BELL, K. K., N. A. MINTON, and B. DOUPNIK, Jr. 1970. Infection of peanut pods by Aspergillus flavus as affected by Meloidogyne arenaria and length of curing time. Phytopathology 60:1284 (Abstr.).

BRODIE, B. B. 1961. Use of 1,2-dibromo-3 chloropropane as a fungicide against Pythium ultimum. Phytopathology 51:798-799.

HATTINGH, M. J., and J. W. GERDEMANN. 1975. Inoculation of Brazilian sour orange seed with an endomycorrhizal fungus. Phytopathology 65:1013-1016.

HIRANO, K., and T. KAWAMURA. 1965. The disease complex caused by nematodes and other micro-organisms. 1. Actions of root-lesion nematodes, Pratylenchus spp., responses to Fusarium spp. in the soil (Japanese text). Ann. Phytopathol. Soc. Jpn. 30:24-30.

HOLDEMAN, Q. L., and T. W. GRAHAM. 1952. The association of the sting nematode with some persistent cotton wilt spots in northeastern South Carolina. Phytopathology 42:283-284 (Abstr.).

LUCAS, G. B., J. N. SASSER, and A. KELMAN. 1955. The relationship of root-knot nematodes to Granville wilt resistance in tobacco. Phytopathology 45:537-540.

MILLER, C. R. 1968. Interaction of Meloidogyne javanica and Phytophthora parasitica var. nicotianae on Nicotiana tabacum 'N.C. 95.' Phytopathology 58:553 (Abstr.).

NIGH, E. L., Jr. 1966. Rhizobium nodule formation on alfalfa as influenced by Meloidogyne javanica. Nematologica 12:96 (Abstr.).

OVERMAN, A. J., and J. P. JONES. 1970. Effect of stunt and root knot nematodes on Verticillium wilt of tomato. Phytopathology 60:1306 (Abstr.).

PITCHER, R. S. 1963. Role of plant-parasitic nematodes in bacterial diseases. Phytopathology 53:35-39.

POLYCHRONOPOULOS, A. G., B. R. HOUSTON, and B. F. LOWNSBERY. 1969. Penetration and development of Rhizoctonia solani in sugar beet seedlings infected with Heterodera schachtii. Phytopathology 59:482-485.

POWELL, N. T. 1971. Interactions between nematodes and fungi in disease complexes. Annu. Rev. Phytopathol. 9:253-274.

TU, C. C., and Y. H. CHENG. 1971. Interaction of Meloidogyne javanica and Macrophomina phaseoli in Kenaf root rot. J. Nematol. 3:39-42.

# 46. Legal, Human, and Environmental Aspects of Nematode Control Chemicals[1]

## John H. Wilson, Jr.

Leader, Joint SON/ASTM E 35.16 Task Force on Human Toxicity and Environmental Effects of Candidate Nematode Control Agents, and extension associate professor, departments of Plant Pathology and Horticultural Science, North Carolina State University, Raleigh, NC 27607

## Legal Aspects

The Federal Insecticide, Fungicide, and Rodenticide Act (FIFRA) as amended in 1972 and 1975 regulates the use of pesticides so as to prevent unreasonably adverse effects on the environment and human health. Testing requirements for assessing the hazards of nematicides (and all pesticides) probably will be more demanding in the future. Hopefully this chapter will stimulate more interest in making nematicide products even safer for humans and the environment.

The amended FIFRA requires that all pesticide users follow the labeled directions except when special regulations allow use at a lower rate, and that the container and any excess be disposed of as specified. The user is responsible for applying the pesticide correctly. The producer must provide all instructions regarding safe, proper use and disposal of the product. The Environmental Protection Agency (EPA) will regulate its use or uses largely through registration of new products and reregistration of all existing products.

The EPA administrator will publish guidelines for registration, which will be revised periodically. Registrations will be approved after the EPA determines whether (i) the composition of the product warrants the proposed claims, (ii) the labeling and other documents submitted comply with FIFRA requirements, and (iii) the product performs its intended function without damaging the environment when used according to common, recognized practice. If these requirements are not met, the applicant will be notified of the reasons for rejection and given 30 days in which to correct the problems before registration is denied.

As a part of the registration-reregistration process, the EPA administrator will classify each pesticide for general use or for restricted use. In some cases, a pesticide may have both general and restricted uses. Private and public workers using restricted-use pesticides must be certified according to a plan submitted by the governor of their state and approved by the EPA.

A private user works with a restricted-use pesticide on property he owns, rents, or sprays for another person, without compensation other than trading personal services. All other users of restricted-use pesticides are certified as commercial (public) applicators. Most states require commercial applicators to pass written tests for certification or licensing or both. Federal law does not require that private applicators be tested for competency.

As the science develops, criteria for evaluating pesticide hazards will improve. Data requirements vary according to the nature of the pesticide (eg, toxicity, persistence, tendency to accumulate in living tissue). Data probably needed to support registration include relatively short-term tests such as acute oral and dermal $LD_{50}$, acute dermal $LD_{50}$, and aquatic organism acute $LC_{50}$. Long-term tests (12–48 months) that may be required include teratogenicity, reproduction, oncogenicity, chronic feeding trials, and foliar residue and exposure studies.

Many fumigant and nonfumigant nematicides probably will be classified for restricted use. Some nematicides may be classified as general or restricted, depending on type of use (eg, commercial, private, homeowner), type of formulation (eg, liquid, granular, dust), method of application, and history of accidents or environmental problems or both.

Human and environmental effects of pesticides will greatly influence the cost and legal implications of developing, producing, and using nematicides in the immediate future.

## Human Safety Aspects

Human health and environmental problems associated with the use and misuse of pesticides can be classified in four categories: (i) acute impact on human health, (ii) potential long-term health effects, (iii) localized environmental effects, and (iv) widespread environmental effects.

Acute health and localized environmental effects have been due largely to improper handling and use of pesticides by applicators. Most nematicides now in use have oral, dermal, and inhalation toxicities that require more than casual attention. Education in the proper handling of pesticides can overcome many health and environmental problems. Certification of restricted-use applicators under amended FIFRA regulations will help to ensure that pesticides be used safely, with minimum risk.

Some pesticides have characteristics (eg, mobility, persistence, and bioaccumulation) that can have long-term health and environmental effects associated with widespread use. Nematicides have not been serious problems in this area. Careful attention must be given to these characteristics while developing future products.

Toxicity (oral, dermal, inhalation) tests are valuable in assessing the hazards of a pesticide. The inherent toxicity of a product, however, may not reflect its actual hazard. For example, the dermal toxicity of a granular product may be high in a laboratory test in which the

[1]A portion of this chapter was presented at the Symposium on Suggested Guidelines and Test Procedures for Nematicide Evaluation, August 16, 1976, held at the 15th Annual Meeting of the Society of Nematologists, Daytona Beach, FL.

product stays on the skin. In the field, however, the same product may not be hazardous simply because it may not adhere to clothing or skin. On the other hand, in laboratory tests the dermal toxicity of an emulsifiable concentrate may closely approximate its hazards in the field. Similarly, in confined wildlife tests, birds may eat the toxicant in their food, but avoid it in the field where they have more choice.

Other basic toxicologic data also can be overemphasized when considering health risks. For example, chloropicrin, a fumigant, has an acute vapor toxicity of 20 ppm, while another common fumigant, methyl bromide, has an acute toxicity of 200 ppm. This tenfold difference is not an indicator of relative toxicity to man. A small quantity of chloropicrin, however, serves as a valuable warning agent in odorless methyl bromide.

Nematicides generally can be classified as fumigants, nonfumigants, or systemics. Fumigants have high vapor pressure, move as a gas, and generally kill the pest on contact. Nonfumigants have low vapor pressure, move as liquids (usually in the soil water), and kill by dermal or oral contact. Systemics are translocated in plants from the place at which the material is applied to the place at which the nematode is feeding.

Fumigant nematicides generally have high acute vapor inhalation toxicities ranging from 100 to 2,000 ppm. Since the oral and dermal toxicities of fumigants tend to be lower than those of most insecticides and nonfumigant nematicides, they are classified as moderately hazardous (acute oral $LD_{50}$ of 50–500 mg/kg and acute dermal $LD_{50}$ of 200–2,000 mg/kg). Volatility affects both toxicity and hazard. Irritation and odor can reduce exposure. As previously stated, a small amount of chloropicrin, a lacrimator, in odorless methyl bromide acts as a warning agent. Many accidents with fumigants, however, are not due to inhalation but to dermal or ocular contact.

Nonfumigant and systemic nematicides, especially high active-ingredient concentrates, generally are highly hazardous chemicals (oral $LD_{50}$ of less than 50 mg/kg and dermal $LD_{50}$ of less than 200 mg/kg). Their low active-ingredient granular formulations, however, can be only moderately hazardous (oral $LD_{50}$ above 50 mg/kg and dermal $LD_{50}$ above 200 mg/kg).

Every attempt should be made to reduce the hazard from the oral, dermal, ocular, and inhalant toxicities of nematicides. Formulation, packaging, warning agents, proper equipment, protective clothing, and good instructions for application can be useful. For example, dry formulations are an excellent way to minimize human toxicity hazard and most nonfumigant nematicides can be used in this form. Certain environmental effects such as overdoses and accidental spill are likewise reduced. Reports in states in which nematicide use is high show far fewer accidental poisonings with granular than with liquid formulations.

Small packages of the safe pesticides tend to prevent overuse and decrease accident, storage, and disposal problems. Large containers of the hazardous pesticides tend to reduce availability for home and garden use. Packaging quality is also important. Highly toxic materials should not be placed in containers that are easily broken or punctured. Some containers deteriorate with age and release the pesticide under storage conditions. Ease of dispensing and low attractiveness of the container for reuse can also help to reduce hazards. Technology in the container safety field needs to be upgraded.

Protective clothing and equipment are needed for safe application of most nematicides. Future labeling will specify the necessary protective clothing (eg, neoprene gloves) and equipment (eg, respirators). Hopefully this information will be specific. Producers, distributors, dealers, and educators should promote safe clothing and equipment. Special clothing or equipment should be sold where the applicator purchases the product. Workers handling and using pesticides, however, are reluctant to use protective clothing because it is often hot, binding, heavy, and uncomfortable. Since impervious protective clothing under humid conditions can lead to heat exhaustion and complicate the hazards of pesticide use, more comfortable protective clothing and equipment is needed. Even with protective clothing, the applicator may not always realize when he is getting pesticide on his skin, in his lungs, or in his mouth. The use of harmless warning agents in pesticides—dyes (skin), gases (nose), bitter substances (mouth)—would be helpful.

Method of application to various sites is important to human and environmental safety. Nematicides are primarily injected into the soil, applied to the soil, and immediately incorporated or applied under a plastic cover. Adverse effects of hazardous pesticides are fewer with soil applications than with use on foliage, on soil surface, on animals, or in open air. Likewise, a nematicide applied to the soil in an open field is less hazardous to the applicator than when applied in a greenhouse or other confined area.

Proper equipment is essential to apply pesticides correctly and safely. Available equipment and its cost might steer a pesticide producer (or user) toward a particular formulation. If equipment for liquid but not granular materials is available, the applicator will probably use the liquid formulation. On the other hand, it might be safer for a worker to use liquid concentrates with proper equipment than to apply a "safer" granular material by hand or with a hand-operated spreader.

The class of applicator will be considered in restricting and recommending nematicides as well as other pesticides. Under the amended FIFRA, applicators will be classified into three groups: (i) commercial/public, (ii) private, certified for restricted use, and (iii) noncertified (general use pesticides only). The training and competence requirements of the three groups will be substantially different. Commercial (public) applicators will be trained, tested, and licensed in most states. Restricted-use private applicators will be trained or tested or both, but general-use applicators will receive no mandatory training. Tremendous progress in the safe use of pesticides can be made if modes of distribution to and use of pesticides by these groups can be controlled.

## Environmental Safety Aspects

The effects of nematicides on the environment are as important as their human health aspects. Many of these effects were mentioned above, since they cannot be separated from human health aspects. Pesticides with significantly adverse effects on the environment will not be tolerated in the future. Presently used nematicides have been reasonably clean in their effects on the environment. Additional points to consider in evaluating nematicides for environmental aspects are discussed below.

Most nematicides are moderately to highly toxic to warm-blooded animals. Safe use requires that they not come into direct contact with humans, domestic animals, wildlife, bees, and other nontarget animals or plants. In addition, the environmental impact of nematicides is generally restricted because (i) the time of use is restricted, (ii) fumigants are usually injected into the soil and kill pests in the gaseous state, and (iii) nonfumigants are usually incorporated into the soil and kill on contact.

The time and manner in which nematicides are used also diminishes the risk of environmental hazards. For example, nematicides are usually applied only once a year. Most fumigant materials require a waiting period before planting, which makes crop and nontarget damage unlikely. Crop, and possibly human and animal injury, usually occurs only with misuse or under unusual environmental conditions such as application to cold or wet soils and trapping the fumigant due to mechanical soil compaction. Phytotoxicity with nonfumigant nematicides is not likely with labeled crops if directions for use are followed. Most nonfumigant nematicides can be applied at or near planting time. If recommended rates per volume of soil are exceeded, however, phytotoxicity is likely. For example, if 6 lb of an active ingredient is required to control the nematode population in an acre of soil 6 in. deep, the actual rate is doubled if the nematicide is incorporated only 3 in. deep. Likewise, if a 6-lb broadcast rate is applied in a 6-in. band on a row 36 in. wide, the rate is six times the recommended rate, and phytotoxicity would be expected. The degree of injury depends on the sensitivity of the crop, the pesticide used, soil type, soil condition, the waiting period, amount of rainfall before planting, and other factors.

Another favorable factor for nematicides is that most nematicides are applied to the soil. They are either incorporated or held into the soil with a soil seal or a gas-tight cover. Therefore, nematicides are not as likely to affect the environment through movement, direct contact, or accidental spraying of nontarget areas as are materials applied above ground. They are, however, subject to runoff through the soil water, soil particles, or vapor drift if conditions are adverse.

Protection of the worker is essential when applying nematicides. Highly volatile fumigants such as methyl bromide and chloropicrin, and to a lesser extent, other liquid fumigants, require protective clothing, because pressure systems are used and hose breakage or rupture could shower the applicator with pesticide. Vapors of most fumigant chemicals can be fatal if inhaled. Burns and severe damage to the eyes can result from contact with volatile fumigants.

The moderately volatile fumigants such as dichloropropenes, ethylene dibromide, and dibromo-chloropropane are applied as liquids. Their inhalation toxicity is not as great as the highly volatile fumigants such as methyl bromide. They are moderately hazardous by ingestion or inhalation, but usually only slightly hazardous by dermal absorption. Long-term effects may result from dermal contact or inhalation of certain nematicides.

Nonfumigant nematicides (eg, ethoprop, fensulfothion, carbofuran, and aldicarb) are highly toxic if ingested orally or absorbed through the skin. Inhalation toxicity is usually relatively low. As previously mentioned, granular formulations are safer to handle than liquids and are equally as effective as nematicides and insecticides. For these reasons, granular formulations of nonfumigant nematicides are preferred to liquid formulations. In addition to being nematicides, some of these materials are used as foliar contact and soil systemic insecticides and nematicides. Foliar uses of nonfumigant nematicides (usually as insecticides) in turn require maximum safety measures to avoid drift and other exposure hazards.

With the exception of a few relatively safe products, most nematicides should be applied only by certified or licensed professional applicators. Exacting application and safety equipment are essential.

Nematicides should be stored in a dry, cool, well-ventilated building. The storage place should be vapor proof, fire resistant, well lighted, and ideally, ventilated with an exhaust fan. The building should be posted and locked. All pesticides should be kept in their original containers and never stored near nutrients, seed, or animals.

Purchasing excess amounts of nematicides should be avoided. Any excess nematicides should be disposed of according to directions on the label. Two good methods are incineration at high temperatures to break down the pesticide into environmentally safe materials or burial in a landfill specially designed for pesticides. Empty containers should be rinsed thoroughly and burned, buried, or placed in an approved sanitary landfill. Pesticide containers should never be reused. Many containers may be returned to the manufacturer or distributor.

Future nematicide candidates must be carefully evaluated to make sure they (i) do not accumulate or become biologically concentrated in plants or animals, (ii) break down under normal use in a reasonable time, (iii) break down rapidly in water and are not highly toxic to fish and other wildlife, (iv) can be handled safely with protective clothing, (v) are used in the safest possible formulation, and (vi) are used by trained pesticide applicators.

## Acknowledgments

I acknowledge the help in preparing this chapter of C. W. Averre, K. R. Barker, and H. E. Duncan, North Carolina State University; Eldon I. Zehr, Clemson University, Clemson, SC; E. M. Wilson, Environmental Protection Agency; P. H. Schroeder, Union Carbide Corporation; A. J. Huvar, Mobil Chemical Company; Steven Hunter, Dow Chemical Company; and W. E. Rader, Shell Development Company.

# Index

Adams, P. B., 61
*Agrostis palustris,* 25, 85
*Allium* spp., 61
*Alternaria* sp., 27, 51
American Society for Testing and Materials, 99–100
*Anguina tritici,* 129
*Aphanomyces euteiches,* 129
*Aphelenchoides* sp., 115
  *ritsema-bosi,* 129
*Apium graveolens* var. *dulce,* 68
Apple
  cedar-apple rust, 41, 43
  fire blight, 46
  flyspeck, 41, 43
  fruit rots, 41, 43
  fungicide tolerance on, 18
  mite suppression on, 36
  powdery mildew, 32–33, 35–40, 43
  quince rust, 41, 43
  scab, 40, 43
  sooty blotch, 41, 43
  timing of sprays on, 42, 46
Arabis mosaic virus, 130
*Arachis hypogaea,* 89
*Aspergillus* spp., 22
  *flavus,* 129
  *ochraceous,* 129
Avocado
  anthracnose, 78
  Cercospora spot, 78
Bactericide
  application methods, 48–49
  timing of application, 49
Barker, K. R., 114
Barua, G. C. S., 72
Barua, K. C., 72
Beer, S. V., 46
Beet ring spot virus, 130
*Belonolaimus* sp., 115, 128
Bergeson, G. B., 128
Bertrand, P. F., 59
*Beta vulgaris,* 23
Bird, G. W., 108, 114
*Botrytis* sp., 22, 51
  *cinerea, 129*
Brodrick, H. T., 78, 80
Carnation ring spot virus, 130
Celery late blight, 68–69
Cellophane-transfer technique, 15–18
*Cephalosporium* sp., 27
*Cercospora* spp., 27
  *apii,* 18, 19
  *arachidicola,* 18, 19
  *beticola,* 19
  *purpurea,* 78
Cetas, R. C., 63
Cherry
  tart, 51–53
  sweet, 55
  leaf roll virus, 130
Chiarappa, L., 54
Chrysanthemum
  Ascochyta blight, 86–88
  Fusarium wilt, 30–31
Citrus
  Alternaria leaf spot, 76

greasy spot, 73–75, 77
melanose, 73, 75–77
scab, 73, 76–77
*Citrus*
  *jambhiri,* 76
  *paradisi,* 74
  *sinensis,* 74
*Cladosporium carpophilum,* 20
*Coccomyces hiemalis,* 51
*Coffea arabica,* 70, 72
Coffee
  coffee berry disease, 70, 71
  leaf rust, 71–72
Cole, H., Jr., 22, 25, 85
*Colletotrichum*
  *coffeanum,* 70
  *gloeosporioides,* 78, 80
Corn, 89
*Corynebacterium*
  *faciens,* 129
  *insidiosum,* 129
  *michiganense,* 129
  *tritici,* 129
*Coryneum carpophilum,* 21
Cotton, 89
Cotton Disease Council, 95
*Criconemoides* sp., 115, 119
  *xenoplax,* 129, 130
*Curvularia trifolii,* 129
*Cylindrocarpon radicicola,* 129
*Cytospora leucostoma,* 59
*Dactylaria* sp., 27
  *humicola,* 28
Davis, R. A., 108
*Diaporthe* sp., 97
  *citri,* 73, 75
Dickson, D. W., 114
*Diplocarpon rosae,* 82
DiSanzo, C. P., 101
Disease complexes, 127, 128–134
Disease control, evaluation of, 26, 33, 36, 38–40, 42, 43, 45, 52, 56, 66, 70–71, 72, 74–75, 76–77, 79, 81, 84, 88, 93–94, 96
Disease indexing, 59–61, 66, 94, 103
Disease severity, 59–61, 66
*Ditylenchus* sp., 115, 122
  *dipsaci,* 128, 129
*Dolichodorus* sp., 115
Dooley, H. L., 32, 57, 63, 68, 82
Dunn, R. A., 114
*Elsinoe fawcetti,* 73, 76
  *mangiferae,* 80
Elson, J. E., 1
Engelhard, A. W., 30, 86
English, H., 54, 59
*Erwinia*
  *amylovora,* 46–50
  *carotovora,* 27
  *chrysanthemi,* 27
  *raponticum,* 129
*Erysiphe cichoracearum,* 18
*Exobasidium vexans,* 72
Experimental controls, 4, 31, 32, 43, 47, 55, 58, 64, 75, 76, 83, 88, 90, 96, 97
Feldmesser, J., 101
Fisher, K. D., 99, 114
Fry, W. E., 63

Fumigation chamber, 22
Fungicide
  application equipment, 36, 42, 44, 56, 58, 62, 64, 83, 89–90
  application techniques, 23–26, 32, 42, 56, 62, 74, 87, 89–90, 96
  bioassays, 15–18
  detection of systemic properties, 17
  effects on quality, 52, 67
  effects on spore production, 70, 75, 76
  fungicidal versus fungistatic properties, 15–16
  laboratory screening, 21, 54, 70, 72
  preliminary field tests, 74
  rates, 42, 89
  registration requirements, 1, 2
  residue detection, 15–18, 79
  spray drift, control of, 35, 42, 51, 74
  tenacity, 15–18
  thermal powder and tablet formulations, 22
  timing of sprays, 64–65, 73–74, 75, 78, 86
  tolerance, 18–20
  use of adjuvants, 78, 81
*Fusarium* spp., 22, 23, 27, 89, 128, 129
  *moniloforme,* 129
  *oxysporum* f. sp. *chrysanthemi,* 30
  *oxysporum* f. sp. *cubense,* 129
  *oxysporum* f. sp. *lycopersici,* 129
  *oxysporum* f. sp. *pisi,* 128, 129
  *oxysporum* f. sp. *vasinfectum,* 129
  *roseum,* 129
*Fusicladium effusum,* 20
*Gloeodes pomigena,* 41, 43
*Glomerella* spp., 51
  *cingulata,* 43
  *gossypii,* 129
Glossary of nematologic terms, 125
*Glycine max,* 89, 97
*Gossypium hirsutum,* 89, 95
Grapefruit, 74–75
Grapevine fanleaf virus, 130
Griffin, G. D., 114, 128
*Gymnosporangium*
  *clavipes,* 41, 43
  *juniperi-virginianae,* 41, 43
Hansing, E. D., 88
Harder, H. H., 21
Heald, C. M., 126
Heidrick, L. E., 63
*Helicotylenchus* spp., 115, 118, 128
  *dihystera,* 130
*Hemicycliophora* sp., 115
*Heterodera* spp., 115, 119, 120, 121, 122, 123
  *glycines,* 119, 128, 129
  *rostochiensis,* 128
  *schachtii,* 105, 128
  *tabacum,* 128
*Helminthosporium* sp., 27, 89
*Hemileia vastatrix,* 71
Hickey, K. D., 37
*Hoplolaimus* sp., 115, 118, 128
Humidity chamber, 85
Inoculation, 26, 29, 31, 32, 36–37, 38, 41, 48–49, 55, 65–66, 69, 82, 86, 87, 93

Inoculum
  concentration, 48, 65, 87
  maintenance of, 28, 32, 41, 47, 51–52, 65, 82, 87
  preparation of, 26, 28, 30, 47, 65, 82, 85, 86, 87, 93, 97
  sources, 74, 82, 87
Johnson, A. W., 106
Johnson, P. W., 108
Kerr, E., 126
Kinloch, R., 126
Knauss, J. F., 27
Leach, L. D., 23
Leach, S. S., 63
Lembright, H. W., 128
Lemon, 76
Lettuce ring spot virus, 130
Lewis, F. H., 40, 51
*Longidorus* spp., 115, 118
  *attenuatus,* 130
  *elongatus,* 130
  *macrosoma,* 130
Lownsbery, B. F., 128
Luepschen, N. S., 21, 35
*Lycopersicon esculentum,* 92–95, 101–102
MacDonald, D., 108
MacDonald, J. D., 23
*Macrophomina phaseoli,* 129
MacSwan, I. C., 57
Mai, W. F., 114
Malek, R. B., 108, 114
*Malus sylvestris,* 18, 32, 35, 37, 40, 43, 46
*Mangifera indica,* 80
Mango
  anthracnose, 80
  bacterial spot, 80
  powdery mildew, 80
  scab, 80
Manji, B. T., 54
Manzer, F. E., 63
McKenry, M., 108
*Meloidogyne* spp., 115, 118, 120, 122, 123, 129, 130
  *arenaria,* 103, 129
  *hapla,* 129, 130
  *incognita,* 101, 103, 129, 130
  *javanica,* 129
Michell, R. E., 114, 126
*Microthyriella rubi,* 43
Minton, E. B., 95
*Monilinia* spp., 18
  *fructicola,* 51, 54–56
  *laxa,* 54–56
Murad, J. L., 114
*Mycosphaerella*
  *citri,* 73, 74, 77
  *ligulicola,* 87
Myers, R. F., 101
*Nacobbus* sp., 115
Neely, D., 15
Nelson, L. A., 2
Nematicide
  application methods, 102, 104, 107
  environmental effects, 107, 136–137
  eradicant properties, 102, 103–104
  experimental controls, 102–103, 106
  greenhouse test procedures, 101–103
  guidelines for testing, 99–100
  harmful effects, 130, 132
  hatching and emergence, 105
  human safety aspects, 135–136
  insect control, 130, 131
  legal aspects, 135
  mycorrhizal effects, 132
  plant responses, 126–127
  registration requirements, 1, 2, 135
  root dips, 102
  site selection, 108–113
  soil properties and, 111–112, 127
  systemic properties, 102

test equipment, 105
  timing of application, 106
Nematodes
  assessment of populations, 106, 111, 114–125
  bioassay procedures, 122
  culture, 104
  distribution in soil, 111
Nematodes, extraction procedures
  cysts, 120–121
  nematodes from soil, 119–120
  nematode eggs, 122
  nematodes from plants, 121–122
  inoculation, 102, 104, 106
  inoculum preparation, 101–102, 104
  root staining procedures, 103
  sampling procedures, 116, 117
  soil infestation, 102
  storing samples, 117–118
Norris, M., 126
Norton, D. C., 114, 128
Ogawa, J. M., 54
*Oidium mangiferae,* 80
Okioga, D. M., 70, 71
O'Melia, F. C., 101
Onion, white rot of, 61–63
Oostenbrink, M., 114
Orange, 74
Ornamental tropical plants, 27–30
Orr, C. C., 108, 126
Osborne, W. W., 126
*Paratrichodorus* spp., 130
*Paratylenchus* sp., 115
  *hamatus,* 129
Pea early browning virus, 130
Peach
  brown rot, 54–57
  leaf curl, 57–58
Peanut, 89
Pear
  fire blight, 46
  timing of sprays, 49
*Penicillium* sp., 22
  *martensii,* 129
Pepper, root-knot nematode, 104
*Persea americana,* 78
*Phoma betae,* 23
*Phragmidium mucronatum,* 82
*Phytophthora* spp., 27, 97
  *cinnamomi,* 128–130
  *citrophthora,* 130
  *infestans,* 63
  *parasitica,* 129, 130
Phytotoxicity
  kinds of symptoms, 26, 52, 67, 86, 88, 91, 93, 94, 98, 126
  measurement, 26, 67, 83, 88, 91, 93
Plug-mix plantings, 92–95
*Podosphaera*
  *leucotricha,* 18, 32, 37, 41, 43
  *oxyacanthae,* 51
Potato late blight, 63–68
Powell, W. M., 108, 114
*Pratylenchus* spp., 115, 117, 120, 129
Prune
  brown rot, 55
  cystospora canker, 59–61
*Prunus*
  *amygdalus,* 55
  *armeniaca,* 21, 55
  *cerasus,* 51, 55
  *domestica,* 55, 59
  *persica,* 55, 57
  *salicina,* 55
*Pseudomonas* spp., 27
  *caryophylli,* 130
  *mangiferaeindicae,* 80
  *solanacearum,* 130
  *syringae,* 56, 60, 129
*Pyrus communis,* 46
*Pythium* spp., 89, 97

*aphanidermatum,* 23, 85, 92, 93, 128
  *arrhenomenos,* 92, 93, 130
  *debaryanum,* 23, 129
  *graminicola,* 129
  *myriotylum,* 92, 93, 129
  *ultimum,* 23, 85, 128, 129, 130
*Radopholus* sp., 115
Ranney, C. D., 95
Raspberry leaf curl virus, 130
Regional cottonseed treatment program, 95
Reidel, R. M., 101, 108
*Rhizobium japonicum,* 130
*Rhizoctoni* sp., 89, 128, 129
  *solani,* 23, 27, 92, 93, 97, 128, 129, 130
Richards, B. L., 63
Riggs, R., 126
Rohde, R. A., 114
*Rosa* spp., 82
Rose
  black spot, 82–84
  powdery mildew, 82–84
  rust, 82–84
Rough, D., 54
*Rotylenchus* sp., 115
  *reniformis,* 129
Ruehle, J. L., 108–114
Sall, M. A., 54
Sanders, P. L., 22, 25, 85
*Schizothyrium pomi,* 41
Schultz, O. E., 63
*Sclerotium*
  *cepivorum,* 61
  *rolfsii,* 27
*Scutellonema* sp., 115
Seed treatment procedures
  cultivar selection, 89
  fungicide application, 89, 90, 96, 97
  fungicide formulations, 90, 96
  fungicide performance, 24, 97
  fungicide rates, 97
  plot design, 91, 96, 97
  replication, 91
  seed infestation, 24, 89
  site preparation, 93
  site selection, 91, 97
Seedling emergence, 23–24
*Septoria apiicola,* 68
Smart, G. C., Jr., 114
*Solanum tuberosum,* 63
Sonoda, R. M., 92
Sorghum, 89
*Sorghum vulgare,* 89
Southards, C. J., 114
Soybean, 89, 97
*Sphaerotheca pannosa,* 82
Statistical procedures
  data analysis, 10, 11, 39, 43, 49, 53, 62, 112–113
  experimental controls, 4, 5
  in site selection, 3, 108–113
  interplot interference, 61–62, 64
  interpretation of results, 11–13, 112, 122–123
  planning experimental designs, 2–10, 112
  replication, 6–7, 62
  sampling procedures, 8–10
Steele, A. E., 101, 105
Strawberry latent ring spot virus, 130
Sugar beet
  cyst nematodes, 105
  seed treatment, 23–25
  seedling diseases, 23
*Taphrina deformans,* 57
Taylor, P. L., 114
Tea-blister blight, 72–73
Test site selection, 3, 28, 46, 51, 54, 61–62, 64, 73, 76, 78, 97, 108–113
Thames, W. H., Jr., 114

*Thielaviopsis basicola,* 130
Thirugnanam, M., 103
Thomason, I. J., 114
*Tilletia foetida,* 89
Tobacco rattle virus, 130
Tobacco ring spot virus, 130
Tomato, 92–95
   root-knot nematode, 101–102
Tomato black ring virus, 130
Tomato ring spot virus, 130
Townshend, J. L., 108–114
*Trichoderma harzianum,* 129
*Trichodorus* spp., 115, 118, 130
   *christiei,* 129
   *cylindricus,* 130
   *primitivus,* 130

   *similus,* 130
   *viruliferus,* 130
*Triticum aestivum,* 89
Turfgrass
   Pythium blight, 25, 85–86
   Rhizoctonia brown patch, 25
   Sclerotinia dollar spot, 25
*Tylenchorhynchus* spp., 115, 129
   *claytoni,* 129
   *dubius,* 129
   *martini,* 129
*Tylenchulus* sp., 115
*Tylenchus agricola,* 129
*Venturia inaequalis,* 18, 19, 20, 40, 43
*Verticillium dahliae,* 128, 129
Wade, E. K., 63

Warren, C. G., 85
Weingartner, D. P., 63
Wheat, 89
Whitney, N. G., 97
*Xanthomonas* sp., 27
   *dieffenbachiae,* 27
*Xiphinema* spp., 115, 118
   *americanum,* 130
   *coxi,* 130
   *diversicaudatum,* 130
   *index,* 130
   *italiae,* 130
Yates, W. E., 54
Yoder, K. S., 18
*Zea mays,* 89
Zehr, E. I., 43